Pharmaceuticals in the Environment

Pharmaceuticals in the Environment: *Current knowledge and need assessment to reduce presence and impact*

Edited by
Benoit Roig

IWA Publishing
London • New York

Published by **IWA Publishing**
Alliance House
12 Caxton Street
London SW1H 0QS, UK
Telephone: +44 (0)20 7654 5500
Fax: +44 (0)20 7654 5555
Email: publications@iwap.co.uk
Web: www.iwapublishing.com

First published 2010

Typeset in India by OKS Prepress Services.

British Library Cataloguing in Publication Data
A CIP catalogue record for this book is available from the British Library

Library of Congress Cataloging-in-Publication Data
A catalog record for this book is available from the Library of Congress

ISBN 10: 184339314X
ISBN 13: 9781843393146

Contents

Acknowledgements

The scientific content of the book was attained during the EU-project KNAPPE (contract n° 036864) from February 1st, 2007 to September 30th, 2008 which was financially supported by grant obtained from the EU Commission within the Global Change and sustainable development Program of the Sixth Framework Program.

I would like to thank Evelyne Touraud and Valentine Viernes for their valuable and excellent support during the whole project. All the persons listed below have been involved in the KNAPPE project, and contributed to the outcome of this book.

Partners of the KNAPPE project:

Armines – Ecole des Mines d'Ales
Valentine Viernes, Delphine Bonnin, Sebastien Leroy, Evelyne Touraud, Ingrid Bazin, Sandrine Bayle, Benoit Roig

Federal Institue of Hydrology (BfG), Koblenz, Germany
Alexa Sadezky, Michael Schlüsener, Dirk Löffler, Thomas Ternes

Silesian University of Technology, Environmental Biotechnology Department, Gliwice, Silesia, Poland
Dagmara Buntner, Sebastian Żabczyński, Korneliusz Miksch

Ecologic – Institute for International and European Environmental Policy, Berlin, Germany
Eleftheria Kampa, Rodrigo Vidaurre, Cornelius Laaser, Thomas Dworak

University of York, York, Heslington, United Kingdom
Louise Summerton, Alistair Boxall, James Clark, Edd Smith

University of Portsmouth, Portsmouth, Hampshire, United Kingdom
Richard Greenwood, Graham Mills

Cemagref, Lyon, France
Marina Coquery, Cecile Miege, Jeanne Garric, Jean-Philippe Besse

Bureau de Recherche Géologique et Minière, BRGM, Orléans, France
Anne Togola, Laurence Amalric

CSIC – Spanish Council for Scientific Research, IIQAB-Instituto de Investigaciones Quimicas y Ambientales de Barcelona, Barcelona, Spain
Damia Barcelo, Mira Petrovic

Associated member of the KNAPPE executive committee:

Association of British Pharmaceutical Industry, London, United Kingdom
Michael Murray

Committee for Sustainable Development in Health (C2DS), Paris, Fance
Véronique Molières

European Federation of Pharmaceutical Industries and Association, Brussels, Belgium
Marianne Poulemaire, Bernard Ferre

European Patients' Forum, Brussel, Belgium
Roxana Radulescu

French Agency of sanitary and health product (AFFSAPS), Paris, France
Paul Houeto

French Agency of food safety, Paris, France
Gwenn Regnault

Ineris, Paris, France
Valéria Dulio

International Federation of Animal Healthcare, Brussels, Belgium
Bernd Halling, Elena Micelli

Janssen Pharmaceuticals, Global Pharmaceutical Supply Group – EMEA Unit of Janssen Pharmaceutica NV, Beerse, Belgium
Birgit Mertens

Liverpool John Moores University, Liverpool, United Kingdom
Mark Cronin

Merck Sharp & Dohme-Chibret, St Germain Laprade, France
Pascal Michoux

SANOFI AVENTIS, Lyon, France
Daniel Gouy, Romain Journel

SAUR, Paris, France
Hélène Darras, Anne Gresle, Corinne Madec, Alain Pericaud

WCA Environment, Faringdon, Oxfordshire, United Kingdom
David Taylor

Contributors

Besse Jean-Philippe
Laboratory for Ecotoxicology
Freshwater systems, ecology and pollution research unit
Cemagref Lyon, 3 bis quai Chauveau CP 220 69336
Lyon Cedex 09, France
jp.besse@yahoo.fr

Boxall Alistair
EcoChemistry Team, York University/CSL
Sand Hutton, York YO41 1LZ, UK
alistair.boxall@fera.gsi.gov.uk

Buntner Dagmara
The Silesian University of Technology
Environmental Biotechnology Department
Akademicka 2, 44100 Gliwice, Poland
dagmarabuntner@post.pl

Clark James
Green Chemistry Centre of Excellence
Department of Chemistry, University of York
Heslington, York, YO10 5DD, UK
jhc1@york.ac.uk

Daughton Christian G., PhD
Chief, Environmental Chemistry Branch
National Exposure Research Laboratory
U.S. Environmental Protection Agency
944 East Harmon Avenue, Las Vegas,
Nevada 89119, USA
Daughton.Christian@epamail.epa.gov

Deswarte Fabien
Green Chemistry Centre of Excellence,
Department of Chemistry, University of York,
Heslington, York, YO10 5DD, UK
fabien.Deswarte@fco.gov.uk

Dworak Thomas
Ecologic Institute
Auhofstrasse 4/7, A-1130 Vienna, Austria
thomas.dworak@ecologic.eu

Garric Jeanne
Laboratory for Ecotoxicology
Freshwater systems, ecology and pollution research unit
Cemagref Lyon, 3 bis quai Chauveau CP 220 69336
Lyon Cedex 09, France
jeanne.garric@cemagref.fr

Greenwood Richard
Professor of Environmental Science School
of Biological Sciences, University of Portsmouth
King Henry Building, King Henry 1 Street,
Portsmouth PO1 2DY, UK
Richard.Greenwood@port.ac.uk

Kampa Eleftheria, Dr.
Ecologic Institute, Berlin
Pfalzburger Strasse 43/44,
D-10717 Berlin, Germany
eleftheria.kampa@ecologic.eu

Laaser Cornelius
Ecologic Institute
Pfalzburger Strasse 43/44, D-10717 Berlin, Germany
cornelius.laaser@ecologic.eu

Le Bot Barbara
LERES – Laboratoire d'Etude et de Recherche
en Environnement et Santé
École des Hautes Etudes en Santé Publique
Avenue du Professeur-Léon-Bernard • CS 74312 • 35043,
Rennes Cedex, France
Barbara.LeBot@ehesp.fr

Löffler Dirk
Federal Institute of Hydrology
Department Water Chemistry
Am Mainzer Tor 1, 56068 Koblenz, Germany
Loeffler@bafg.de

Miksch Korneliusz
The Silesian University of Technology
Environmental Biotechnology Department
Akademicka 2, 44100 Gliwice, Poland
korneliusz.miksch@polsl.pl

Graham Mills
Professor of Environmental Chemistry
School of Pharmacy & Biomedical Science
University of Portsmouth,
White Swan Road, Portsmouth PO1 2DT, UK
Graham.Mills@port.ac.uk

Mompelat Sophie
LERES – Laboratoire d'Etude et de Recherche
en Environnement et Santé
École des Hautes Etudes en Santé Publique
Avenue du Professeur-Léon-Bernard • CS 74312 • 35043,
Rennes Cedex, France
Sophie.Mompelat@ehesp.fr

Roig Benoit
LGEI Centre, Ecole des Mines d'Alès,
6, Avenue de Clavières, F-30319 Alès, France
Benoit.Roig@mines-ales.fr

Ruhoy Ilene Sue, MD, PhD
Assistant Professor, Basic and Clinical Sciences
Touro University Nevada,
College of Osteopathic Medicine
Henderson, NV 89014, USA
Ilene.Ruhoy@tun.touro.edu

Sadezky Alexa
Federal Institute of Hydrology
Department Water Chemistry
Am Mainzer Tor 1, 56068 Koblenz, Germany
Asadezky@yahoo.com

Schlüsener Michael
Federal Institute of Hydrology
Department Water Chemistry
Am Mainzer Tor 1, 56068 Koblenz, Germany
Schluesener@bafg.de

Summerton Louise
Green Chemistry Centre of Excellence,
Department of Chemistry
University of York
Heslington, York, YO10 5DD, UK
ls25@york.ac.uk

Ternes Thomas
Federal Institute of Hydrology
Department Water Chemistry
Am Mainzer Tor 1, 56068 Koblenz, Germany
Ternes@bafg.de

Thomas Olivier
LERES – Laboratoire d'Etude et de Recherche
en Environnement et Santé
École des Hautes Etudes en Santé Publique
Avenue du Professeur-Léon-Bernard • CS 74312 • 35043,
Rennes Cedex, France
olivier.thomas@ehesp.fr

Touraud Evelyne
LGEI Centre, Ecole des Mines d'Alès
6, Avenue de Clavières,
F-30319 Alès, France
Evelyne.Touraud@mines-ales.fr

Vidaurre Rodrigo
Ecologic Institute, Berlin
Pfalzburger Strasse 43/44,
D-10717 Berlin, Germany
rodrigo.vidaurre@ecologic.eu

Żabczyński Sebastian
The Silesian University of Technology
Environmental Biotechnology Department
Akademicka 2, 44100 Gliwice, Poland
sebastian.zabczynski@polsl.pl

Foreword

Pharmaceuticals in the environment – Why should we care?

Christian G. Daughton and Ilene S. Ruhoy

"Take only pictures. Leave only footprints." Perhaps the ultimate expression for the concepts of sustainability and the "ecological footprint," this credo of the hiker and spelunker reflects the collective importance of the seemingly innocuous, minuscule impacts that can accrue from each individual's isolated actions. In reality, however, this credo is simply not possible to uphold, as humans unavoidably leave behind indelible trails of telltale fingerprints. Wherever we live or travel, we impart unique chemical signatures on the environment in the form of minute residues of pharmaceuticals and personal care products that we excrete, wash from our bodies, or discard to sewerage or trash. Even the literal fingerprints we leave behind can contain residues of drug ingredients secreted through our sweat. While the residues originating from each individual may be insignificant by themselves, the combined contributions from all individuals, as well as from medicated animals, can reach measurable levels in surface and ground waters and on land receiving treated sewage residuals.

Yet another acronym for one of the many menageries of chemicals that without invitation take up residency in the environment is PiE – Pharmaceuticals in the Environment. With continually escalating attention devoted by the technical literature and popular press, pharmaceuticals as pollutants in the environment comprise a large, diverse class of thousands of active pharmaceutical ingredients (APIs). These chemicals serve as a widely recognized archetype for the much larger galaxy of so-called "emerging contaminants" (ECs). The ECs aren't necessarily new to the environment – rather it's the attention we have begun to devote to their ubiquitous presence – where they occur as interlopers at best, or troublemakers at worst.

Why do (or should) we devote such attention to the PiE issue, especially if we stop to consider that APIs reside in the ambient environment at concentrations that not long ago were considered infinitesimally low – from parts-per-billion to sub-parts-per-trillion (sub-nanograms per liter, or picomolar) and lower? Roughly a decade ago, these levels were extraordinarily difficult for chemists to routinely detect for any chemical in an environmental matrix and were also far below any levels of known toxicological concern. At the least, however, exposure to PiE incrementally contributes to the overall burden of chemical stressors with which each organism must maintain its homeostasis.

The mere fact that APIs gain unintended, unexpected, and usually unwelcomed entry to the environment has probably served to better engage the public with a graphic realization of the intimate, immediate, and inseparable connection that every individual has with the environment – PiE makes us acutely aware of the chemical sea that surrounds us, composed of the universe of chemicals that we create as well as those naturally formed in the biosphere. Perhaps more so than for any other class of environmental contaminants, PiE has also made us focus on the much larger world of unregulated contaminants, where quite a variety of linguistic shortcuts have evolved to describe these chemicals with which we coexist. Some of the many terms and acronyms used in the literature are shown in Table 1; these chemicals include not

©2010 IWA Publishing. *Pharmaceuticals in the Environment: Current knowledge and need assessment to reduce presence and impact.* Edited by Benoit Roig. ISBN: 9781843393146. Published by IWA Publishing, London, UK.

just anthropogenic chemicals but also naturally occurring xenobiotics. The coining of each term or acronym has generally resulted from a need to categorize chemical stressors in different ways to suit different purposes.

Table 1 Terminology of chemical pollutant classifications.[1]

Class or grouping	Grouped according to
EDC/EAC (Endocrine Disrupting [or Active] Chemical) **CMR** (Carcinogenic, Mutagenic, toxic to Reproduction)	Toxicological mode of action or endpoint
PBT (Persistent, Bioaccumulative Toxic) **vPvB** (very Persistent, very Bioaccumulative) **POP** (Persistent Organic Pollutant)	Environmental properties
Micro-pollutants; micro-constituents **Trace pollutants/residues** **TOrC** (Trace Organic Compounds)	Frequency/level of occurrence
OWCs[2] (Organic Wastewater Contaminants)	Location of occurrence
PPCPs[3] (also PhACs, PACs, and PiE; these each comprise a subset of PPCPs, which includes personal care products)	Type of intended usage
Priority pollutants	Regulation
Unregulated pollutants	Absence of regulation
"Chemical weeds"[4]	Public perception ("out of place chemicals")
Emerging contaminants/pollutants[5] **ECCs** (emerging chemicals of concern) **EPOCs** (emerging pollutants of concern) **ESOCs** (emerging substances of concern) **COCs** (chemicals of concern)[6] **COPCs** (chemicals of potential concern) **CECs** (chemicals of emerging concern)	Novelty, fad, timeliness, or new concern
Xenobiotics, exotics **Xenobiotic organic compounds** (**XOCs**)	Foreign versus endogenous
Toxicants, toxins, toxics, perturbogens (agonists, antagonists, activators, repressors, inhibitors, regulators, modulators)	Overall toxicity (note: "toxins" comprise a special subset of toxicants – namely, those that are naturally synthesized, primarily proteins; "toxics" is jargon for "toxicants")
HPV[7] (high production volume) chemicals	Quantity (manufactured/imported in US $>$ 1 million pounds/year)
PDPs or **HDPs** (population- or human-derived pollutants/constituents) **POHO** (pollutants of human origin)	Source or origin

While the published literature on unregulated and emerging contaminants is enormous, the literature devoted to PiE alone is exponentially expanding. Are we doing a sufficient job at mining these new data and synthesizing them into a meaningful knowledge base? If not, we risk wasteful duplications of effort and gross inefficiencies in targeting limited research resources (see: http://dx.doi.org/10.1897/09-138.1). Even worse, we risk not realizing that the answers to

[1] Adapted and extended from: *Renewable Resources Journal*, 2005, 23(4): 6–23.

[2] Coined by Kolpin et al., *Environmental Science & Technology*, 2002, 36(6): 1202–1211.

[3] "Pharmaceuticals and personal care products" (coined by Daughton and Ternes, Environmental Health Perspectives, 1999, 107[suppl. 6]:907–938); PhACs: "pharmaceutically active compounds" (coined by Sedlak, Gray, and Pinkston, Environmental Science & Technology, 2000, 34:508A–515A); PiE: "pharmaceuticals in the environment" [came into use around 2001, when PhRMA (Pharmaceutical Research and Manufacturers of America) created its internal PIE Task Force].

[4] Doined by Daughton, *Renewable Resources Journal*, 2005, 23(4): 6–23.

[5] Term first used prominently by the National Research Council ("Identifying Future Drinking Water Contaminants: based on the 1998 Workshop on Emerging Drinking Water Contaminants," National Academy Press, Washington, DC, 1999).

[6] Came into use in early 1990s.

[7] See the U.S. EPA's The High Production Volume Information System (HPVIS): http://www.epa.gov/hpvis/index.html

certain questions may already exist. How do we best define what outcomes are desired from further research, especially outcomes that benefit as many stakeholders and beneficiaries as possible?

Moreover, without a firm idea as to whether PiE is an important issue, why should we continue to invest resources? The answer to this question is partly provided by some of the very questions that PiE has provoked. What are the ramifications as chemists continue to improve their analytical armamentarium for detecting and identifying ever-lower concentrations of pollutants? How will society respond to the expanding knowledge of increasing numbers of xenobiotics occurring in our waters and foods simply because the already minuscule concentrations that are detectable are driven ever lower? Can risk be understood or effectively communicated in the face of chemical mixtures at such low concentrations that the statistical power and abilities of toxicology today are exceeded? How will the new paradigm of confronting the possibility of subtle environmental effects be effectively communicated to the public so that it is perceived in a meaningful manner and that trust in drinking water supplies can be maintained? Can any risk possibly contributed by PiE be placed in a meaningful context of overall risk posed by the galaxies of other anthropogenic and naturally occurring xenobiotics to which humans and the environment are continually exposed? Perhaps the major question is how do we ensure the sustainability (and minimize the exposure hazards) of a chemical-based, chemical-centric society in the most cost-effective manner?

Even though there seem to be very few definitive answers to what might appear to be too many questions, these questions alone can justify continued examination of the PiE issue. But there are a number of other reasons that PiE should hold our attention as scientists and as consumers. PiE makes obvious the connection that should ordinarily exist between the practice of medicine and the study and protection of the condition of the environment. The two are intimately tied but little recognized as such. The two share many commonalities and connections. Just consider the processes of data collection, epidemiology, diagnosis, mitigation/treatment, prognosis, determination of vulnerability, and pollution/disease prevention. Each of these plays a critical role in both health care and environmental protection – in the ecology of health and in the health of ecology. Of key significance in determining how to allocate limited resources to ever-growing numbers of priorities, improvements in one can often leverage unintended, collateral improvements in the other. By implementing any number of the myriad ways to minimize PiE, significant improvements in the quality and cost of healthcare can possibly be achieved.

For the aquatic environment, major unknowns include the consequences of chronic (sometimes mutigenerational) exposure to very low levels of multiple pharmaceutical residues. This exposure sometimes involves receptors that differ from those in humans, and mechanisms of action can change as the exposure levels are reduced. The potential for effects can increase with exposure to multiple like-acting drugs, as a result of concentration (or dose) addition. Unpredictable and perhaps surprising adverse effects can result, as witnessed by the decimation of certain vulture populations from scavenging carcasses from domestic animals that had been treated with diclofenac, or demonstrated in the collapse of fish populations by exposure to low parts-per-trillion of 17α-ethynylestradiol.

Exposure for humans to environmental API residues, compared with aquatic exposure, is probably lower because feral residues occur in drinking water at greatly reduced levels; but foods grown on sewage- or manure-amended acreage may contain substantially higher concentrations – a consequence of the preferential sorption of certain APIs to sewage sludge. The occurrence of even minute residues of APIs in drinking waters, especially waters created for immediate reuse, poses special problems with regard to risk perception as they constitute unexpected, unintentional, unwelcome exposure that traces directly back to a water's origin from sewage.

In contrast to aquatic exposure, additional unknowns arise regarding human exposure.

How do we assess the significance of unintended exposure due to:

- chronic exposures to APIs designed for short-term therapeutic use;
- exposure routes that differ from the intended clinical routes of administration (e.g., oral ingestion of APIs intended exclusively for dermal use);
- simultaneous exposure to low-levels of multiple APIs, especially those that are contraindicated (this could be particularly problematic for APIs present below purported no-effects levels but which share common modes of action, making the effective dose the sum of the individual doses);
- unintended, unexpected exposure of certain sub-populations to APIs that they should actively avoid (e.g., drugs contraindicated during pregnancy). With regard to the last item, exposure during critical times of developmental vulnerability can yield subtle but lasting functional defects; this concern was codified in the "Faroes Statement" on "Human Health Effects of Developmental Exposure to Chemicals in Our Environment" (http://dx.doi.org/10.1111/j.1742-7843.2007.00114.x), which emphasizes that beyond the traditional view of toxicity – where the "dose makes the poison" – exists a second dimension where the "timing amplifies the poison".

An important aside is the way in which risk communication has often been formulated in attempts to convey the extremely low risk currently assessed for the trace levels of APIs in drinking water. Attempts to place trace concentrations into context by using analogies (such as a part-per-trillion roughly equating to but a single drop in an Olympic-size pool) may only inject further confusion – or at worst serve to misdirect. The concept of concentrations not only can add confusion to risk communication, it can also serve to distract from pertinent facts or even pose a degree of self-contradiction. For example, statements are often made that concentrations in the part-per-trillion (ng/L) range are vanishingly low. But at the same time, it is common knowledge in the water and beverage industries that certain ubiquitous organic chemicals interact with biological receptors and result in obvious biological responses at levels below the lowest currently measurable levels for APIs in water. For example, objectionable taste or odor can be imparted to water at concentrations down into the tens of picograms per liter (nearing a part per quadrillion, pg/L). Examples of chemicals that can elicit such taste or odor effects at sub-ppt levels include: geosmin, guaiacol, 2, 4, 6-trichloro- and 2, 4, 6-tribromo-anisole, and 2-isobutyl- and 2-isopropyl-3-methoxypyrazine. This incongruity in attempting to convey a lack of significance in low concentrations of APIs can only lead to confusion.

Finally, even if concentrations might currently be considered extremely low, continued reductions in water supply (from extended drought) and increasing incidence of effluent-dominated receiving waters, punctuated with raw sewage overflow events, could inevitably lead to higher concentrations of all trace contaminants (because of less dilution). The problem of diminished stream flows is further exacerbated when sewage escapes untreated such as in locales where straight-piping (direct discharge without treatment) of raw sewage is still practiced.

If we acknowledge that it may not be possible to sustain a chemical-by-chemical approach to regulating or even for monitoring environmental contaminants, then we need to examine if there are simple ways in which to at least make continual progress in lessening, and preferably minimizing, exposures. Even in the absence of understanding the extent and magnitude of the actual chemical-exposure universe, certain actions aimed at reducing exposure to APIs could have unrelated collateral benefits, greatly increasing their cost-effectiveness.

Most of the pollution prevention measures that could be implemented to reduce the introduction of ingredients from pharmaceuticals to the environment stand a good chance of also reducing the introduction of other substances, some of which we may not even be currently aware. A reachable goal would be to educate the public to the fact that ALL of their actions, activities, and behaviors can impact our shared environment. This major learning point can be reinforced by providing guidance on the environmentally sound and safe disposal of all consumer items. The immediate outcome sought is to redirect consumer wastes down the avenues with the smallest ecological footprints. The ultimate outcome is to continually reduce unneeded, imprudent consumption.

Likewise, all treatment technologies (none of which was ever specifically designed for removing APIs from wastewater, sewage sludge, or drinking water) act by broad mechanisms that span many chemical classes. By controlling one class, control of others will also occur. Money spent in removing one class of pollutants is highly leveraged because it can also facilitate the collateral removal of other contaminants – some of which, including certain APIs, undoubtedly continue to elude the detection prowess of analytical chemists.

Examining the life cycle of a medication perhaps reveals the most important aspect of why we should care about trace levels of APIs in the environment. PiE actually serves to dramatize wasted healthcare resources and lost opportunities to achieve therapeutic objectives. Our focus to date has tended to dwell on discrete aspects of the life cycle of APIs – limited primarily to establishing environmental occurrence and studying source control (waste and drinking water treatment) and the potential for aquatic effects. The actual origin of the problem and possible solutions, however, have garnered significantly less attention, with the one exception being the design of approaches for dealing with unwanted, leftover medications; in the U.S., the first federal guidance for consumer disposal of unused drugs was issued in February of 2007 by the White House Office of National Drug Control Policy and subsequently elaborated upon by the US FDA (http://www.fda.gov/Drugs/ResourcesForYou/Consumers/BuyingUsingMedicineSafely/EnsuringSafeUseofMedicine/default.htm). Important to note, however, is that the magnitude of the contribution of APIs to the environment via disposal (versus excretion) is unknown and probably varies dramatically across APIs (as a function of the pharmacokinetics, delivery routes, and compliance rates for the respective medications). While the proper disposal of unwanted drugs might be important for reducing the unnecessary entry of at least certain drugs to the environment, it may never achieve dramatic reductions in the overall magnitude of PiE. Prudent disposal is perhaps more important for minimizing the very real problem of human morbidity and mortality due to diversion of drugs from accumulated stockpiles of medications and those medications imprudently discarded to trash, and the consequent poisonings of humans and pets.

Beyond the proper disposal of unwanted drugs, the ultimate focus with regard to pollution prevention is perhaps better placed on the way in which drugs are marketed, prescribed, and dispensed. Healthcare systems could be re-designed so that only the optimal medications are prescribed (for correct and therapeutically justified purposes) in minimal doses individualized for each patient, and dispensed in quantities and for durations to facilitate their full consumption; the potential for environmental impact can also be considered in selecting medications. The ideal outcome would be a profound reduction in the types and quantities of leftover drugs requiring disposal; see http://dx.doi.org/10.2165/0002018-200831120-00004. Little recognized is that redesign of prescribing and dispensing practices could also lessen the quantities of APIs that gain entry to the environment as a result of their intended use – e.g., simply by reducing excretion as a result of lower doses or shorter courses. Many additional factors beyond the prescribing and dispensing chains play roles, such as patient compliance and direct-to-consumer advertising (in the US and Australia).

The end result of a greener healthcare system could be not just a cleaner environment, but also more efficient usage of healthcare resources, reduced healthcare costs, improved therapeutic outcomes, and reduced incidence of purposeful abuse and accidental poisonings from diversion of stockpiled or improperly disposed drugs. The health of humans is indeed intertwined with that of the environment. Perhaps the healthcare community should develop a credo analogous to the spelunker's – something akin to "Prescribe and dispense only what's needed – no more. Leave no medication behind."

NOTICE

The US Environmental Protection Agency (EPA), through its Office of Research and Development (ORD), funded and performed the research described. This manuscript has been subjected to the EPA's peer and administrative review and has been approved for publication.

Chapter 1

Introduction

Benoit Roig and Evelyne Touraud

1.1 PHARMACEUTICAL IN THE ENVIRONMENT: BACKGROUND

At a time when sustainable development is an increasingly important consideration, questions of safety and environmental protection are becoming more and more embedded in all industrial activities. Decision No. 2455/2001/EC in accordance with the Water Framework Directive established a list of "priority substances selected from amongst those that present a significant risk to, or via, the aquatic environment". This list is based on the toxicity, persistence, bioaccumulation potential, human health risk and the monitored and modeled concentration of each substance in the aquatic environment. Some other substances (emerging pollutants) can be considered as a complementary list likely to be of relevance to the WFD in the future. In 2002, United States Geological Survey (USGS) released the first nationwide study of the occurrence of pharmaceuticals, hormones and other organic wastewater contaminants in US streams (Kolpin *et al.*, 2002). 95 emerging contaminants potentially associated with human, industrial and agricultural waste have been monitored from 139 streams across a wide range of setting in 30 states. Since 2002, USGS has published more than 160 reports that document and demonstrate the existence of these substances in US waters, the sources of these substances, the assimilation of some of these by organisms (Kinney *et al.*, 2008) and adverse ecological health effect (Vajda *et al.*, 2008).

Many of these compounds may be found at low concentrations in treated wastewater and sometimes in drinking water. Although this is a relatively novel area of research, there are some concerns that exposure to emerging pollutants may increase cancer risks (Sharpe and Skakkebaek, 1993; Cantor *et al.*, 2006, Richardson *et al.*, 2007) as well as physiological changes in humans and animals (Xu *et al.*, 2005; Ghisari *et al.*, 2005; MacLusky *et al.*, 2005; Sawai *et al.*, 2003; Hong *et al.*, 2004; Biau *et al.*, 2007). Of particular concern, is human and animal exposure to so-called Pharmaceutical and Personal Care Products (PPCPs). These compounds are defined by Daughton and Ternes (1999) as a very broad, diverse collection of thousands of chemical substances, including prescription and over-the-counter therapeutic drugs, fragrances, cosmetics, sun-screen agents, diagnostic agents, nutraceuticals, biopharmaceuticals, and many others. This broad collection of substances refers, in general, to any product consumed by individuals for personal health or cosmetic reasons.

Pharmaceuticals in the Environment is no longer a new issue but recently it has become a priority concern, particularly for politicians and the general public. Indeed, the occurrence of active pharmaceutical ingredients in water bodies is an undesired side effect of their normal use. In order for the substances to develop their intended effects in the human body, sufficient intact molecules must reach the diseased cell before they are degraded into a multitude of metabolites by the body's biochemical processes. In order to achieve this goal, pharmaceuticals are optimized for stability. This has two consequences: on one hand, the active ingredients are not metabolized completely in the human

 Pharmaceuticals in the Environment: Current knowledge and need assessment to reduce presence and impact. Edited by Benoit Roig. ISBN: 9781843393146. Published by IWA Publishing, London, UK.

body but excreted primarily via urine and thus reach domestic wastewater; on the other hand, the desired stability of the molecules hinders their biological degradation in conventional sewage treatment plants. Many of the active ingredients studied so far become only partially removed, others not at all. On reaching the rivers and lakes in the effluent of the sewage treatment plants, the stable molecules can then make their way into the groundwater and finally – via drinking water – back to humans.

Contrary to the initial belief, domestic wastewater (human activity) is the main source of occurrence of active pharmaceutical ingredients in the environment. Alongside, pharmaceutical production sites and hospitals or other medical facilities play a lesser role. Whereas – at least in Europe – residues are generally largely removed from contaminated production wastewaters by means of costly technology on-site, the contribution of hospitals is maximum 10 to 20 per cent of the total quantity of pharmaceuticals released into the environment each year.

Substantial amount of work has already been done since the 1990s both in Europe and other parts of the world. Figure 1.1 shows the exponential expansion of scientific literature in the topic of pharmaceuticals in the environment since 1991.

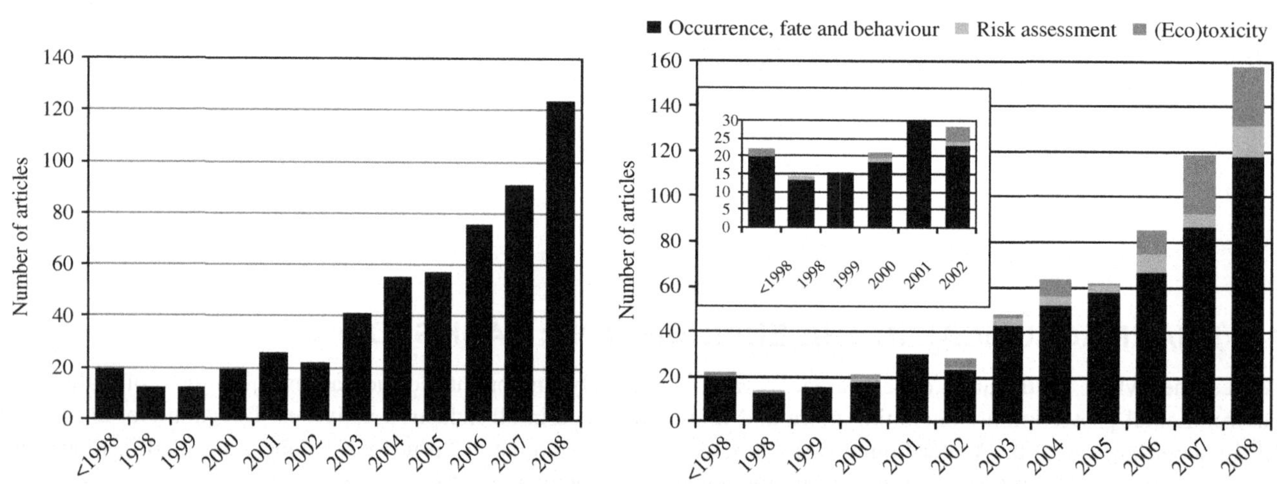

Figure 1.1 Evolution of the scientific production concerning pharmaceuticals in the environment between 1991 and 2008 (around 550 articles).

Regarding the tackled fields, 80% of the articles deal with one of the following aspects, occurrence, fate and behaviour of pharmaceuticals within treatment plants and in environmental waters. Since the last five years, studies regarding the impact of pharmaceuticals in the environment (eco-toxicity) and the assessment of their potential environmental risk are growing. However, this research has been fragmentary, resulting in a weak connection between, the different data.

During the Knappe project, an integrated approach enabling a holistic understanding of the issue of pharmaceutical products in the environment as a whole has been implemented. The approach is based on the whole life cycle of a pharmaceutical product (from the manufacture to the presence in the environment and the exposure of aquatic and terrestrial organisms) by considering the available data and the actors of each level of the cycle. Knappe project has initiated discussions with the concerned groups of experts (pharmaceutical industry, water managers, healthcare community, patients, regulatory institutions) and provided the information to all stakeholders (Figure 1.2).

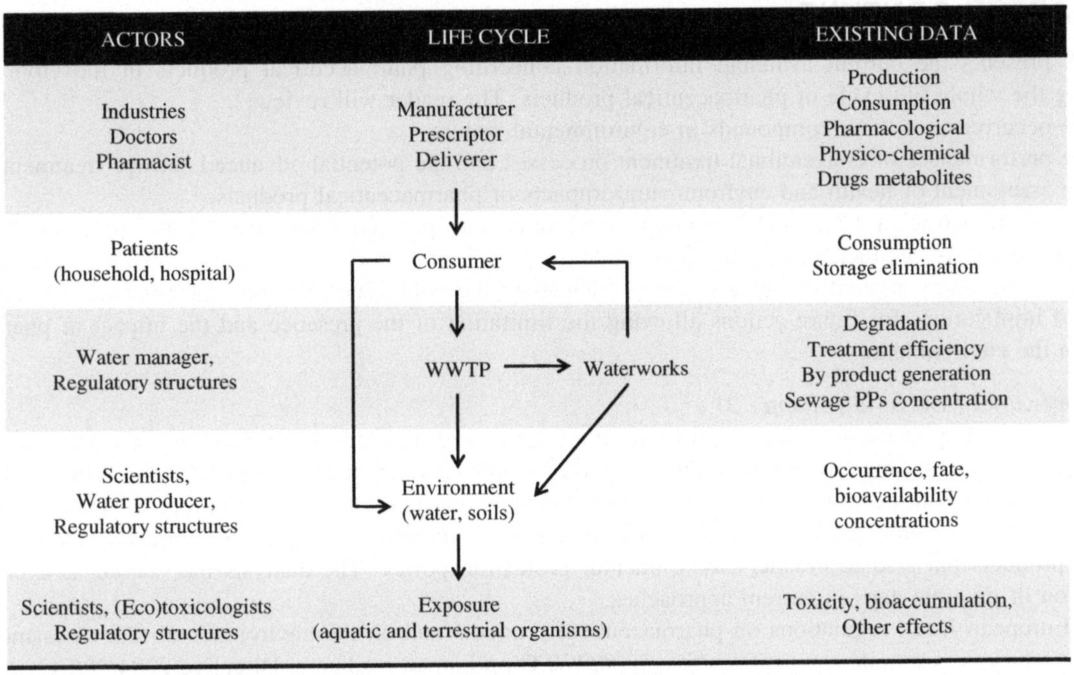

Figure 1.2 Integrated approach for pharmaceutical products in the environment.

1.2 OBJECTIVES OF THIS BOOK

The objectives of this book is to review the existing information concerning the presence of pharmaceutical products in the environment and to put emphasis on questions deserving attention by integrating results of previous and on going EU projects and published data from both governmental sources and scientific literature. In particular the following question deserved specific attention:

- What is the list of most relevant pharmaceutical products (PPs) for the aquatic environment? Which indicators for supporting environmental managers, health authorities?
- What is the efficiency of urban and industrial sewage treatment plants over a year? What is the fate and behaviour of PPs in sewage treatment plants? If receiving waters are used for potable water supplies, does the presence of these compounds represent a potential hazard to human health?
- Could we solve some problems by environmental or cleaner technologies?
- What regulatory approaches, incentives, prevention actions can be implemented in order to lower PPs concentration in the environment? Does a European practical guidance can be developed?
- Can the impacts of PPs on the environment be reduced through the use of eco-pharmacostewardship approaches including the use of clean synthesis, classification and labelling, and better communication of methods of 'good practice'?
- How can we better monitor the environmental impact of a pharmaceutical once it has received a marketing authorisation?
- How can we manage the discharge of PPs in the environment.

This work was implemented in a European Specific Support Action (KNAPPE, n°036864) supported by EU Commission (DG Research) as part of the 6th Framework Program. During 20 months, the project has identified and compiled current knowledge, information and legislation concerning protection of the environment from emitted residues of pharmaceutical products. With the collaboration of the main actors of the pharmaceutical products life cycle (industrials, consumers, medical actors, regulatory organisations, water managers, ecotoxicologists, environmentalists), it allows to discuss the main issues and to propose a couple of recommendation to lower the presence and impact for the environment.

The following chapters are extracted from the Knappe project deliverables and make a global synthesis of the available data concerning in particular occurrence, detection, fate, behaviour, removal treatments, environmental and health impacts. Finally, some recommendations will be proposed to the reader who, as scientist or simple consumer, can act at different level to improve the consideration of the problematic.

1.3 CHAPTER CONTENT

This book presents the current available information concerning pharmaceutical products in the environment by considering the whole life cycle of pharmaceutical products. The reader will review:

- the occurrence of these compounds in environmental waters,
- the performances of conventional treatment processes and the potential advanced sewage treatment plants,
- the assessment of health and environmental impacts of pharmaceutical products,
- Eco-pharmacostewardship and eco-pharmacovigilance as pro-active approaches for limiting discharge of pharmaceutical products in the environment.

All these aspects are covered in the seventh first chapter of the book. The remainder of this book will focus on the outputs and implications for future actions allowing the limitation of the presence and the impact of pharmaceutical products in the environment.

Policy framework at EU level (Chapter 2)

Key existing policy instruments and approaches at EU level and in selected Member States, which are relevant to limiting the discharge of pharmaceutical products into the water environment are reviewed. Policies treated in this state of the art addressed issues of authorization of pharmaceutical products, pollution prevention, wastewater treatment as well as monitoring of environmental quality. The review paid attention to regulations directly relevant to human medicinal products but also to broader environmental protection policy. The analysis has served as a basis for the identification of possible gaps in current approaches.

At the European level, regulations on pharmaceutical products and on the environmental risk assessment in their authorization process are quite recent: Directive 2004/27/EC on human medicine, Directive 2004/28/EC on veterinary medicine. Although separate regulations exist for "human medicinal products" and "veterinary medicinal products", these two categories of pharmaceuticals are addressed in consecutive regulations which are based on the same principles and are highly similar in content. According to Directive 2004/27/EC on human PPs, for all new authorizations of pharmaceuticals, the environmental effects must be examined and this assessment must accompany the authorization application. Prior to 2004, detailed environmental risk assessment was only carried out in exceptional cases for human medicine. However, the granting of a marketing authorization of human medicine cannot be refused using the environmental impact as criterion.

For veterinary compounds, the situation is different. Contrary to human pharmaceuticals, the granting of a marketing authorization of a veterinary medicine can be denied due to an unacceptable risk for the environment. The 2004 EU regulatory amendments indeed introduced certain key changes in the authorization of veterinary medicines. The criterion of environmental safety has been given the same weight as consumer safety in the concluding risk-benefit assessment and, therefore, can decide about the authorisation or non-authorization of a new veterinary pharmaceutical product.

Concerning, *Environmental protection regulations and/or EU policies or initative*, no pharmaceutical products are included on the list of substances of particular interest to control and monitor. The main reason for this can be seen in the selection process for the first list of priority substances, which dates back almost 10 years and was based on already existing official lists of pollutants. Pharmaceutical products, however, are often referred to as emerging pollutants, which means that their presence in and impact on the aquatic environment is just being discovered and researched.

Exposure bases on life cycle (Chapter 3)

Many tons of pharmacologically active substances are used every year to abort diseases, to prevent minor illness, or to face deeper hurt. Following proper therapeutic use the pharmaceuticals are often excreted via urine and feces (either unchanged or metabolized), directly washed off after dermal application or improper discharge and subsequently released into waste water. During sewage treatment many pharmaceuticals are not (totally) eliminated and end up in surface water. Thus, there is a multitude of input sources in environment with no evident quantitative data available (concerning the relative distribution of pharmaceuticals from all emission sources). Finally, drinking water production may be influenced when surface or bank filtrated water is used as raw water source. Because pharmaceuticals have been designed to have a specific physiological effect on humans or animals, the compounds may pose a potential environmental threat. Therefore the calculation and or the measurement of pharmaceutical emission in the aquatic environment is of special interest. In this context, the chapter will present the main modes of measurement and calculation currently used to assess the real exposure of aquatic and terrestrial population.

Real situation: occurrence of the main investigated PPs in water bodies (Chapter 4)

An extensive data compilation on the environmental occurrence of PPs was created, including 58400 measurements of 178 pharmaceutical products in 22 countries in wastewater treatment plant influents, and effluents, surface water, groundwater, bank filtrate, marine water and drinking water. For each country, average mean environmental concentrations (A-MECs) were calculated from the compiled data for each molecule, regarding the seven aquatic matrices, when measurements were available. Additionally, consumption data of drugs for five European countries (France, Germany, Poland, Spain and the UK) have been used to predict their environmental concentrations following the approach used in the EU-Project Poseidon.

A comparison revealed that predicted and averaged measured environmental concentrations matched well for wastewater treatment plant influents. For effluents and surface waters, the agreement of the data was limited to some extent. Variability of the STP treatment efficiency and transformation processes in surface waters for example were not covered by the prediction model and can explain the differences.

The data compilation was used to establish a set of indicator substances for the determination of the wastewater share in surface waters and for determining whether a water body is influenced by poorly or raw wastewater. The presence of these indicator substances which are readily (bio)degradable in wastewater treatment, in surface waters indicate the input of non or poorly treated wastewater with potential consequences for water quality and water usage, concerning pollutants and pathogens. No pharmaceutical products could be classified as indicator substances specific for the input from hospitals or veterinary medicine. Potential indicator substances for hospital input either lacked the specificity of only being used and excreted in a hospital environment or their environmental concentrations were too low for a reliable quantification.

Due to the different registration situation in the various countries, with deviating and/or overlapping use in human and veterinary medicine, no pharmaceutical compounds are in general fully attributable to veterinary purposes only.

Currently, the available data on the consumption and the occurrence of pharmaceuticals in the aquatic environment on a European level is of very limited comparability. Efforts should be made to achieve a common book keeping system on the usage of all pharmaceutical products within the EU, including prescribed drugs, OTC drugs, and all compounds used for diagnosis and in hospitals.

Performance of conventional treatment processes (Chapter 5)

The efficiency of elimination of pharmaceuticals by conventional treatment processes and advanced treatment are assessed. In order to highlight the differences of the crucial parameters impact on the removal of these molecules, literature data was described according to the following factors: (i) Sludge retention time (SRT), (ii) Hydraulic retention time (HRT), (iii) Reactor configuration, (iv) Advanced technologies.In this frame, pharmaceutical compounds are tackled as therapeutic groups: antibiotics, anticonvulsants, anti-inflammatories (analgesic), β-blockers, hormones, tranquilizers, X-ray contrast media and lipid regulators.

According to the available data concerning wastewater treatment it can be stated, that sewage treatment plant (STP) (e.g. conventional activated sludge processes, membrane assisted bioreactors) configuration is not a factor, which have the highest impact on the drugs removal. It is rather connected with parameters, such as sludge retention time (SRT) and hydraulic retention time (HRT).

Biological monitoring and end points (Chapter 6)

Medicines play an important role in the treatment and prevention of disease in humans and animals. Even though the side effects on human and animal health have been widely documented, only recently have the potential environmental impacts of the manufacture and use of medicines been considered. The data review showed that there is now a large body of data that has been generated over the past decade on the effects of pharmaceuticals on aquatic organisms. Using this data, generally, pharmaceuticals show low acute toxicity to fish, daphnids and algae and while the majority of pharmaceuticals also show low toxicity in standard chronic studies, a number of pharmaceuticals are highly toxic in standard chronic tests. A comparison of standard acute and chronic test endpoints with available monitoring data indicate that, with a few exceptions, pharmaceuticals do not pose an unacceptable risk to the environment. The effects of a number of pharmaceuticals have also been assessed using non-traditional ecotoxicity tests and endpoints. The effect endpoints for the majority of these novel tests are many orders of magnitude lower than the traditional endpoints and existing uncertainty factors do not reflect these differences. Additionally, most of the novel endpoints can be linked to important ecological functions so can be regarded as ecologically relevant. However, even when novel data are included in the risk characterization, most pharmaceuticals seem to pose a low risk to ecosystem health. There are however some exceptions that may need further assessment or control in the future.

Review of environmental risk assessment and prioritization strategies for pharmaceuticals (Chapter 7)

Prioritization of human pharmaceuticals is necessary due to the high number of pharmaceuticals used, which hinders the possibility to assess the eco-toxicity of every compound. To implement a relevant prioritization strategy, there is a need to accurately assess the environmental exposure and the environmental effects. From the review of ERAs and prioritization strategies conducted in the last ten years, it is possible to highlight the following parameters and to propose some concluding remarks:

Regarding exposure assessment:

- The use of simple models, as EMEA model, to calculate PECs for surface water is in general in good agreement with field measurements.
- Accurate, consumption amounts are essential, but data are sometimes unavailable, depending on the country.
- Metabolism data and excretion rates are essential but data are often incomplete or unavailable.
- STP removal rates are lacking which is a major limitation of the accuracy of ERA.
- PEC for other compartments than water column is not well assessed.
- Bioaccumulation of pharmaceuticals remains to be more studied.

Regarding effect assessment:

- Chronic eco-toxicological data are lacking.
- Due to gaps in ecotoxicological data for pharmaceuticals (due to the high number of pharmaceuticals), other ways of assessing the effects of pharmaceuticals have to be investigated and validated.
- Pharmacological data can be useful to estimate the biological effects on aquatic organisms however, (i) the access to such data is sometimes not possible, (ii) the relevance of such data for environmental considerations remains to be confirmed.
- Investigation of the evolutionary conservation of drug targets is important information that can help for a relevant use of pharmacological data, and for targeting sensitive species in bioassays.
- QSAR models should be investigated with regard to pharmaceuticals, which are compounds with specific mechanisms of action.

Thus, it becomes necessary to harmonize the different prioritization strategies and models now available in European countries. As pharmacological and toxicological data are not easily available, the ERA of old pharmaceuticals at least, could greatly benefit from the development of a database on existing pharmacological and ecotoxicological data.

Review of available information and discussion with the main actors of the life cycle of medicine showed that the presence of pharmaceutical products in the environment is not inevitable but can be limited and mitigated by different anticipating and/or reacting actions. The involvement of all is required to achieve these objectives. The two last chapters of this book seek to explore ways in which the impact of pharamaceuticas on the environment can be reduced at all stages in their lifecycle (production, prescription, use and disposal) through the use of eco-pharmacostewardship and eco-pharmacovigilance approaches.

Recommendations on research and development (Chapter 8)

In order to assess a risk to human beings or the environment it is necessary – in the classical approach – to establish a causal relationship between an agent and an observed harmful effect. In risk assessment of chemicals generally and particularly in the case of pharmaceutical products, this is manifestly possible only in rare cases, especially due to the lack of some scientific and technical data. A number of areas have been identified where it is difficult to obtain the information necessary to carry out a reliable risk assessment. One fundamental area is the lack of information on volumes of some compounds used in some countries. Reasonably, reliable information can be obtained for prescribed drugs that are still within patent, but it becomes increasingly difficult for generic drugs, and over the counter products in most European countries, with the exception of Sweden where pharmacies are state controlled. Work is needed in a number of areas to provide reliable representative information on (i) drug development, (ii) behaviour (removal efficiency, partitioning between water and solids) in STPs, (iii) potential for transfer from sludge to soil to ground and surface waters, (iv) behaviour in surface and ground waters, (v) toxicity to aquatic and soil organisms, and related potential for environmental harm. In order to obtain a complete picture of the life cycle of a medicine it is necessary to obtain reliable and representative information on levels of medicines in various divisions of the pathway to and within the aquatic environment, including manufacture and distribution, end use, STPs, waste water, surface and ground waters, drinking water, sludge, soil, soil and aquatic biota. It is not feasible to obtain detailed information on all of the medicines in use, or to monitor them, even only in surface waters. It will therefore be necessary to assess potential risk on the basis of available information in order to prioritize research and monitoring efforts.

Recommendations on communication and education measures (Chapter 9)

Contrary to other product areas, environmental aspects have previously hardly played a role in prescribing or purchasing, use or disposal of medicinal products. In the scenario of the substance life cycle, however, the phase of handling drugs also provides approaches for effective options of action for reducing the contamination of water bodies. In general, these can be implemented at two points: On the one hand in reducing and/or modifying the consumption of pharmaceuticals, and on the other hand in the avoidance of medicinal wastes that may be improperly disposed of via domestic wastewater (or in ensuring proper disposal of medicinal products, respectively).

The use of medicinal products is associated with high and undisputed individual and social benefit. Moreover, a medical decision for a medicinal therapy generally depends on a number of factors that must be weighed against each other. Here, in each individual case, priority is given to the best possible strategy for curing or palliation of a disease. When in doubt, the rule is "health protection before environmental protection". At the same time, it must also be assumed that further regulations in an already heavily regulated system that are aimed at an objective that is perceived to be secondary will encounter opposition particularly with physicians. Measures that are intended to make the handling of drugs more environmentally compatible must therefore satisfy at least three prerequisites: (i) they must not result in a quality loss in prevention and therapy; (ii) they must be integrable into the complex professional routine of physicians and pharmacists; (iii) they should co-operate with already societal desirable plans for reform in the healthcare system.

Behavioral changes that should contribute to a reduction in contamination of water bodies, however, require a corresponding awareness, communication and education of the problem, particularly if they are to be implemented voluntarily.

REFERENCES

Biau, S., Bayle, S., de Santa Barbara, P. and Roig, B. (2007) The Chick embryo: an animal model for hormonal compound effect detection. *Analytical and Bioanalytical Chemistry* **387**(4), 1397–1403.

Cantor, K.P., Ward, M.H., Moore, L.E. and Lubin, J. (2006) Water Contaminants. In *Cancer Epidemiology and Prevention*, 3rd edition (eds D. Schottenfeld, J.F. Fraumeni Jr), Oxford University Press, New York.

Daughton, C.G. and Ternes, T.A. (1999) Pharmaceuticals and personal care products in the environment: agents of subtle change? *Environ. Health Perspect.* **107**(suppl 6), 907–938.

Ghisari, M. and Bonefeld-Jorgensen, E.C. (2005) Impact of environmental chemicals on the thyroid hormone function in pituitary rat GH3 cells. *Mol. Cell Endocrinol.* **244**(1/2), 31–41.

Hong, C.C., Shimomura-Shimizu, M., Muroi, M. and Tanamoto, K. (2004) Effect of endocrine disrupting chemicals on lipopolysaccharide-induced tumor necrosis factor-alpha and nitric oxide production by mouse macrophages. *Biol. Pharm. Bull.* **27**(7), 1136–1139.

Kinney, C.A., Furlong, E.T., Kolpin, D.W., Burkhardt, M.R., Zaugg, S.D., Werner, S.L., Bossio, J.P. and Benotti, M.J. (2008) Bioaccumulation of pharmaceuticals and other anthropogenic waste indicators in earthworms from agricultural soil amended with biosolid or swine manure. *Environ. Sci. Technol.* **42**(6), 1863–1870.

Kolpin, D.W., Furlong, E.T., Meyer, M.T., Thurman, E.M., Zaugg, S.D., Barber, L.B. and Buxton, H.T. (2002) Pharmaceuticals, hormones, and other organic wastewater contaminants in U.S. streams, 1999–2000: a national reconnaissance? *Environ. Sci. Technol.* **36**, 1202–1211.

MacLusky, N.J., Hajszan, T. and Leranth, C. (2005) The environmental estrogen bisphenol a inhibits estradiol-induced hippocampal synaptogenesis. *Environ Health Perspect* **113**(6), 675–679.

Richardson, D., Plewa, M.J., Wagner, E.D., Schoeny, R. and DeMarini, D.M. (2007) Occurrence, genotoxicity, and carcinogenicity of regulated and emerging disinfection by-products in drinking water : a review and roadmap for research. *Mutation Research* **636**, 178–242.

Sawai, C., Anderson, K. and Walser-Kuntz, D. (2003) Effect of bisphenol A on murine immune function: Modification of interferon, IgG2a, and disease symptoms in NZB × NZW F1 mice. *Environ. Health Perspect.* **111**(16), 1883–1887.

Sharpe, R.M., Skakkebaek, N.E. (1993) Are oestrogens involved in falling sperm counts and disorders of the male reproductive tractTM? *Lancet* **341**, 1392–1395.

Vajda, A.M., Barber, L.B., Gray, J.L., Lopez, E.M., Woodling, J.D. and Norris, D.O. (2008) Reproductive disruption in fish downstream from an Estrogenic wastewater effluent. *Environ. Sci. Technol.* **42**(9), 3407–3414.

Xu, L.C., Hong, S., Chen, J.F., Bian, Q., Qian, J., Song, L. and Wang, X.R. (2005) Evaluation of androgen receptor transcriptional activities of bisphenol A, octylphenol and nonylphenol *in vitro*. *Toxicology* **216**(2/3), 197–203.

Chapter 2

Policy framework at EU level

Eleftheria Kampa, Rodrigo Vidaurre, Cornelius Laaser and Thomas Dworak

2.1 INTRODUCTION

This chapter reviews key existing policy instruments and approaches at EU level, which are relevant to limiting the discharge of pharmaceutical products (PPs) into the aquatic environment. EU policies treated in this review address issues of authorisation of PPs, pollution prevention, wastewater treatment as well as monitoring of environmental quality. The analysis serves as a basis for the identification of potential gaps in current approaches. Future policy action might build on this assessment in order to reduce the risk of harmful impacts to the environment.

The chapter starts off with an overview of the EU policy framework relevant to the presence of PPs in the environment. The review puts PPs into the broader context of European environmental protection policies but also pays particular attention to regulations directly relevant to medicinal products. In this context, it seeks to establish links to important water and other natural resource protection policies, such as the EU Water Framework Directive, the Groundwater Directive or the European Soil Strategy. The presentation of EU medicinal regulations, including environmental risk assessment requirements for PPs, is kept concise and refers to the extensive publications of other scholars on this topic. Policies on animal medicine are treated less extensively and only in so far as they are relevant to the water environment.

In the final section, an integrated discussion of the current policy framework is provided, based on the findings of the present review as well as existing expert literature and discussion papers.

2.2 POLICY FRAMEWORK AT EU LEVEL

The following outlines the key elements of the EU policy framework, which are relevant to current discussions on the presence of PPs in the aquatic environment. Firstly, the development of regulations for the authorisation of medicinal products is described; progressively these regulations have given more emphasis to the environmental risk assessment of PPs. Secondly, the section explores existing and potential links of PPs and current EU policies on environmental and water protection.

The status of the information presented on the EU policy framework is as of April 2008 and might be subject to change according to new EU policy developments.

2.2.1 Medicinal regulations & policies

2.2.1.1 Product authorisation and environment risk assessment

In the EU, an environmental risk assessment (ERA) for animal and human pharmaceuticals during product authorisation was first established in regulation in the early 1990s. Currently, ERA is regulated by Directives 2004/27/EC on human medicine, 2004/28/EC on veterinary medicine and the new regulation EC/726/2004 for medicine containing or consisting of genetically modified organisms. These directives mainly address environmental risk in the consumption phase of PPs. The minimisation of environmental impacts during the production phase of PPs is governed by the EU IPPC Directive 96/61/EC on Integrated Pollution Prevention Control from industrial sources.

According to Directive 2004/27/EC on human PPs, environmental effects must be examined for all new authorisations of PPs and this assessment must accompany the authorisation application. The environmental risk assessment typically involves the generation of data on the environmental exposure and ecotoxicity of PPs. Prior to 2004, detailed environmental risk assessment was only carried out in exceptional cases for human medicine (Holzmann, 2005). However, the granting of a marketing authorisation of human medicine cannot be refused using the environmental impact as criterion.

For veterinary PPs regulated under 2004/28/EC, the situation is different. Contrary to human PPs, the granting of a marketing authorisation of a veterinary medicine can be denied due to an unacceptable risk for the environment (Koschorrek and Apel, 2006). The 2004 EU regulatory amendments indeed introduced certain key changes in the authorisation of veterinary medicines. The criterion of environmental safety has been given the same weight as consumer safety in the concluding risk-benefit assessment and, therefore, can decide about the authorisation or non-authorisation of a new veterinary PP (Holzmann, 2005).

In addition to the legal requirements set in the directives listed above, specific guidelines are available which recommend the scope and the actual procedure for the environmental assessment of medicinal products. Currently, the ERA of human pharmaceuticals is based on the Guidelines of the European Agency for the Evaluation of Medicinal Products (EMEA)[1], while the environmental impact assessment (EIA) of veterinary pharmaceuticals is based on the VICH guidance (International Cooperation on Harmonisation of Technical Requirements for Registration of Veterinary Products).[2]

Also in the US, an environmental risk assessment strategy is being implemented by the regulatory agency FDA (Food & Drug Administration) to assess the risk of human pharmaceuticals (see FDA, 1998). The ERA methodology of the FDA is roughly similar to that of the EMEA. The FDA methodology is not treated further in this chapter, since the focus is on the EU policy framework.

The current EMEA guidance document for human medicinal products came into effect in the end of 2006. According to this guidance, an ERA is required for all new drug marketing authorisation applications as well as for applications requesting major changes or extensions to existing authorisations which would result to increase of the environment's exposure to the drugs in question. An ERA is not required for renewals of authorisations or applications requesting minor changes to existing marketing authorisations.[3] There may also be cases in which the absence of an ERA could be justified, e.g. marketing authorisation applications for generic medicinal products. Vitamins, electrolytes, amino acids, peptides, proteins, carbohydrates and lipids are exempted because they are unlikely to result in significant risk to the environment. Similarly, vaccines and herbal medicinal products are also exempted due to the nature of their constituents. In case of marketing authorisation applications for radio-pharmaceutical precursors for radio-labelling and radio-pharmaceuticals, additional requirements on emission standards for radiation set by Council Directives 96/29/Euratom and 97/43/Euratom should be taken into account (CHMP, 2006).

In the EMEA guidance document, the ERA follows a 2-phased structure of assessment (Figure 2.1).

[1]Guideline on the environmental risk assessment of medicinal products for human use. European Medicines Agency. London, 01 June 2006. Doc. Ref. EMEA/CHMP/SWP/4447/00.

[2]Guidance for industry: Environmental Impact Assessment (EIAs) for veterinary medicinal products (VMPs). VICH International Cooperation for on Harmonisation of Technical Requirements for Registration of Veterinary Medicinal Products. Phase I (VICH GL6) and Phase II (VICH GL38).

[3]Minor changes to a drug marketing authorisation are in fact notifications, i.e. applicants just inform regulatory authorities and implement the changes without delay or approval. Examples are change in the name or address of the marketing authorisation holder or the manufacturer, change in the name of the medicinal product, change in the name of the active substance, minor change in the manufacturing process of the active substance, etc.

Examples of major changes to a drug marketing authorisation include change in the manufacturing process of the active substance that could affect efficacy, deletions in security information, deletions of contraindications, changes in adverse reactions, etc.

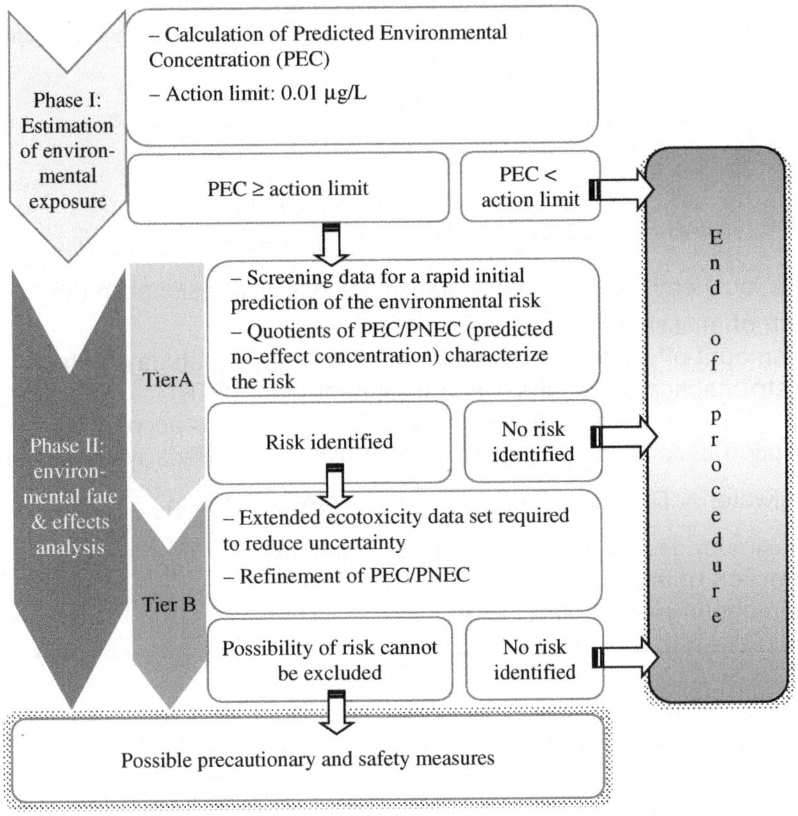

Figure 2.1 Environmental risk assessment for human pharmaceutical products (based on CHMP, 2006).

In Phase I, the drug concentration expected to occur in the aquatic environment is calculated (predicted environmental concentration ($PEC_{surfacewater}$)). If the PEC is below a defined action limit of 0.01 µg/l, it is assumed that this specific medicinal product is unlikely to represent a risk for the environment. Thus, the assessment procedure does not continue. If the calculated PEC is equal or above this action limit, a Phase II environmental fate and effects analysis is required. In some cases, the action limit may not be applicable. Some drug substances may affect the reproduction of vertebrate or lower animals at concentrations lower than 0.01 µg/l. These substances have to enter Phase II and a tailored risk assessment strategy must be followed that addresses its specific mechanism of action.

Phase I:

$$PEC_{surfacewater} = \frac{DOSEai \times Fpen}{WWinhab \times Dilution}$$

DOSEai: Maximum daily dose consumed per inhabitant (mg/hab/day)
Fpen: Percentage of market penetration (default value = 0.01)
WWinhab: Amount of wastewater per inhabitant per day (l/hab/day; default value = 200)
Dilution: Dilution factor from sewage treatment plants to surface water (default value = 10)

In Phase II, information on the physical, chemical and toxicological properties are obtained and assessed in relation to the extent of the environmental exposure. Phase II is split into Tier A and Tier B. A screening information base set of data for which Tier A is investigated allows for a rapid prediction of the environmental risk linked to the use of the product. Quotients of PEC/PNEC characterise the risk for the aquatic compartment and for microorganisms. PNEC is the predicted no-effect concentration, which is derived from the lowest no observed effect concentration (NOEC) or dose-response result of a set of laboratory ecotoxicity studies representing the aquatic or terrestrial compartment. Certain additional assessment criteria, such as half-life biodegradation, are given with fate data. If at Tier A no risk is identified, the assessment is complete. If at Tier A risk is identified, the subsequent Tier B level requires an extended ecotoxicity data set, which should allow for a reduction of uncertainty and addressing data gaps. Tier B is the ultimate step in the risk assessment and resembles an iterative process of PEC and PNEC refinement. $PEC_{surfacewater}$ can be refined as:

Phase II:

$$PEC_{surfacewater} = \frac{Elocalwater \times Fstp}{WWinhab \times CAPACITYstp \times Factor \times Dilution}$$

Elocalwater: local emission to wastewater of the relevant residue (mg/d)
Fstp: fraction of emission directed to surface water
WWinhab: amount of wastewater per inhabitant per day (l/hab/day; default value = 200)
CAPACITYstp: capacity of local sewage treatment plant (STP)
Factor: factor taking the adsorption to suspended matter into account
Dilution: dilution factor from sewage treatment plants to surface water (default value = 10)

where Elocalwater = DOSEai × Fexcreta × Fpen × CAPACITYstp

DOSEai: maximum daily dose consumed per inhabitant
Fexcreta: fraction of the active substance excreted
Fpen: percentage of market penetration

As concerns the specific environmental compartments targeted by the ERA for human medicine, it should be noted that in Phase I, the PEC calculation is restricted to the aquatic compartment. In Phase II (Tier A), standard long-term toxicity tests on fish, daphnia and algae are carried out to determine the $PNEC_{water}$. An exposure assessment for groundwater is also required, since entry into the groundwater via bank filtration is considered. Furthermore, the fate of the substance in the sewage treatment plant (STP) is investigated via a biodegradability test. The fate of substances which are not readily biodegradable have to be investigated in a water-sediment study. It is assumed that a substance with high adsorption coefficient is retained in the STP and may reach the terrestrial compartment with land spreading of sewage sludge. In this case, an environmental assessment of the drug substance in the terrestrial compartment has to be conducted, unless the substance is readily biodegradable. The terrestrial risk assessment complements the aquatic risk assessment and does not replace it (CHMP, 2006).

In general, in Tier B of Phase II, a base set of tests investigating biodegradation in soil, toxicity to soil invertebrates and acute effects on terrestrial plants and microorganisms is conducted. If a substance is not ready biodegradable and if the results from the water-sediment study demonstrate significant shifting of the drug substance to the sediment, effects on sediment organisms have to be investigated (CHMP, 2006).

In the end of the ERA procedure, if the possibility of environmental risk cannot be excluded, specific arrangements to limit risk should be envisaged on a case-by-case basis (possible precautionary and safety measures). In any event, in the case of human medicine, environmental impacts can not constitute a criterion for refusal of a marketing authorisation according to Directive 2004/27/EC.

More details on the ERA process of human medicine can be found in the EMEA 2006 guideline (CHMP, 2006) as well as in the literature (Kümmerer, 2004; Koschorreck and Apel, 2006; Umweltbundesamt, 2005).

For veterinary medicinal products, first draft EU guidelines on their EIA existed already in the early 1990s. The VICH guideline on Phase I assessment currently in force dates back to 2001. In addition, in 2005, the VICH Phase II guidelines came into force in the EU. According to the present guidelines for veterinary medicine, an EIA is mandatory for all new applications, independent of the application procedure (central or national marketing authorisation) and type ("full", "generic" etc.), and is therefore required for all marketing authorisations submitted in the EU irrespective of the underlying legal basis. In respect to renewals, the legal provisions require "a re-evaluation of the risk-benefit balance". Further guidance on the interpretation of the data requirements, in particular in respect to marketing authorisation applications for generics, extensions and variations, as well as for renewals, is under development (CVMP, 2007).

The EIA for veterinary medicine follows the same 2-phased structure (Phase I and Phase II with Tiers A and B) as the ERA for human medicine (Figure 2.1). However, some key differences exist due to the different nature and areas of application of human and veterinary medicine.

For instance, in Phase I, the EIA for veterinary medicine is distinguished into an aquatic and a terrestrial branch, depending on whether the medicine is used to treat aquatic or terrestrial species. Distinct PECs are thus calculated for soil and water, while ecotoxicity studies are related respectively to aquatic species and terrestrial species (earthworms, microbes and plants). In the aquatic branch, the EIA continues beyond Phase I only if the environmental introduction concentration of the medicine ($EIC_{aquatic}$) is calculated to be more than 1 μg/L. In the terrestrial branch, the EIA continues beyond Phase I only if the predicted environmental concentration of the medicine in the soil (PEC_{soil}) is calculated to be more than 100 μg/Kg.

The Phase II guidance for veterinary medicine contains different sections for each of the following major branches: aquaculture, intensively reared terrestrial animals, and pasture animals. Exposure to both the terrestrial and aquatic compartment may be applicable to a particular medicine, depending on its route of environmental introduction. For instance, veterinary medicine administered to intensively reared animals has the potential to impact terrestrial non-target species directly, and non-target species indirectly in surface waters due to transport via water, including when adsorbed to soil particles and organic matter. In this case, due to possible leaching of active ingredients into groundwater, it is necessary to calculate PECs both for surface and groundwater. Veterinary medicine used to treat pasture animals may also impact terrestrial non-target species, as well as non-target species in surface waters. Therefore, also in the case of medicine for pasture animals, it is necessary to calculate PECs for both surface and groundwater.

In general, if there is evidence that there will be no exposure to a particular compartment (i.e. water, soil/sediment and dung), then it may be possible to waive studies for that compartment. However, sound scientific evidence must be presented in the marketing application dossier supporting the omission of these studies.

In the end of the EIA procedure for veterinary medicine, doubts about the acceptability of environmental risk can lead to changes of the characteristics of the product or of the area of application of the product, or even the refusal of marketing authorisation by the authorities.

2.2.1.2 Drug take-back schemes

EU medicinal legislation requires the set up of take-back schemes for unused and expired medicine in all Member States. These requirements are set out in Directive 2004/27/EC for human medicinal products (Article 127b). Reference to these collection systems is to be made on the labelling or package leaflet. These systems had to be set up by the end of October 2005. As concerns veterinary products, the requirement to have appropriate collection systems dates back to 2001 and was set out by Directive 2001/82/EC.

Take-back schemes are, on the one hand, a means to reduce the increased presence of pharmaceuticals in the environment. If not disposed of in a sound way, unused and expired medicine can enter the water environment either directly when flushed down the toilet or indirectly via landfill leachate when disposed of in domestic solid waste (landfill leachate may contaminate groundwater if the landfill is not properly sealed). On the other hand, take-back schemes also target the problem of unsafe pharmaceutical storage practices that can result in accidental poisonings and drug abuse. The extent of the establishment and the degree of effectiveness of take-back schemes for human drugs is quite different among European countries. In-depth assessments of scheme effectiveness (estimated recovery rate of unused/expired drugs) are missing for most countries. Individual available assessments often provide different and contradictory figures, pointing to the need for a more consistent approach that should be used in such evaluations. High levels of public awareness and education on the environmental consequences of the disposal of unused/expired drugs seem to be essential for the success of such schemes (Clark *et al.*, 2007).

2.2.2 Environmental protection regulations & policies

The following introduces key EU water policies, but also other EU environmental policies and initiatives which are related or could become related to the issue of PPs in the (aquatic) environment of Europe.

2.2.2.1 Water policies

Water Framework Directive (WFD)

The Directive 2000/60/EC (WFD) requires reaching good status in all EU waters by 2015. For surface water, "good status" is defined as "the status achieved by a surface water body when both its ecological status and its chemical status are at least good". In general, "good status" is defined as "slight changes in the values of the relevant quality elements as compared to the values found at closely undisturbed conditions". The requirement of the WFD for good chemical status of water refers to the concentration of substances that are harmful to the ecosystem.

During production, distribution, use, and through undeliberate loss, many chemical substances enter the aquatic environment. Depending on their physico-chemical properties, they can impair the ecosystem and pose a risk to human health. The WFD established two paths to minimize negative impacts of chemical substances on water bodies:

1. *Definition of a list of priority substances* that are considered to be particularly harmful to the aquatic environment on a European scale due to their toxicity, persistence, and bioaccumulation potential. Candidate substances for the list were

selected on a "list-based" approach, according to which the candidates were selected from official lists and monitoring programmes.[4] In summary, 658 substances were compiled (Klein *et al.*, 1999). More than 300 of them were subject to risk assessment and extensive consultations before a final list of priority substances was published (Decision 2455/2001/EC), which then became Annex X of the Directive. Annex X of the WFD lists 33 priority substances (mostly organic pollutants (e.g. pesticides), but also some heavy metals and others), for which the Commission has to define environmental quality standards according to Article 16(7). Member States must apply measures to progressively reduce the concentration of priority substances below the defined quality standards. In case of priority hazardous substances, measures have to aim at cessation or phasing-out of discharges.

At present, no pharmaceutical products are on the list of priority substances, with the exception of a few substances that have a broad spectrum of application and are also being used in the research and/or manufacturing of PPs.[5] The selection process for the first list of priority substances dates back almost 10 years and was based on already existing official lists of pollutants. PPs, however, are often referred to as emerging pollutants, which means that their presence in and impact on the aquatic environment is just being discovered and researched. According to WFD Article 16(4), the European Commission is asked to review the list of priority substances every four years. By definition, PPs that could be nominated for the list must exhibit toxic, persistent, and eventually bio-accumulative characteristics, and therefore have relevance on a supra-national scale. Moreover, there is still considerable uncertainty regarding the risk of long-term uptake of presumably not harmful substance traces. In this context, the WFD puts emphasis on the precautionary principle, stating that especially in identifying priority hazardous substances, any potential adverse effects of the product should be taken into account and should lead to a scientific assessment of the risk (recital 11 and 44, WFD).

In recent discussions on the proposed Directive of environmental quality standards amending the WFD,[6] three pharmaceutical substances (carbamazepine, diclofenac and iopamidol) were proposed as "subject to identification as possible priority substances".[7] However, in the final compromise reached between European Parliament and Council of Ministers, these pharmaceutical substances as well as several other proposed substances were dropped from the finally agreed list of substances to be subject to environmental quality standards under the new Directive.

2. *Requirement to identify all other pollutants that are discharged in significant quantities in the river basin or sub-basin.* Member States have to set quality standards for such river basin-specific pollutants, and take action to meet those quality standards by 2015 as part of good ecological status (Article 4, 11 and Annex V, WFD). This refers predominantly to substances listed in Annex VIII of the WFD which provides an indicative list of the main pollutants and inter alia covers *"substances and preparations, or the breakdown products of such, which have been proved to possess carcinogenic or mutagenic properties or properties which may affect steroidogenic, thyroid, reproduction or other endocrine-related functions in or via the aquatic environment."* (Annex VIII, No. 4, WFD).

If PPs that enter the aquatic environment in significant quantities pose a risk to a water body due to their physico-chemical properties, they have to be covered by this requirement. The prioritisation to identify Annex VIII substances is an on-going process, with the possibility of further substances being added (and their corresponding standards for environmental quality proposed). The process uses criteria on occurrence, persistence, toxicity and bioaccumulation in order to identify new substances. Oestradiol and ethinyl oestradiol have both been identified by the Environment Agency as *potential* UK specific pollutants. At the time of writing, work was underway to investigate whether PNECs (predicted no effect concentrations) can be developed for these substances. Based on the outcomes of this work and peer reviews, these substances may or may not become UK specific pollutants (pers. comm. Environment Agency UK, January 2008).

Consistently implemented, the two above-mentioned paths of the WFD provide the means to track down and minimise the effects of harmful chemical substances discharged into the aquatic environment. However, it should be mentioned that the comprehensive WFD mechanisms for maintaining and improving the overall quality of all waters in Europe can only work if the pressure on the aquatic environment and the related risk have been detected. Regarding

---

[4]List I and II of Council Directive 76/464/EEC, Annex 1A and 1D of the Third North Sea Conference (3. NSC), Priority lists 1–3 identified under Council Regulation No 793/93, OSPAR list of individual candidate substances, HELCOM lists of priority substances, Pesticides prioritised under Council Directive 91/414/EEC, Monitored substances not mentioned on any of the lists above (based on the monitoring data obtained from Member States) (Klein *et al.*, 1999, p. 7).

[5]E.g. trichloromethane (chloroform), trichloroethylene, tetrachloromethane, nonylphenol, naphtalene, mercury, dicloromethane, cadmium, chlorobenzene, anthracene, dicloroethane.

[6]COM(2006) 398 final: Proposal for a Directive of the European Parliament and of the Council on environmental quality standards in the field of water policy and amending Directive 2000/60/EC.

[7]European Parliament, Committee on the Environment, Public Health and Food Safety 2007: Compromise Amendments 1–3. Proposal for a Directive of the European Parliament and of the Council on environmental quality standards in the field of water policy and amending Directive 2000/60/EC. Draft report (PE 378.719v01-00), Anne Laperrouze. PE 386.558v01-00. 26.3.2007. http://www.europarl.europa.eu/meetdocs/2004_2009/documents/dv/659/659305/659305en.pdf.

pharmaceuticals, this process is currently only beginning and is being complicated by the fact that the presence of pharmaceutical products is in many cases difficult to detect with existing standard analytical methods. On the other hand, one of the strengths of the WFD is that its implementation (identifying pressures and impacts on water bodies, design of monitoring programmes, developing river basin management plans) is designed as an iterative process, and new findings can thus be included in the next management cycle. PPs can be incorporated into this scheme, once more data have become available.

Furthermore, most MS are currently adjusting their monitoring programmes to the main pollutants[8] (e.g. nutrients, heavy metals) and to the requirements set out in the list of priority substances. PPs are mostly not yet on the agenda when designing monitoring programmes. However, in countries like UK, Germany and several Swedish counties, monitoring of PPs is taking place to some extent within investigation programmes of the authorities on the occurrence and risk of PPs in the environment. In Germany, three pharmaceutically active substances (diclofenac, carbamazepin, and ibuprofen) have even been proposed for inclusion in the WFD monitoring (pers. comm., German Federal Environment Agency, January 2008).

Drinking Water Directive

The Drinking Water Directive (DWD) 98/83/EC aims at the protection of human health from harmful influences that might occur through pollution of water intended for human consumption. It also aims at assuring the suitability and purity of water for human consumption. It is strongly linked and synchronised with the WFD. The parameters to be monitored are defined in Article 5 and Annex I of the DWD. It lists 48 parameters (50 for bottled waters) categorised into three distinct groups: microbiological, chemical (meaningful to human health), and indicator parameters (meaningful to supply engineering, aesthetics, and health care). No pharmaceutical products are listed in the DWD. However, some substances used in pharmaceutical manufacturing are included in its list (which also appear in the WFD list of priority substances, see above).

In 2003, the European Commission decided not to include endocrine disrupters in the DWD, due to the lack of sensitive detection methods and reliable data concerning their risk to human health in view of their occurrence in water. However, Member States are free to set values for additional parameters not included in Annex I of the DWD, when necessary for the protection of human health. In practice, despite the lack of legal requirements to regularly monitor PPs so far, research projects frequently address the issue of PP concentrations in drinking water.

Groundwater Directive

The new Groundwater Directive 2006/118/EC complements the WFD in terms of chemical status criteria for groundwater. It clarifies good chemical status criteria and specifies how to identify and reverse pollution trends. The Directive establishes a regime which sets groundwater quality standards for a minimum list of pollutants and introduces measures to prevent or limit inputs of pollutants into groundwater. It requires to establish groundwater quality standards by the end of 2008, to carry out pollution trend studies using existing data and data which are mandatory by the WFD ("baseline level" data obtained in 2007–2008), to reverse pollution trends so that environmental objectives are achieved by 2015 using the measures set out in the WFD, to make operational measures to prevent or limit inputs of pollutants into groundwater so that WFD environmental objectives can be achieved by 2015 and to comply with good chemical status criteria (based on EU standards of nitrates and pesticides and on threshold values established by Member States).

No pharmaceutical products are listed in the Annexes of the Groundwater Directive. However, Member States are required to "amend the list of threshold values whenever new information on pollutants, groups of pollutants, or indicators of pollution indicates that a threshold value should be set for an additional substance" (Article 3(6), Groundwater Directive). The Directive also includes the requirement to identify and reverse significant increasing trends in pollutant concentrations. The repeated review of the status, trends and measures that is synchronised with the WFD cycles is meant to take new pollutants and risks into account and tackle them accordingly.

Bathing Water Directive

The Bathing Water Directive 2006/7/EC is concerned with monitoring and testing bathing water quality in order to protect bathers from health risks and to preserve the environment from pollution. While the first Directive introduced in 1976 required regular monitoring of 19 pollutants or other parameters (for instance water colour), the revised Directive 2006/7/EC reduces this list to just two microbiological indicators of faecal contamination, Escherichia Coli and Intestinal Enterococci. This simplification reflects the fact that due to most other water pollution being covered by the WFD, the primary health threat to bathers comes from faecal material. The issue of pharmaceuticals in the aquatic environment is therefore not dealt with under this Directive.

[8]"Main" in terms of amount introduced to the aquatic environment.

Urban Waste Water Treatment Directive

The Urban Waste Water Treatment Directive (UWWTD) 91/271/EEC concerns the collection, treatment and discharge of urban waste water from urban agglomerations. Its purpose is to stimulate Member States to invest in the collection and treatment of urban waste water; the aim is to reduce the input of organic matter and nutrients into European waters according to specific standards and deadlines set out in the Directive (Kemp, 2002).

The Directive defines three levels of treatment: primary, secondary and tertiary treatment. The basic rule for the level of treatment is secondary (i.e. biological) treatment. Treatment has to be more stringent (tertiary with additional nutrient removal) for discharges in so-called sensitive areas (defined according to UWWTD Annex II).

Although the parameters that the UWWTD aims to reduce are only related to organic material and nutrients, a side-effect of treatment is that a large number of other pollutants are also removed from wastewater, in particular many micropollutants such as heavy metals and pharmaceuticals. The reduction of the concentration of pharmaceuticals in treated wastewater depends on the particular treatment processes applied, and increases with higher levels of treatment. However, because many pharmaceuticals resist biodegradation (ICON Consultants, 2001), they are frequently found in significant concentrations in sewage sludge.

The UWWTD works with an Emissions Limit Value approach for the parameters it takes into consideration, which can be explicitly subject to changes; it is in theory possible that the Emission Limit Values are extended to other parameters, e.g. pharmaceuticals. However, no other contaminants have been taken up in the Directive's annexes in its 17 years of existence. In the same time, the ongoing implementation of the EU WFD poses new challenges to treatment plant operators. Especially the issue of how the treatment plants used to an Emissions Limit Value approach will respond to the WFD approach of Environmental Quality Standards is receiving more and more attention.

2.2.2.2 Other environmental policies and initiatives

European Strategy for Soil Protection

The status of soil is one of the factors regulating water quality; thus, current European policies for soil protection are relevant to water status and possibly also to the presence of pharmaceutical products in environmental waters. In 2006, the European Commission adopted a Communication for a Soil Thematic Strategy and a proposal for a Soil Framework Directive, with the aim of protecting soils across the EU. The Communication for a Soil Thematic Strategy explains why further action is needed to ensure a high level of soil protection, setting the overall objective of the Strategy, and explains what kind of measures must be taken. The proposal for a Framework Directive sets out common principles for protecting soils across the EU, including among others prevention, inventory, and remediation of contaminated soil. Although pharmaceutical substances are not part of the current EU policy documents under discussion on soil protection, they have been identified as a priority issue for further research by the working groups set up in preparation of the Soil Strategy (Van-Kamp *et al.*, 2004).

Sludge Directive

Sewage sludge used in agriculture is covered by the Sewage Sludge Directive 86/278/EEC. The Directive's aim is to "encourage the correct use of sewage sludge" in agriculture, while preventing "harmful effects on soil, vegetation, animals and man"; these effects could be due to the uptake of contaminants in agricultural products. The restrictions on use of sewage sludge are currently limited to seven heavy metals: cadmium, copper, nickel, lead, zinc, mercury, and chromium.

The Directive does not make explicit mention of pharmaceuticals in sewage sludge. Because the Directive encourages the application of sewage sludge due to its valuable properties, instead of favouring other forms of disposal such as combustion, it actually encourages at present the spreading of pharmaceuticals and their residues in the environment. Pharmaceuticals in sewage sludge applied to fields can also end up being leached into groundwater, and eventually making their way into surface water and/or drinking water. However, the Directive does foresee a process of its own adaptation to technical and scientific progress, so future changes to the Directive could in theory address pharmaceuticals.

Some German federal States (e.g. Bavaria and Nordrhein-Westphalia) have recently passed legislation that restricts the use of sewage sludge in agriculture. The presence of pharmaceuticals in sewage sludge, and as consequence the possibility of them entering environmental waters, was one of the reasons for passing such decisions.

Endocrine Strategy

Some chemicals can act on the endocrine system to disturb the mechanisms of the body or to initiate processes at abnormal times in the life cycle. These so-called endocrine disrupters comprise natural hormones, natural chemicals, man-made chemicals, and pharmaceuticals. Indeed, several pharmaceuticals are designed to be hormonally highly active, e.g. the contraceptive pill and treatments for hormone-responsive cancers. The residues of such pharmaceuticals enter the aquatic environment via various pathways and may have endocrine disruption effects on humans and wildlife if taken up unintentionally.

The European Endocrine Strategy addresses the potential environmental and health impacts of endocrine disruption in short-term, mid-term, and long-term actions. A key activity within the strategy is the identification and evaluation of relevant substances for their role in endocrine disruption. Any pharmaceuticals detected to pose risk to consumers under the endocrine strategy efforts will be dealt with in existing environmental regulation as far as possible. For the aquatic environment, the endocrine strategy refers explicitly to the WFD and its anti-pollution approach (CEC, 2004).

REACH

The REACH Regulation EC No. 1907/2006 deals with the registration, evaluation, authorization and restriction of chemicals. It entered into force on June 1, 2007. It exempts substances "used in" medicinal products from its main provisions, because medicinal products are already regulated within the scope of the Human Use Directive 2001/83, the Veterinary Use Directive 2001/82, and the Regulation 726/2004 (governing the centralized procedure of authorisation and supervision of medicinal products). Some provisions of the Regulation, however, remain applicable, including disclosure of information (on hazardous properties and regarding risk management measures) and restrictions on marketing and use of substances that pose an "unacceptable" health or environmental risk (which will probably rarely be applied to medicinal products) (Bogaert *et al.*, 2007).

IPPC Directive

The IPPC Directive 96/61/EC (on Integrated Pollution Prevention and Control) concerns the minimisation of pollution from various industrial sources. Annex I of the Directive lists industrial installations for which operators are required to obtain an authorisation (environmental permit) from the authorities in the EU countries. The list of installations which have to comply with the terms of the Directive includes "installations using a chemical or biological process for the production of basic pharmaceutical products". New installations of this kind have to apply for a permit and have to bring by evidence that emissions to the environment during production are minimal and accord to Best Available Techniques. Furthermore, Annex III lists main polluting substances which should be taken into account for fixing emission limit values. For water, this Annex lists inter alia substances and preparations which have been proved to possess carcinogenic or mutagenic properties or properties which may affect reproduction in or via the aquatic environment as well as persistent hydrocarbons and persistent and bioaccumulable organic toxic substances.

With the above requirements, the IPPC Directive ensures that emission of PPs into the environment during their production process is minimised. All in all, in Europe, there has been much progress in reducing emissions of PPs in the production phase by respecting relevant legislation and by improving the technologies used in industrial manufacturing. It is thus argued that efforts to further limit the discharge of PPs into the environment should concentrate on steps after but also prior to production (design, consumption and disposal phases).

2.3 DISCUSSION OF CURRENT POLICY INSTRUMENTS

In this section, the current policy framework and its instruments to limit the discharge of pharmaceutical products (PPs) is discussed, in view of its relevance for the protection of the aquatic environment.

In general, environmental policy responses to water pollution are, as a rule, based on certain principles: the precautionary principle, the principle that pollution should be avoided at source, the prevention principle, as well as the polluter-pays principle.

Current EU policies related to risk reduction from the presence of PPs in the environment can be seen in the context of the precautionary and prevention principles, but also in the context of end-of-pipe solutions (mainly wastewater treatment aspects). Having said that, it should be kept in mind that, in current EU environmental policy, precautionary and preventive approaches instead of end-of-pipe solutions are being promoted at all decision and policy levels.

2.3.1 Policy Framework in the light of the precautionary and prevention principles

The precautionary principle emphasises that, where evidence of a threat to the health of the environment exists, scientific uncertainty must not be allowed to delay reasonable forms of management action. Therefore, given that concerns exist that pharmaceuticals in natural waters could have subtle, long-term effects on the reproduction, development, and/or behaviour of aquatic species, uncertainty should not be a reason to allow environmental contamination by pharmaceuticals to continue unabated. This urges the scientific and policy community to consider and possibly to implement management strategies to mitigate the release of PPs to the environment (Doerr-MacEwen and Haight, 2006).

On the other hand, there are arguments that the interpretation and possible application of the precautionary principle need to be adapted to the conditions of pharmaceutical risk management. The precautionary principle has evolved in the setting of environmental policy making, a field with significant differences to the characteristics of pharmaceutical risk management (Callréus, 2005). In a recent review of expert stakeholders' views on the management of pharmaceuticals in the environment, concerns were expressed about the precautionary principle. These concerns were related mainly to the proportionality of management action, the balancing of risks and socioeconomic considerations, and the level of scientific evidence required to engage in precautionary management action. Despite concerns, however, a number of risk management strategies for PPs in the environment such as drug take-back schemes were considered by most as sensible approaches (Doerr-MacEwen and Haight, 2006).

All in all, environmental concerns are now more strongly integrated in EU regulations on PPs than in the past, at least with respect to the marketing authorisation of PPs. As already mentioned, since September 2005, each new human medicine must undergo environmental risk assessment, while in the case of veterinary medicine authorisation, environmental security has gained the same weight as consumer security in the risk-benefit analysis. Also, detailed guidelines are now available for the environmental risk assessment procedure of both human and veterinary medicine.

Nevertheless, certain gaps still exist in the medicinal policy framework, especially in the practical aspects of its implementation; these should be targeted in future efforts addressing the presence of PPs in the environment. In this context, the authors would like to emphasise the following issues:

- **Environmental safety of "old medicine".** So far, environmental risk is not assessed for so-called "old medicine". "Old medicine" includes products authorised before the adoption of EU legislation requesting environmental risk assessment (in 1993 for human PPs and in 1990 for veterinary PPs) and the adoption of relevant ERA guidelines (in 2006 for human PPs and in 1998 for veterinary PPs). "Old medicine" indeed constitutes a large proportion of medicinal products authorised and available on the market, and there is currently no system for their environmental evaluation and control. It is argued that, in analogy to similar European regulatory provisions for other groups of substances like industrial chemicals, biocides and pesticides, the environmental safety of "older medicine" (both human and veterinary) should also be assessed. Priority lists for an "old medicine" programme should be drafted, and applied at the European level to minimise the environmental evaluation work load (SRU, 2007).
- **Authorisation of veterinary medicine and risk mitigation.** Doubts about the acceptability of the environmental impact of veterinary medicine can lead to changes of the characteristics of the product, of the area of its application, or even to the refusal of marketing authorisation by the authorities. An identified environmental impact may also be mitigated to an acceptable level by special precautions included in the information that accompanies the product in labeling and packaging. Such precautions can include, for example, the application of manure coming from treated animals at a minimum distance from surface waters or restrictions to the direct access of treated animals to surface waters. Those addressed by such precautionary measures are livestock owners and those responsible for handling manure.

Montforts (2005), however, identified several constraints which decrease the technical and legal effectiveness of such measures. On the one hand, measures taken on the basis of the relevant medicinal Directive (2001/82/EC) are not legally binding for veterinary doctors and farmers (the consumers). Therefore, there seems to be a need for complementary regulations with regard to PP application instructions and their addressees in national legislation in a Europe-wide harmonised manner. On the other hand, precautionary measures are only acceptable under the medicinal Directive 2001/82/EC, if their effect can be demonstrated using risk assessment methodologies. For measures completely prohibiting the release of treated animals or the use of contaminated manure, it is quite easy to demonstrate that they have an effect. However, the impact of temporary storage of contaminated manure or of keeping a security distance of treated animals from surface waters cannot be easily quantified due to the lack of standardised methods. Therefore, to render these precautionary measures effective for risk mitigation (and hence suitable for labelling and packaging), risk assessment methodologies must be further developed and applied.

- **Authorisation of human medicine and risk mitigation.** Although approval of authorisation for veterinary medicines can be refused on the basis of risks to the environment, this is not the case for human medicine. The principle therein is that the therapeutic benefit for sick humans is to be considered beyond possible damages to the environment. This principle can doubtless be followed; however, a possible future approach could for instance require that the therapeutic benefit be superior to any alternatives and that a broad, little-controllable application (e.g. by recipe-free circulation for self-medication) of pharmaceuticals that pose an environmental risk be prevented. Thus, although it is unrealistic to prohibit the authorisation of human PPs on environmental grounds, different alternatives should be considered in the authorisation process to reduce environmental exposure as far as possible (SRU, 2007). If alternative compounds are available promising the same success in therapy, but posing at the same time a lower environmental risk, a refusal of authorisation should not be excluded per se (Knacker *et al.*, 2006).

It should also be mentioned that, in general, only little experience for human medicine in the risk management area has been gathered so far (Koschorrek and Apel, 2006). In Germany, although the federal environment agency has been successful in a significant number of cases in combining authorisation of veterinary drugs with practicable environmental risk mitigation measures, in the field of human pharmaceuticals, there is a need for further practical efforts (Holzmann, 2005; Schlimm, 2005). At the moment, there is no concrete proposal and no binding regulations for mitigation measures relevant to human medicine (SRU, 2007). It is however expected that the 2006 publication of the EMEA guideline on the ERA of human PPs will contribute to the broader use of risk mitigation measures for human PPs.

- **Drug take-back schemes.** The extent of establishment and the degree of effectiveness of take-back schemes for drugs is quite different among European countries. In-depth assessments of scheme effectiveness (estimated recovery rate of unused/expired drugs) are missing to a great extent. High levels of public awareness and education, firstly regarding the environmental consequences of the disposal of unused/expired drugs and, secondly, on the operation of such schemes seem to be key for their success. Problems in the implementation of take-back schemes are largely related to lack of awareness, leading to erroneous consumer behaviour. There is thus need to improve the labelling of pharmaceuticals and consumer information, so as to assure the return of unused pharmaceuticals to pharmacies.

As far as the EU policy framework beyond medicinal regulations is concerned, at present it does not adequately take account of the precautionary and prevention principles on the issue of pharmaceuticals in the environment; this is partly due to implementation deficiencies and partly due to legislation gaps. All in all, it is not argued here that additional directives are needed. A suitable EU policy framework seems to be in place already, but some modifications to existing directives may be needed:

- **Water Framework Directive.** The WFD provides an appropriate policy framework to deal with all sources of chemical pollution in European waters if implemented properly.

Firstly, it establishes a list of priority chemical substances for which measures must be progressively applied to reduce their concentration below defined quality standards. It is argued that in future revisions of the list of priority substances, certain PPs should be considered.

Secondly, the WFD (Annex VIII) requires to identify all other pollutants that are discharged in significant quantities in the river basins and to set relevant quality standards that should be met by 2015 as part of good ecological status. The definition of target substances and standards is up to the Member States and to specific basin authorities. For instance, in the UK, two PPs are under discussion for addition to the list of UK specific pollutants according to WFD Annex VIII.

However, all in all, the respective WFD mechanisms and the Directive's precautionary principle are not yet used fully to deal in a focused way with the issue of PPs in the aquatic environment. The issue is currently not politically discussed, which is partly due to its complexity, uncertainties, and lack of knowledge (as described in the next section).

A first important step in order to better use the WFD implementation process for filling in knowledge gaps on PPs in the aquatic environment is to include, where considered necessary, PPs in the "characterisation of river basin districts and review of environmental impact of human activity" (WFD Art. 5). On the basis of "river basin district characterisation" results, decisions for future monitoring of substances and relevant necessary water management measures can be taken.

Appropriate and adequate monitoring of pharmaceutical substances in the environment is in general missing. It is recommended to include the monitoring of specific pharmaceutical substances – those which are of high environmental relevance due to the effects, persistence or mobility – in existing monitoring programmes like those for the WFD implementation (SRU, 2007).

- **Drinking Water & Groundwater Directives.** PPs are not listed yet in the Annexes of these two key water directives that refer to substances to be monitored. Nevertheless, with respect to drinking water (whose source is often groundwater), there is frequently concern on possible health impacts due to the presence of PPs. Firstly, public demand for drinking water of good quality cannot be overlooked by water managers. Secondly, the precautionary principle is often called for in this case, when bearing in mind uncertainties such as the following: the increasing spread of antibiotic-resistant bacteria can pose a health risk for humans; there is lack of knowledge on risk to human health (via drinking water) from chronic exposure to PPs that are designed for short-term use.
- **Bathing Water Directive.** The recent amendment of the Bathing Water Directive may be seen as a step forward when discussing this Directive in the overall framework of new EU water policies. However, due to several implementation gaps of the overall framework and especially the fact that PPs are currently not widely discussed in WFD implementation, there is the risk that the issue of PPs in bathing waters will not receive adequate attention in policy implementation.

- **Other environmental policies and initiatives**. The following refers to European soil policy, the Sewage Sludge Directive, the EU Endocrine Strategy, the REACH Regulation, and the IPPC Directive:
  - PPs are not yet part of current policy documents under discussion on ***European soil protection***. However, the possible relevance of European soil protection for PPs has been identified on a policy level by placing these substances among the priority issues for further research in this field (in terms of their quantification and risk evaluation methods).
  - Despite the fact that PPs' presence in sewage sludge could lead to the unintended uptake of relatively high concentrations of PPs via food consumption (when food is grown on fields where sewage sludge is applied), there are no limits for PP concentrations yet in the ***Sewage Sludge Directive***. Another possible fate for pharmaceuticals and other pollutants contained in sewage sludge is their leaching into groundwater, and eventually making their way into surface water and/or drinking water. In fact, in some EU countries, due to preoccupation with the issue of contaminants in sludge, including drugs, the application of sewage sludge on agricultural fields has been prohibited.
  - An identification and evaluation process of substances with potential environmental and health impacts of endocrine disruption continues and improves in the context of the ***Endocrine Strategy***. Any pharmaceuticals detected under the Endocrine Strategy efforts will be dealt with in existing environmental regulation as far as possible. For the aquatic environment, the endocrine strategy also refers to the WFD and its strategy against pollution.
  - PPs are also exempt from the main provisions on registration, evaluation, authorization, and downstream use of chemicals in the EU ***REACH Regulation***. Some provisions of REACH remain applicable, including disclosure of information (on hazardous properties and regarding risk management measures) and restrictions on marketing and use of substances that pose an "unacceptable" health or environmental risk. However, these provisions will probably rarely be applied to medicinal products.
  - The ***IPPC Directive***, which aims at pollution prevention and control in industrial production processes, only applies to the production of PPs. This Directive ensures that emission of PPs into the environment during production is minimised. There has been much progress in reducing emissions of PPs in the production phase by respecting relevant legislation and by improving the technologies used in industrial manufacturing. In consequence, efforts to further limit the discharge of PPs into the environment should concentrate on steps beyond (but also prior to) production.

2.3.2 Policy framework and end-of-pipe solutions

Once PPs are released by consumers into sewage, either via urine or after disposal of unused/expired drugs down the drain, only an end-of-pipe solution remains to limit the presence of PPs in the natural aquatic environment.

Although precautionary and pollution preventive policy approaches should be a priority, the policy framework could also be improved in terms of PP removal in the phase of wastewater treatment. The most relevant EU policy is the Urban Wastewater Treatment Directive (UWWTD) which focuses on the reduction of organic material and nutrients. The removal of other pollutants, including pharmaceuticals and heavy metals, has so far only been a side-effect of wastewater treatment.

Firstly, the full implementation of this Directive in all EU Member States would have positive impacts on the levels of occurrence of PPs in the water environment. Secondly, a future amendment of the UWWTD could be envisioned to include emission limits on pharmaceuticals. Requirements to deal with pharmaceutical loads in wastewater may also result for plant operators from the WFD. Given the new approach of the WFD on water quality, that combines emission limit values with environmental water quality standards, operators of urban wastewater treatment plants may be forced to upgrade their technologies. In any event, technologies which could improve the degree of removal of PPs from wastewater, like ozonation and membrane filtration, are already available.

The wastewater treatment policy framework could also be modified to intensify efforts for installing special (pre-) treatment facilities for hospital wastewater, where this is assessed to be a good management option. In some countries, hospitals are considered hotspots for pharmaceuticals emissions because their emission route accounts for a high percentage of the total load of human pharmaceuticals in the environment (e.g. see Roorda *et al.* (2008) for an illustration of the situation in the Netherlands). In consequence, it is suggested that treating hospital wastewater at source could reduce the treatment load in the large municipal treatment facilities. Certainly, though, a detailed analysis for selecting the most cost-effective alternative (municipal treatment or on-site hospital treatment) should be made for each specific case or on national level.

In this entire discussion on wastewater treatment options, a crucial issue is the cost recovery for investments in new technologies. Next to possible policy modifications, economic instruments (e.g. ear-marked water taxes) could be used to financially support the upgrade of treatment. In this context, it is important to clarify the role of the polluter (industry or consumer) in order to develop an appropriate approach for implementing the polluter-pays principle.

2.3.3 What Complicates further eu policy development to limit discharge if PPs into waters?

Based on the discussions above, it is apparent that there are still certain policy gaps (either legislative gaps or gaps in the implementation of legislation) when it comes to the issue of PPs in the aquatic environment. Taking further action to close some of these gaps is, however, still difficult, due to lack of relevant information and knowledge as well as uncertainties on PPs in the environment.

It becomes clear that we need **better understanding, better data and further research** on PPs in the aquatic environment. Some of the key issues to be clarified include:

- **Knowledge on the fate and effects of PPs in the environment.** So far, knowledge on the fate and effects of PPs in the environment has been sparse. It is hoped, however, that given the recent adoption of EU guidelines on the ERA of PPs, more data on the fate and effects of PPs will be generated in the near future (Koschorreck and Apel, 2006).
- **Knowledge on low-level effects and chronic effects of PPs.** The environmental risk assessment for PPs should guarantee that possible low level and long-term effects of PPs are not overlooked. The strict application of an action limit, which terminates any further risk assessment for substances not exceeding the PEC in surface water (0.01 μg/L for human medicine), may lead to the disregard of such effects, considering that experience with the ecotoxicological effects of human PPs is still very limited.

Concerning chronic effects, more data need to be made available that allow their assessment in ecotoxicological studies, especially for products that have been on the market for a long time.

Concerning low-level effects, there is a strong opinion that pharmaceutical compounds do not pose a large risk because they are present in such low concentrations (ng/l), with most effects only seen in the mg/l range. However, it should be kept in mind that disease resistance to pharmaceuticals is favoured by low concentration exposure and compounds such as hormones have effects at very low levels. The effects of active compounds in the low (ng/l range) cannot be excluded, as experience with pesticides has shown, impacts can be significant at low levels (Stuer-Laurisden *et al.*, 2000).

- **Cumulative effects of multiple PP substances.** Although in the environment organisms are exposed to a variety of substances, the potential impact of mixtures of individual substances in the environment is recognised as an unresolved issue within the currently implemented ERA schemes. Especially in the case of pharmaceuticals from similar medicinal classes, it can be expected that they have similar modes of action under environmental conditions, and hence, additive or synergistic effects are likely to occur (Knacker *et al.*, 2006). So far, only individual products undergo an environmental risk assessment and thus, the risk assessment process is lacking in cumulative assessments of potential environmental risk from substances of comparable structure (SRU, 2007).
- **General data availability and accessibility.** At present, the limited quality of data used in ERA and the difficulty to access ERA information used in marketing authorisations remain key problems for several interested stakeholders. There is also great unbalance on data availability for newly authorised PPs versus PPs brought on the market prior to ERA regulations (SRU, 2007). In addition, ERA data and information are only open to the respective national competent authorities (with the exception of Sweden where ERA results are made public online on a voluntary basis). The UK Environment Agency argues for open access to non-commercially sensitive information to facilitate effective environmental assessments. It has thus called upon the EMEA and the MHRA (the English equivalent) to put in place a suitable system making environmental information on pharmaceuticals easily available and accessible (Environment Agency, 2003).

Another important data limitation concerns figures for medication consumption. In most countries, and for both human and veterinary products, no reliable figures for pharmaceutical use are available. This is an important stumbling block when trying to determine the environmental impacts of the use of human and veterinary pharmaceuticals. A possible way of addressing this issue is to increase the requirements regarding the documentation of use (SRU, 2007).

Besides the scientific and data gaps highlighted above, the following issue further complicates policy discussions on PPs in the environment. Namely, difficulties in setting regulatory limits on the release of PPs into water from different sources (the public, wastewater treatment plants, industry) and in applying economic instruments (e.g. taxes to finance pollution-reduction measures and the "polluter-pays-principle") result from the question "who is the polluter". Different definitions of the polluter (consumers of medicine; industry; doctors as prescribers; hospitals; pharmacists; or even public agencies issuing PP marketing authorisations) can be used, with varying implications. It is thus important to clarify who the polluter is and to carry out research on the costs and benefits of the different options.

All in all, the development of policy options and instruments to limit the discharge of PPs into the aquatic environment certainly requires an open consultation and dialogue process involving key public and private stakeholders. The need to amend existing policies or to develop new policies has to be based on a balanced consideration of the precautionary and prevention principles of environmental policy making, the polluter-pays principle as well as new scientific knowledge on the risks and impacts of the unintended release of PPs into water.

Finally, it should be kept in mind that the pool of optional approaches to limit the discharge of PPs into the aquatic environment goes also beyond EU regulations and policies. In this context, the following actors can play a role in various ways: the industry (e.g. by better communication of the environmental risk of drugs via environmental classification systems or by introducing environmental criteria in the early steps of product development), the public domain (e.g. by promoting and ensuring safe waste disposal with incineration), the health care system, doctors as well as patients (e.g. by responding to communication efforts on the environmental risk of PPs and making better use of drug take-back schemes).

REFERENCES

Bogaert, P., Kingham, R., Castle, G. and García Molyneux, C. (2007) Reach And Its Impact On Pharmaceuticals. Covington & Burling LLP.

Callréus, T. (2005) The precautionary principle and pharmaceutical risk management. *Drug Safety*, **28**(6), 465–471.

CEC – Commission of the European Communities (2004) Commission Staff Working Document on implementation of the Community Strategy for Endocrine Disrupters – a range of substances suspected of interfering with the hormone systems of humans and wildlife. COM (1999) 706.

CHMP – Committee for Medicinal Products for Human Use (2006) Guideline on the environmental risk assessment of medicinal products for human use. CPMP/SWP/4447/00. London: European Medicines Agency.

Clark, J., Summerton, L., Smith, E., Kampa, E. and Touraud, E. (2007) Discussion Document on Ecopharmacostewardship of Pharmaceutical Products. KNAPPE Deliverable available at: http://www.knappe-eu.org/fichiers/47-D5.1%20Ecopharmacostewardship%20discussion%20document.pdf.

Council Directive of 12 June 1986 on the protection of the environment, and in particular of the soil, when sewage sludge is used in agriculture (86/278/EEC). Official Journal L 181, 4/7/1986.

Council Directive of 21 May 1991 concerning urban waste water treatment (91/271/EEC). Official Journal L 135, 30/5/1991.

Council Directive 96/29/EURATOM of 13 May 1996 laying down basic safety standards for the protection of the health of workers and the general public against the dangers arising from ionizing radiation. Official Journal L 159, 29/6/1996.

Council Directive 96/61/EC of 24 September 1996 concerning integrated pollution prevention and control. Official Journal L 257, 10/10/1996.

Council Directive 97/43/Euratom of 30 June 1997 on health protection of individuals against the dangers of ionizing radiation in relation to medical exposure, and repealing Directive 84/466/Euratom. Official Journal L 180, 9/7/1997, pp. 22–27.

Council Directive 98/83/EC of 3 November 1998 on the quality of water intended for human consumption. Official Journal L 330, 5/12/1998.

Council Regulation (EC) No 726/2004 of the European Parliament and of the Council of 31 March 2004 laying down Community procedures for the authorisation and supervision of medicinal products for human and veterinary use and establishing a European Medicines Agency. Official Journal L 136, 30/4/2004.

CVMP – Committee for Medicinal Products for Veterinary Use 2007: Guideline on environmental impact assessment for veterinary medicinal products – in support of the VICH guidelines GL6 and GL38. EMEA/CVMP/ERA/418282/2005. London: European Medicines Agency.

Directive 2000/60/EC of the European Parliament and of the Council establishing a framework for the Community action in the field of water policy. Official Journal L 327, 22/12/2000.

Directive 2001/82/EC of the European Parliament and of the Council of 6 November 2001 on the Community code relating to veterinary medicinal products. Official Journal L 311, 28/11/2001 pp. 1–66.

Directive 2001/83/EC of the European Parliament and of the Council of 6 November 2001 on the Community code relating to medicinal products for human use. Official Journal L 311, 28/11/2001 pp. 67–128.

Directive 2004/27/EC of the European Parliament and of the Council of 31 March 2004 amending Directive 2001/83/EC on the Community code relating to medicinal products for human use. Official Journal L 136, 30/4/2004 pp. 34–57.

Directive 2004/28/EC of the European Parliament and of the Council of 31 March 2004 amending Directive 2001/82/EC on the Community code relating to veterinary medicinal products. Official Journal L 136 , 30/04/2004 pp. 58–84.

Directive 2006/7/EC of the European Parliament and of the Council of 15 February 2006 concerning the management of bathing water quality and repealing Directive 76/160/EEC. Official Journal L 64, 4/3/2006.

Directive 2006/118/EC of the European Parliament and of the Council of 12 December 2006 on the protection of groundwater against pollution and deterioration Official Journal L 372, 27/12/2006, pp. 19–31.

Doerr-MacEwen, N.A., Haight, M.E. (2006) Expert stakeholders' views on the management of human pharmaceuticals in the environment. *Environmental Management* **38**, 853–866.

Environment Agency 2003: Human Pharmaceuticals: Position Statement. http://www.environment-agency.gov.uk/aboutus/512398/289428/698017/?lang = _e.

FDA (Food and Drug Administration) 1998: Guidance for Industry – Environmental Assessment of Human Drugs and Biologics Applications, Revision 1. FDA Center for Drug Evaluation and Research, Rockville.

Holzmann, T. (2005) 2004 – ein wichtiges Jahr für die Umweltsicherheit von Arzneimitteln. Grußwort. in: Umweltbundesamt (Ed.) 2005: Arzneimittel in der Umwelt – Zu Risiken und Nebenwirkungen fragen Sie das Umweltbundesamt. Texte 29-05. UBA. Dessau. 5–8.

ICON – I C Consultants 2001: "Pollutants in Urban Waste Water and Sewage Sludge." Final Report for DG Research.

Kemp, R. (2002) Treatment of an EU directive: UWWTD, Compliance with the Urban Waste Water Treatment in Germany, Spain, England and Wales, and The Netherlands.
Klein, W. (1999) Revised Proposal for a List of Priority Substances in the Context of the Water Framework Directive (COMMPS Procedure. With co-operation of S. Denzer, M. Herrchen and P. Lepper *et al.* Fraunhofer-Institute for environmental chemistry and ecotoxicology.
Knacker, T., Liebig M. and Moltmann J.F. (2006) Environmental Risk Assessment. In *Human Pharmaceuticals, Hormones and Fragrances. The challenge of micropollutants in urban water management* (ed. T. Ternes, A. Joss), pp. 121–147, IWA Publishing. London.
Koschorreck, J. and Apel, P. (2006) A brief overview on the legal background and the regulatory instruments of the environmental risk assessment for Pharmaceuticals in the EU, USA, Japan, Australia and Canada. In *Human Pharmaceuticals, Hormones and Fragrances. The challenge of micropollutants in urban water management* (ed. T. Ternes, A. Joss), pp. 107–119, IWA Publishing. London.
Kümmerer, K. (Ed.) (2004) Pharmaceuticals in the Environment. Sources, Fate, Effects and Risks. 2nd edition. Springer. Berlin, Heidelberg, New York.
Montforts, M. (2005) Vorsorge vor Umweltrisiken in der Produktinformation von Veterinärpharmaka: Juristische Fallstricke und Methodologische Herausforderungen. In Umweltbundesamt (Ed.): Arzneimittel in der Umwelt – Zu Risiken und Nebenwirkungen fragen Sie das Umweltbundesamt. Texte 29-05. UBA. Dessau.
Regulation (EC) No 726/2004 of the European Parliament and of the Council of 31 March 2004 laying down Community procedures for the authorisation and supervision of medicinal products for human and veterinary use and establishing a European Medicines Agency. Official Journal L 136, 30/4/2004.
Regulation (EC) No 1907/2006 of the European Parliament and of the Council of 18 December 2006 concerning the Registration, Evaluation, Authorisation and Restriction of Chemicals (REACH), establishing a European Chemicals Agency, amending Directive 1999/45/EC and repealing Council Regulation (EEC) No 793/93 and Commission Regulation (EC) No 1488/94 as well as Council Directive 76/769/EEC and Commission Directives 91/155/EEC, 93/67/EEC, 93/105/EC and 2000/21/EC. Official Journal L 396, 30/12/2006.
Roorda, J., Derksen, A. and Wortel, N. (2008) Challenging conventional sanitation options for health care units. Presentation at the International Conference on New Sanitation Concepts and Models of Governance [19–21.05.2008], Wageningen.
Schlimm, J. (2005) Umweltbewertung von Arzneimittel in Deutschland – ein kurzer Abriss. In Arzneimittel in der Umwelt – Zu Risiken und Nebenwirkungen fragen Sie das Umweltbundesamt, (ed. Umweltbundesamt), Texte 29-05. UBA. Dessau. 5–8.
SRU – Sachverständigenrat für Umweltfragen (Ed.) 2007: Arzneimittel in der Umwelt. Stellungnahme. Nr 12, April 2007. SRU. Berlin.
Stuer-Lauridsen, F., Birkved M., Hansen L.P., Holten LützhØft, H.C. Halling-Sørensen, (2000) Environmental risk assessment of human pharmaceuticals in Denmark after normal therapeutic use. *Chemosphere* **40**, 783–793.
Umweltbundesamt (Ed.) 2005: Arzneimittel in der Umwelt – Zu Risiken und Nebenwirkungen fragen Sie das Umweltbundesamt. Texte 29-05. UBA. Dessau.
Van-Camp, L., Bujarrabal, B., Gentile, A.-R., Jones, R.J.A., Montanarella, L., Olazabal, C. and Selvaradjou, S.-K. (2004) Reports of the Technical Working Groups Established under the Thematic Strategy for Soil Protection. EUR 21319 EN/6, 872 pp. Office for Official Publications of the European Communities, Luxembourg.

Chapter 3

Exposure based on life cycle

Sophie Mompelat, Barbara Le Bot, Olivier Thomas and Benoit Roig

3.1 EXPOSURE ROUTES

There is a multitude of input sources in environment with no evident quantitative data available (concerning the relative distribution of PPs from all emission sources (Figure 3.1).

Although humans and animals treated with PPs constitute the main contamination source of water resources (surface, ground and drinking waters), PPs are qualitatively, quantitatively, spatially and temporally shared out into different routes depending on whether the drug or medicine is for veterinary (land- or water-based) or human consumption, and in this last case depends on whether patients are located in private households or in hospitals and other places (e.g., schools, residential facilities like retirement facilities). For example, for dispensed medicines for human consumption, the major route of entry to the aquatic environment following dispensing is through excretion and the waste water cycle. However, old medicines or those remaining upon the end of the course of treatment are often disposed of in ways that lead to their appearance in environmental water. In a study by Bound and Voulvoulis (2005), household disposal pathways and characteristics in the United Kingdom were reported for various drug types. The largest disposal pathway was trash/bins, while sink and toilet represented between 3.6 and 16.7% for painkillers, antihistamines antibiotics and beta-blockers. Only 14 to 33% of respondents to the survey took their surplus medicines back to the pharmacy. No hormones, lipid regulators or antidepressants were disposed of *via* waste water. This pattern of disposal is similar to those reported elsewhere (Bound and Voulvoulis, 2006a). However, many countries have made efforts to change this (e.g., take-back schemes have been successful in increasing the proportion of waste medicines that is disposed of through pharmacy outlets). On the other hand, prescribed drugs in hospitals are rather designated to treat heavier pathologies than in households. As an example, antineoplastics (also named cytotoxics) used in cancer chemotherapy are exclusively prescribed and principally (intravenously or orally) administered in hospitals. Few years ago, they were released mainly in hospital effluents between 5–500 g/L (Kummerer, 2001), but nowadays, 75% of antineoplastics are given in outpatient departments with a huge trend for oral administration at home (Johnson *et al.*, 2008). Improper disposal of unused or expired drugs, which are directly thrown in toilets or end up in landfill, and pharmaceutical residues from manufacture spill accidents (Reddersen *et al.*, 2002) can also be regarded as other significant local points of potential contamination.

Thus, following use, significant proportions of some pharmaceuticals are excreted unchanged whilst some are metabolised and then either parent compound or detoxification products are released through the urine and faeces into waste water systems or, for countryside households directly released into septic tanks (Carrara *et al.*, 2008). Both, waste water treatment plant (WWTP) and septic system are not designed to eliminate highly polar micropollutants like PPs, but degradation during the waste water treatment process (or removal with sewage sludge), will depend on the concentration levels of the medicine of interest, its physicochemical characteristics and stability, the type of waste water treatment plant involved and its efficiency. Consequently, the elimination rates cover the whole range between 0 to 100% (Radjenović *et al.,* 2007; Drewes, 2007). Among PPs not retained in WWTP, only 10 to 20% of carbamazepine is removed (Radjenović *et al.,* 2007; Perez and Barcelo, 2007), and then found with other not removed PPs in treated wastewater (WW) discharges in surface water (SW) (i.e., rivers, lakes, streams, estuaries). For dispersed households, septic tanks are also susceptible to release PPs, but this time, into ground water (GW) (Carrara *et al.,* 2008). Differently, direct release of veterinary pharmaceuticals in environment may occur via application in aquaculture (i.e., fish farming), but also indirect release by way of animals topically treated, and mostly *via* run-off and leaching through fields from manure spreading to agricultural fields and livestock wastes (Boxall *et al.,* 2003; 2004; Khan *et al.,* 2007; Sarmah *et al.,* 2006; Kemper, 2008).

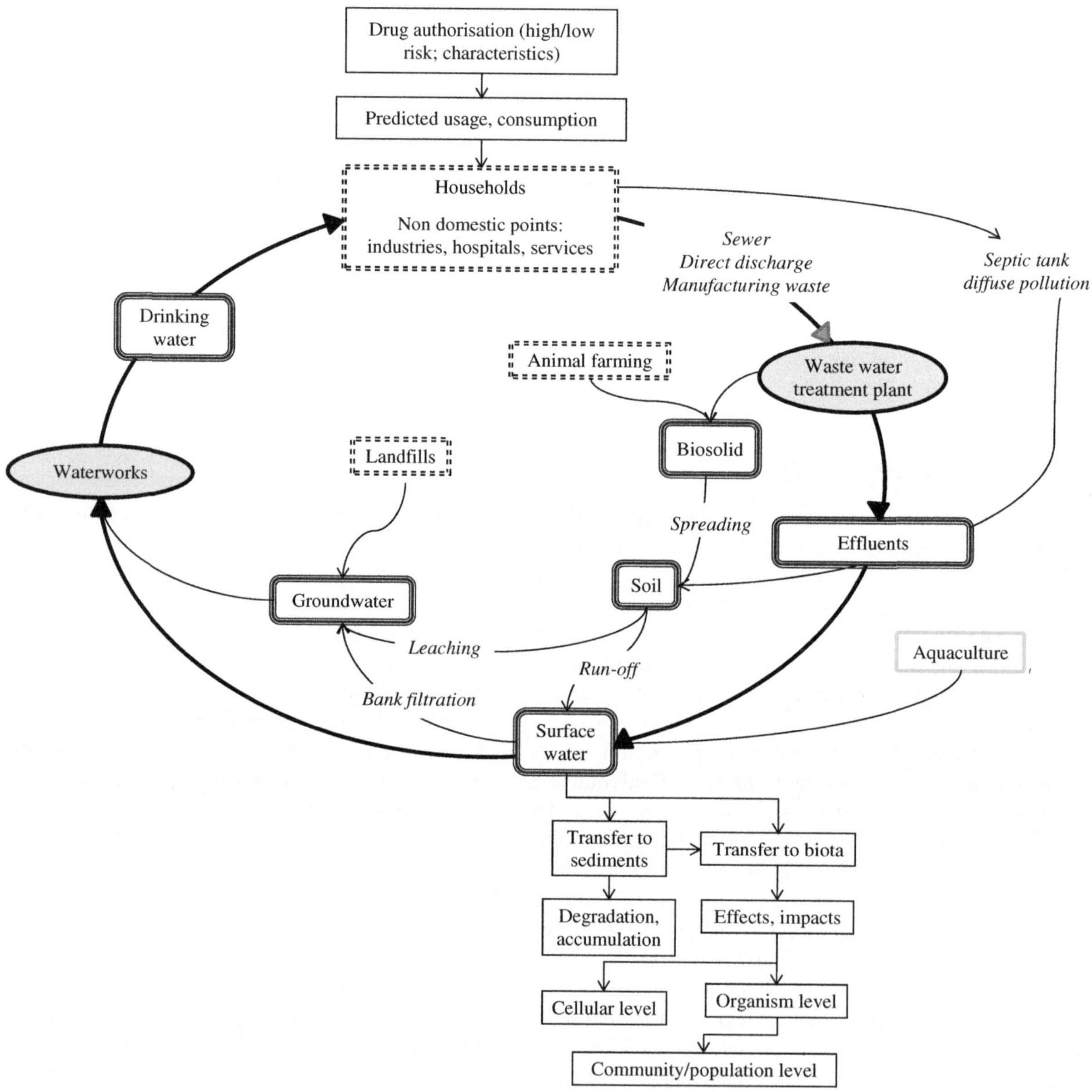

Figure 3.1 Exposure routes of pharmaceuticals from the aquatic environment to human based on their simplified life cycle (adapted from Mompelat *et al.*, 2009).

In some cases, biotic or abiotic degradation pathways may lead to the formation of metabolites or transformation products at concentrations that may be toxicologically relevant. Disposal of sewage sludge via farming applications (sludge/manure application) may increase the possible threat of ground water and surface water contamination. Indeed, GW and SW are physically closely linked, and therefore, these compartments can contaminate one another, e.g. GW pollution from artificial recharge of polluted SW or of treated WW reuse (despite soil or bank filtration) and exfiltration from GW to connected SW (Jux *et al.,* 2002).

Once released into surface waters, the fate of pharmaceuticals, including their transfer into and movement within sediments, and persistence will depend on their physicochemical characteristics (e.g., hydrophobicity, volatility, reactivity). This will, in turn, dictate the exposure of biota to the drug its potential environmental impact. It will also affect the degree of contamination of drinking water.

In order to model these systems and to obtain realistic risk assessments both good quality field data and reliable estimates of inputs are needed. As shown in Figure 3.1, accurate information is required to predict or estimate pharmaceutical usage, its pattern of use (e.g., over the counter sales, hospital-dispensed drugs, and any seasonality in the use), its volume of production, dispensing in the community and usage (based for example on market penetration factors). This information is critical in the prediction of environmental concentrations and therefore the risk of these substances in the aquatic environment.

3.2 MODELLING OF EXPECTED CONCENTRATIONS IN VARIOUS MATRICES

There is still no evidence that the PPs presence in drinking water, although at trace level, can be exclude from the potential risk of chronic adverse effects on human beings. However, there are, currently, no specific norms to rule on contamination limits of PPs in drinking water and no regulatory guidance to monitor them.

There is a tiered approach to environmental risk assessment for medicines for both human (Jones *et al.*, 2002) and veterinary (CHMP, 2006). The first phase of the risk assessment at the registration/authorisation stage based principally on checks for persistence for non-polar (log Kow > 4.5) compounds, and the calculation of predicted concentrations (PEC) in surface waters. However, the calculation is based on assumptions such as an evenly distributed usage over time and space), and these may introduce a large uncertainty into the calculation. Other assumptions such as no metabolism, biodegradation or retention of the drug lead to conservative estimates of risk.

Under the EMEA risk assessment guidelines for medicines for human use, a basic risk assessment of exposure of surface water can be undertaken by predicting environmental concentrations, principally in surface waters. Predicted surface water concentrations can be obtained using the following equations (EMEA, 2006; Bound and Voulvoulis, 2004)

$$PEC_{Surface\ water} = \frac{DOSEai \times F_{pen}}{WASTEWinhab \times DILUTION \times 100} \quad (3.1)$$

With:

$PEC_{Surface\ water}$	=	Predicted environmental concentration (M/V^3)
DOSEai	=	maximum daily dose of drug consumed per inhabitant (M $inhabitant^{-1}$ T^{-1})
F_{pen}	=	market penetration of the drug (%)
WASTEWinhab	=	volume of wastewater per inhabitants (V $inhabitant^{-1}$ T^{-1})
DILUTION	=	dilution factor from effluent to surface water

This relationship assumes that the drug is used uniformly over the time period and target area; that no metabolism occurs in the patient and that WWTPs, where there is no degradation or retention of the compounds, are the main point of entry to the system (Bound and Voulvoulis, 2004)

Using knowledge of the amount of drug excreted by the patient and the amount removed from the system by WWTPs processes (via adsorption and degradation), the $PEC_{Surface\ water}$ is revised as shown in Eq. (3.2). A separate calculation with a newly derived PEC for any metabolites is also performed. The inclusion of these factors requires a significant increase in data.

$$PEC_{Surface\ water} = \frac{Elocal_{Water} \times F_{WWTPs}}{WASTEWinhab \times CAPACITY_{WWTPs} \times FACTOR \times DILUTION} \quad (3.2)$$

With:

F_{WWTPs}	=	fraction of the drug that enters surface waters (efficiency of the WWTP)
$CAPACITY_{WWTPs}$	=	capacity of the local wastewater treatment plant ($inhabitants^{-1}$)
FACTOR	=	Degree of sorption to suspended matter
$Elocal_{Water}$	=	local emission to wastewater given by Eq. (3.3)

$$ELOCAL_{Water} = \frac{DOSEai \times F_{excret} \times F_{pen} \times CAPACITY_{WWTPs}}{100} \quad (3.3)$$

With:

F_{excret}	=	fraction of drug excreted

The estimation of PEC values requires in turn the estimation of a number of variables as shown above. Often the variability of these variables and the associated uncertainty are poorly defined, and this increases the overall uncertainty associated with the PEC values (see chapter 4).

In many cases safety factors such as assuming no losses during sewage treatment (i.e., influent concentration = effluent concentration) are introduced to provide protection against underestimating the PEC. For the calculation of PEC values for surface waters it is necessary to know the flow of the surface water and the proportion of waste water to surface water.

Often it is considered appropriate to estimate the worst case scenario that is obtained by using low flow conditions with maximum expected effluent concentrations. Loss of pollutants present in the primary sewage during treatment can span orders of magnitude between different plants, and is normalised to the input of a drug into the sewage treatment. Inputs can vary over orders of magnitude depending on the market penetration of the drugs of interest.

3.3 MODELLING TOXICITY

Human health risk assessment (HHRA) relies on model indicators usually designated to the ecotoxicological risk assessment (ERA) on aquatic and terrestrial fauna. Therefore, ERA indicators like predicted environment concentrations (PECs) (Anderson *et al.*, 2004; Bound and Voulvoulis, 2006b), predicted no effect concentrations (PNECs), PEC/PNEC ratio, etc, which are estimated according to various approaches to obtain the most accurate and realistic assessment, may be refined to take into account the human variability and interspecies differences (Dorne *et al.*, 2007). Nevertheless, the scarcity of data needed to evaluate the PECs and their own set limits make the assessment controversial (Jones *et al.*, 2005; Dorne *et al.*, 2007). Besides, PPs consumption, WWTP and DWTP removals, metabolization, and fate in the environment are variables which affect the uncertainty of the predictive factors. Moreover, they are intended to model wide geographical area, despite available data concern, for the majority of them, spatial and temporal isolated points (Fent *et al.*, 2006).

Recent tools such as toxicogenomic with its associated disciplines, and quantitative structure/activity relationships (QSARs) modelling, combined with probabilistic models, constitute efficient means for the comprehension of the human biologic effect and mechanisms (toxicokinetics and toxicodynamics) (Jones *et al.*, 2005; Dorne *et al.*, 2007).

Alternatively to the PEC/PNEC ratio, the I_{70} indicator is also used in the risk assessment with indirect long-term exposure via DW (Webb *et al.*, 2003). Nevertheless, similarly to PEC/PNEC, I_{70} benchmark does not take into account the cytotoxics which have no cut-off limit below which no carcinogenic effects may occur. Besides, chronic effects at low doses assessed with I_{70} indicator (Webb *et al.*, 2003; Kostich and Lazorchak, 2008) is intended for one single PP.

The comparison of measured or predicted environmental concentrations with the toxicologically relevant one is a crucial stage of risk assessment for aquatic systems. While much uncertainty is associated with the measurement or prediction of surface water concentrations, even more uncertainty can be found when trying to estimate safety levels (no observed-effect levels) based on acute or chronic toxicological data. This in turn will vary widely depending on the pharmaceutical(s) of interest and the receptor. A human health risk assessment of 26 pharmaceutical products in US waters was conducted by Schwab *et al.* (2005). Acceptable daily intakes (ADI) were estimated with the following relationship:

$$ADI = \frac{1000 \times POD}{UF_1 \times UF_2 \times UF_3 \times UF_4 \times UF_5} \tag{3.4}$$

where POD is the point of departure (mg/kg/day) often estimated from lowest therapeutic effect dose (from clinical trial prior to registration), and uncertainty factors (UF) related to interspecific variability, intraspecific variability, the extrapolation from a low effect to no effect level, the duration of exposure in toxicological studies, and a general "data quality" factor. Predicted no effect concentrations (PNEC) in drinking water and in fish tissue can then be estimated based on these ADI:

$$PNEC_{DW} = \frac{1000 \times ADI \times BW \times AT}{IngR_{DW} \times BCF \times EF \times ED} \tag{3.5}$$

$$PNEC_{F} = \frac{1000 \times ADI \times BW \times AT}{IngR_{F} \times BCF \times EF \times ED} \tag{3.6}$$

Where BW is body weight (kg), AT is average time (days), *IngR* is adult/child ingestion rate (L/day), EF is exposure frequency, ED is exposure duration and BCF is the bioconcentration factor (accumulation in fish tissue).

These PNEC values were then compared with measured pharmaceutical concentrations in US streams and with maximum measured concentrations in the literature. PEC values were also calculated using the *Ph*ATE programme (Anderson *et al.*, 2004) (assuming low flow and no depletion). Nineteen out 26 human medicines showed measured environmental concentrations (MEC) and PNEC values well below 1 for the three scenarios investigated there. A ratio between 0.1 and 1 was observed for ciprofloxacin (all scenarios), metformin (combined fish/drinking water), ranitidine (combined fish/drinking water) and warfarin (combined fish/drinking water).

By the nature of the tolerance distribution estimates of toxicity, such as the PNEC in the trials (approaching 100% and zero response) are associated with large uncertainties, and hence the need for the use of large safety factors. However, potentially more significant uncertainty is introduced when laboratory-based toxicity measurements using limited test species are used to extrapolate to environmental impact. A wide range of methods is available for biological monitoring, and these are reviewed in detail in Section 6. There is a need for further work in this area to compare available methods, to assess the representativeness of current toxicity test protocols, and to identify the most appropriate test species with low tolerances for various pharmaceutical classes in order to provide a margin of safety based on objective measurements rather than on arbitrary safety factors.

Future trends in HHRA are the evaluation of mixture effects. A single PP may be out of danger on humans meanwhile it may be toxic in the presence of other pollutants. Mostly of the few existing studies about mixture effects are dedicated to ERA (Jones *et al.,* 2005; Sanderson *et al.,* 2007; Cleuvers, 2003). However, a recent study estimates the human exposure to mixture PPs with similar mode of action (Kostich and Lazorchak 2008). It were been highlighted that daily dose is more rapidly reached when PPs are ingested as a mixture in DW (Kostich and Lazorchak, 2008).

Additive effects, synergic effect, i.e., the magnificiance or potentialization of one PP onto one another, inhibition or induction interaction effects (Dorne *et al.,* 2007).

REFERENCES

Anderson, P.D., D'Aco, V.J., Shanahan, P., Chapra, S.C., Buzby, M.E., Cunningham, V., Duplessie, O.M., Hayes, E.P., Mastrocco, F.J., Parke, N.J., Rader, J.C., Samuelian, J.H. and Schwab, B.W. (2004) Screening analysis of human pharmaceutical compounds in U.S. surface waters. *Environmental Science & Technology*. **38**(3), 838–849.

Bound, J.P. and Voulvoulis, N. (2004) Pharmaceuticals in the aquatic environment – a comparison of risk assessment strategies. *Chemosphere* **56**, 1143–1155.

Bound, J.P. and Voulvoulis, N. (2005) Household disposal of pharmaceuticals as a pathway for aquatic contamination in the United Kingdom. *Environmental Health Perspectives* **113**(12), 1705–1711.

Bound, J.P., Kitsou, K. and Voulvoulis, N. (2006a) Household disposal of pharmaceuticals and perception of risk to the environment. *Environmental Toxicology and Pharmacology* **21**, 301–307.

Bound, J.P., Kitsou, K. and Voulvoulis, N. (2006b) Predicted and measured concentrations for selected pharmaceuticals in UK rivers: Implications for risk assessment, *Water Research* **40**, 2885–2892.

Boxall, A.B.A., Fogg, L.A., Kay, P., Blackwell, P.A., Pemberton, E.J. and Croxford, A. (2003) Prioritisation of veterinary medicines in the UK environment. *Toxicology Letters* **142**, 207–218

Boxall, A.B.A., Fogg, L.A., Blackwell, P.A., Kay, P., Pemberton, E.J. and Croxford, A. (2004) Veterinary medicines in the environment. *Rev. Environ. Contam. Toxicol.* **180**, 1–91.

Carrara, C, Ptacek, C.J., Robertson, W.D., Blowes, D.W., Moncur, M.C., Sverko, E. and Backus S. (2008) Fate of pharmaceutical and trace organic compounds in three septic system plumes, Ontario, Canada. *Environ. Sci. Technol.* **42**(8), 2805–2811.

Cleuvers, M. (2003) Aquatic ecotoxicity of pharmaceuticals including the assessment of combination effects. *Toxicology Letters* **142**, 185–194.

Committee for Medicinal Products for Human Use (CHMP). (2006) Guideline on the Environmental Risk Assessment of Medicinal Products for Human Use. European Medicines Agency (EMEA) veterinary medicines and inspections, London. 1–12.

Dorne, J., Ragas A.M.J., Frampton G.K., Spurgeon, D.S. and Lewis, D.F. (2007) Trends in human risk assessment of pharmaceuticals. *Analytical and Bioanalytical Chemistry* **387**, 1167–1172.

Drewes, J.E. (2007) Removal of pharmaceutical residues during wastewater treatment. *Comprehensive Analytical Chemistry* **50**, 427–449.

European Medicines Agency (2006) Guideline on the environmental risk assessment of medicinal products for human use. London, 1st June 2006, CPMP/SWP/4447/00.

Fent, K., Weston, A.A. and Caminada, D. (2006) Ecotoxicology of human pharmaceuticals. *Aquatic Toxicology* **76,** 122–159.

Jones, O.A.H., Voulvoulis, N. and Lester, J.N. (2002) Aquatic environmental assessment of the top 25 English prescription pharmaceuticals. *Water Research* **36**, 5013–5022.

Jones, O.A.H., Lester, J.N. and Voulvoulis, N. (2005) Pharmaceuticals: a threat to drinking water? *Trends in Biotechnology* **23**, 163–167.

Johnson, A.C., Jurgens, M.D., Williams, R.J., Kummerer, K., Kortenkamp, A. and Sumpter, J. P. (2008) Docytotoxic chemotherapy drugs discharged into rivers pose a risk to the environment and human health? An overview and UK case study. *Journal of Hydrology* **348**, 167–175.

Jux, U., Baginski, R.M., Arnold, H.G., Kronke, M. and Seng, P.N. (2002) Detection of pharmaceutical contaminations of river, pond, and tap water from Cologne (Germany) and surroundings. *Int. J. Hyg. Environ. Health* **205**, 393–398.

Kemper, N. (2008) Veterinary antibiotics in the aquatic and terrestrial environment. *Ecological indicators* **8**, 1–13.

Khan, S.J., Roser, D.J., Davies, C.M., Peters, G.M., Stuetz, R.M., Tucker, R. and Ashbolt, N.J. (2008) Chemical contaminants in feedlot wastes: Concentrations, effects and attenuation. *Environment International* **34**(6), 839–859.

Kostich, M.S. and Lazorchak, J.M. (2008) Risks to aquatic organisms posed by human pharmaceutical use. *Science of The Total Environment* **389**, 329–339.

Kummerer, K. (2001) Drugs in the environment: emission of drugs, diagnostic aids and disinfectants into wastewater by hospitals in relation to other sources – a review. *Chemosphere* **45**, 957–969.
Mompelat, S., Le Bot, B. and Thomas, O. (2009) Occurrence and fate of pharmaceutical products and by-products, from resource to drinking water. *Environment International* **35**(5), 803–814.
Perez, S. and Barcelo, D. (2007) Fate and occurrence of X-ray contrast media in the environment. *Anal Bioanal Chem* **387**, 1235–1246.
Radjenović, J., Petrović, M., Barcelo, D. and Petrović, M. (2007) Advanced mass spectrometric methods applied to the study of fate and removal of pharmaceuticals in wastewater treatment. *Trends in Analytical Chemistry* **26**(11), 1132–1144.
Reddersen, K., Heberer, T., Dunnbier, U. (2002) Identification and significance of phenazone drugs and their metabolites in ground- and drinking water. *Chemosphere* **49**, 539–544.
Sanderson, H., Laird, B., Pope, L., Brain, R., Wilson, C., Johnson, D., Bryning, G., Peregrine, A.S., Boxall, A. and Solomon, K. (2007) Assessment of the environmental fate and effects of ivermectin in aquatic mesocosms. *Aquatic Toxicology* **85**, 229–240.
Sarmah, A.K., Meyer, M.T., Boxall, A.B.A. (2006) A global perspective on the use, sales, exposure pathways, occurrence, fate and effects of veterinary antibiotics (Vas) in the environment. *Chemosphere* **65**, 725–759.
Schwab, B.W., Hayes, E.P., Fiori, J.M., Mastrocco, F.J., Roden, N.M., Cragin, D., Meyerhoff, R.D., D'Aco, V.J. and Anderson, P.D. (2005) Human pharmaceuticals in US surface waters: A human health risk assessment. *Regulatory Toxicology and Pharmacology* **42**, 296–312.
Webb, S., Ternes, T., Gibert, M. and Olejniczak, K. (2003) Indirect human exposure to pharmaceuticals via drinking water. *Toxicology Letters* **142**, 157–167.

Chapter 4

Real situation: Occurrence of the main investigated PPs in water bodies

Alexa Sadezky, Dirk Löffler, Michael Schlüsener, Benoit Roig and Thomas Ternes

4.1 INTRODUCTION

For several years the presence of pharmaceutical products (PPs) has been investigated in various aqueous matrices such as raw wastewater (WWTPinfluent), wastewater treatment plant effluents (WWTPeffluent) (Paxeus *et al.,* 2004; Schlusener and Bester 2005; Gros *et al.,* 2006), surface water (SW) (Wiegel *et al.,* 2004; Seitz *et al.,* 2006, Kasprzyk-Hordern *et al.,* 2008, Tamtam *et al.,* 2008), bank filtrate (BF) (Heberer *et al.,* 2004), groundwater (GW), drinking water (DW) (Reddersen *et al.,* 2002; Ternes *et al.,* 2002; Petrovic *et al.,* 2003; von Gunten *et al.,* 2006; Zwiener, 2007) and marine water (MW) (Buser *et al.,* 1998; Weigel *et al.,* 2002; Bruchet *et al.,* 2005). Recent reports have shown the presence of a large variety of PPs in WWTP effluents and in surface waters, with concentrations up to several µg/L (Ternes *et al.,* 1998; Kolpin *et al.,* 2002; Metcalfe *et al.,* 2003; Zuehlke *et al.,* 2004; Roberts *et al.,* 2006; Buerger *et al.,* 2006). More than 180 PPs have been detected in various environmental matrices. This is mainly due to the continuous improvement of analytical techniques which allow to identify and quantify a broad range of substances at concentration levels down to the lower ng/L range. Considerable effort is moreover being made in the development of specific analytical techniques for the determination of PPs in complex matrices (including wastewaters and sludge) and in the assessment of their environmental behaviour (sorption, biodegradability). The majority of the pharmaceutical residues enters municipal WWTPs via domestic and hospital wastewaters or through discharges of the pharmaceutical industry. Up to date the PPs most frequently studied in WWTPs are antibiotics, antiepileptics, antiphlogistics, X-ray contrast media, lipid regulators, betablockers and tranquillizers (Moffat *et al.,* 2004). These compounds vary strongly concerning their chemical structure and their physico-chemical properties.

The presence of human pharmaceuticals can be used to indicate to which extent water bodys such as rivers, ground water and drinking water are influenced by municipal wastewater. In principle, the percentage of wastewater (treated and non-treated) in surface waters can be estimated with concentrations of appropriate pharmaceuticals.

The aim of this study was to evaluate and to interconnect current knowledge (scientific publications and communications, national and international project reports, reports from national water management organisations and data bases), in order to establish an inventory list of PPs most relevant in terms of exposure to the aquatic environment. The environmental relevance of PPs is evaluated on the basis of existing data from occurrence measurements in the aquatic environment, consumption patterns in different European countries, physico-chemical properties, and available data about pharmacokinetics and environmental fate.

The data compilation was further used in this work to predict environmental concentrations of PPs and to verify their agreement with measured data. Another objective was to evaluate whether the presence of PP indicators could be used to determine the presence and proportion of different contaminating sources in rivers, lakes and ground water. Finally all compiled data was used to identify data gaps in the current knowledge on PPs in the environment, which might then be used as a starting point in future studies.

The results of this work can then be used to develop a classification system for PPs in order to prioritize their environmental risks. The classification system is supposed to support water managers, health authorities and special panels involved in the establishment of river basin management plans and strategies to minimize point and diffuse pollution of surface waters. It helps to identify those PPs of most concern, which should then be integrated into future monitoring programs.

4.2 LITERATURE STUDY ON PPS

PPs of concern are those which are administered and sold in European countries in high quantities, are detected in the aquatic environment and/or having a high ecologic risk. For that purpose, the content of relevant national and international scientific literature, reports and data bases regarding these topics were evaluated and relevant literature data were summarized. A data compilation was elaborated containing data on consumption, occurrence, physico-chemical properties, pharmacokinetics (metabolization-, excretion rates) and environmental fate (persistence, transformation, sorption) of PPs.

In order to collect the relevant data on pharmaceutical products, the content of several databases were considered: e.g. Clarke's Analysis of Drugs and Poisons (Moffat *et al.,* 2004); Hazardous Substances Data Bank (HSDB) (2007); AccessMedicin (2007). Data on the occurrence of PPs in the aquatic environment was extracted from a total of 170 references (national reports, Ph.D., Theses, national and international scientific literature) from the years 1998 to 2007 (Sadezki *et al.,* 2007). Literature data were assembled to quantitatively assess the occurrence of pharmaceuticals in seven compartments: treated and raw wastewater, surface water (rivers and lakes), bank filtrate, ground water, drinking water, and marine coastal water. National consumption data for pharmaceutical products (1998 to 2006) was contributed by the respective project partners from the five European countries participating in the KNAPPE project.

Some active compounds used in pharmaceutical products were excluded, e.g. caffeine and quinine, since some substances are additionally widely used as additives in soft drinks (e.g. quinine in concentrations of up to 85 mg/L in "Tonic-Water"). Their impact on the aquatic environment due to their application in pharmaceutical products can be neglected. Furthermore, theophyllin is used in soft-drinks and occurs also naturally at low concentrations in tea and chocolate. Natural hormones were also excluded since these substances are naturally excreted by mammals (e.g. humans). Since most existing analytical methods for the determination of EE2 in other aqueous matrices (e.g. surface water, bank filtrate, groundwater, tap water and marine/coastal water) are not reliable at the relevant environmental concentrations at the pg/L-level, EE2 was totally excluded from the data list.

4.2.1 Consumption

National consumption (prescription/sales) data from 1998 to 2006 was obtained from the five European countries participating in KNAPPE: France, Germany, Poland, Spain and the UK (England and Wales). National consumption data were normalized in metric tons per year for each country and in g/person/year considering the total population of the countries (Table 4.1 and Table 4.2). The complete dataset is given in Table 4.9.

Most of the available consumption data on Pharmaceuticals were published for Germany (BLAC, 2003), France (Besse and Garric, 2007) and UK (Sebastine *et al.,* 2003; Webb, 2004). For Germany, consumption quantities are mainly based on the numbers of prescribed daily doses for the different pharmaceuticals multiplied by the respective amount of a defined daily dose (DDD). The substance specific amounts of the DDDs were obtained from the WHO (WHO, 2006) for the most relevant medicinal application. For England and Wales the consumed quantities of PPs were obtained by summing up all prescribed amounts of each compound considering the various products, packages sizes and formulations. French data was provided from the AFFSAPS (Agence Française de Sécurité Sanitaire des Produits de Santé); for 2004 the AFFSPAPS data (top 100 products most consumed, sales numbers) were recalculated to obtain kg/year by Besse and Garric (2007). For Poland and Spain, only the numbers of sold packages/containers were available. Therefore, consumption was predicted using a simple approach: The quantity of the pharmaceutical in the smallest available package was multiplied by the number of packages sold. As an example, for Poland 38, 687, 702 packages of paracetamol were sold in 2006. The smallest package contains 10 pills with each 500 mg of the active compound. Therefore it can be estimated that in Poland 193, 439 kg of paracetamol were used in 2006, which is comparable with other European countries on an annual per capita basis (Poland 5 g/capita, Germany 8 g/capita, Spain 4 g/capita, UK 7 g/capita).

Table 4.1 Mean annual consumption of selected pharmaceuticals in metric tons for France, Germany, Poland, Spain and the UK (England, Wales). Data available for the period from 1999–2006.

Class	Compounds	France	Germany	Spain	UK	Poland
Analgesic	Acetylsalicylic acid	638	1125	no data	82	no data
	Diclofenac	10	73	2	28	19
	Ibuprofen	203	261	108	149	193
	Paracetamol	2799	367	147	821	186
Antiulcerative	Ranitidine	12	85	20	31	no data
Antiepileptic	Carbamazepine	36	81	no data	40	32
Antibiotic	Amoxicillin	386	99	no data	80	no data
	Sulfamethoxazole	20	54	no data	1	7
Antihypertensive	Troxerutin	444	41	no data	no data	no data
Antidiabetic	Metformin	717	518	no data	309	no data
Urostatic	Allopurinol	54	138	no data	24	no data
Contrast medium	Iopromide	73	65	no data	no data	no data

Table 4.2 Average annual consumption of selected pharmaceuticals in g per person and year for France, Germany, Poland, Spain and the UK. Data available for the period from 1999–2006.

		Annual per capita consumption of PPs (g/cap a)				
Class	**Compound**	**France**	**Germany**	**Spain**	**UK**	**Poland**
Antibiotic	Amoxicillin	6.50	1.20	no data	1.54	no data
	Ciprofloxacin	0.21	0.17	0.09	0.12	0.13
	Clarithromycin	0.25	0.12	0.13	0.08	0.27
	Sulfamethoxazole	0.34	0.65	no data	0.02	0.17
Antiepileptic	Carbamazepine	0.61	0.98	no data	0.77	0.84
Analgesic	Tramadol	0.44	0.30	no data	0.27	no data
	Ibuprofen	3.4	3.2	2.6	2.8	5.04
	Paracetamol	47.14	4.46	3.60	15.68	4.84
Contrast medium	Iopromide	1.24	0.79	no data	no data	no data
Antidiabetic agent	Metformin	12.1	6.3	no data	5.9	no data

The data from France reflects the total amount consumed including those quantities sold without prescription – freely over the counter (OTC-drugs). In contrast, the consumption data from Germany, Poland, Spain and the UK (England and Wales) do not include OTC-drugs. The quantity of a pharmaceutical sold in OTC-products can be a considerable contribution to the total quantity used, in some cases, e.g. ibuprofen or paracetamol, even higher than the amount prescribed. The different procedures of surveying PP consumption in these five countries illustrate that consumption data of pharmaceuticals are merely a qualified estimation.

Apart the therapeutic pharmaceuticals, consumption data for diagnostics, e.g. iodinated contrast media, are hardly available. In total, consumption data was available for 50 PPs for France, 130 PPs for Germany, 31 PPs for Spain, 98 PPs for the UK and 25 PPs for Poland.

4.2.2 Exposure

The compilation of exposure data prepared within the KNAPPE Project includes 58400 records of environmental measurements for 178 PPs and metabolites which were detected in samples of the aquatic environment from 22 countries. The data points comprise about 4300 measurements of raw wastewater (WWTPinf), 15000 data points for treated wastewater (WWTPeff), 28200 for surface water (SW), 1700 for bank filtrate (BF), 6700 for ground water (GW), 1300 for tap water (DW) and 1100 for marine/coastal water (MW) (Table 4.3).

Table 4.3 Numbers of measured occurrence data – data compilation for aquatic matrices in different countries.

Country	WWTPinf	WWTPeff	SW	BF	GW	DW	MW	Sum
Australia					16			**16**
Austria	42	90	31		42			**205**
Belgium	3	3						**6**
Brazil		12	2					**14**
Canada	714	1477	4865			30		**7086**
China	15	15	246			15	180	**471**
Croatia	135	135						**270**
Czech Rep.			72					**72**
Finland	202	202	128					**532**
France	463	1026	1606	31	60	133	232	**3551**
Germany	1137	8888	15133	1689	6537	744	17	**34145**
Greece		135						**135**
Italy		436	544		12	52		**1044**
Japan	66	66						**132**
Korea	141	278	120					**539**
Norway	25	40					24	**89**
Romania			64					**64**
Spain	540	462	432					**1434**
Sweden	69	282						**351**
Switzerland	64	431	280					**775**
UK	150	248	925					**1323**
USA	559	807	3764		36	313	667	**6146**
Sum	**4325**	**15033**	**28212**	**1720**	**6703**	**1287**	**1120**	**58400**

(WWTPinf = WWTPinfluent, WWTPeff = WWTPeffluent, SW = surface water, BF = bank filtrate, GW = groundwater, DW = drinking water, MW = marine water)

From the data included, WWTP effluent and surface water samples represent almost 75% of all measurements. Only very few data was available for the other aquatic compartments. WWTP influent samples seem under-represented, as probably not all analytical research groups were able or willing to measure in highly matrix loaded STP influents. This might be due to the fact, that analytical methods for surface water or groundwater frequently have to be improved to be applicable for raw sewage. The availability of occurrence data for PPs in drinking water and marine water is also very limited. It should be emphasized that a significant share of data on PPs in water, concerning especially measurements of PPs in drinking water samples, remains non-published or confidential, and was thus not available for this data evaluation. Moreover, for reliable measurements in tap water and marine water, analytical procedures have to be improved or newly developed to enhance the limits of quantification of PPs into the picogramm per liter range.

Table 4.3 also shows an overview of the number of measurements for several countries. More than half of all measured data available originated from Germany. For all other countries, especially within the EU, significantly less occurrence data was available.

Mean, maxima, medians, 90-percentiles were considered for WWTP influents and effluents, surface water, bank filtrates, ground water, tap water and marine/coastal waters.

To enable a comparison of the literature data, an *Average Measured Environmental Concentration* (A-MEC) was defined characterizing the "mean" concentration of a PP in each of the compartments. A-MECs were calculated for all countries for which occurrence data were available. The A-MEC of a pharmaceutical compound in a specific compartment was calculated as the weighted mean of all data points, reported as statistical median. In those cases where no median was available the mean was used for the calculation of the A-MEC instead. For statistical reasons all values below the limit of detection (LOD) or below the limit of quantification (LOQ) were set to 0. The mathematical expression for the calculation of the weighted mean follows Eq. 4.1:

$$A-MEC = \frac{\sum_{i=1}^{n}(\text{Number of samples}_{\text{subset}} \times \text{Median Concentration}_{\text{subset}})}{\sum_{i=1}^{m} \text{Number of samples}_{\text{total}}} \quad (4.1)$$

The resulting A-MEC values and the maximum measured concentrations for PPs in the seven compartments of the aquatic environment are summarized in Table 4.10.

Figure 4.1 shows the distribution of studied samples according to therapeutic classes of PPs. Most frequently measured were antibiotics, followed by antiphlogistics (analgesics, anti-inflammatory drugs).

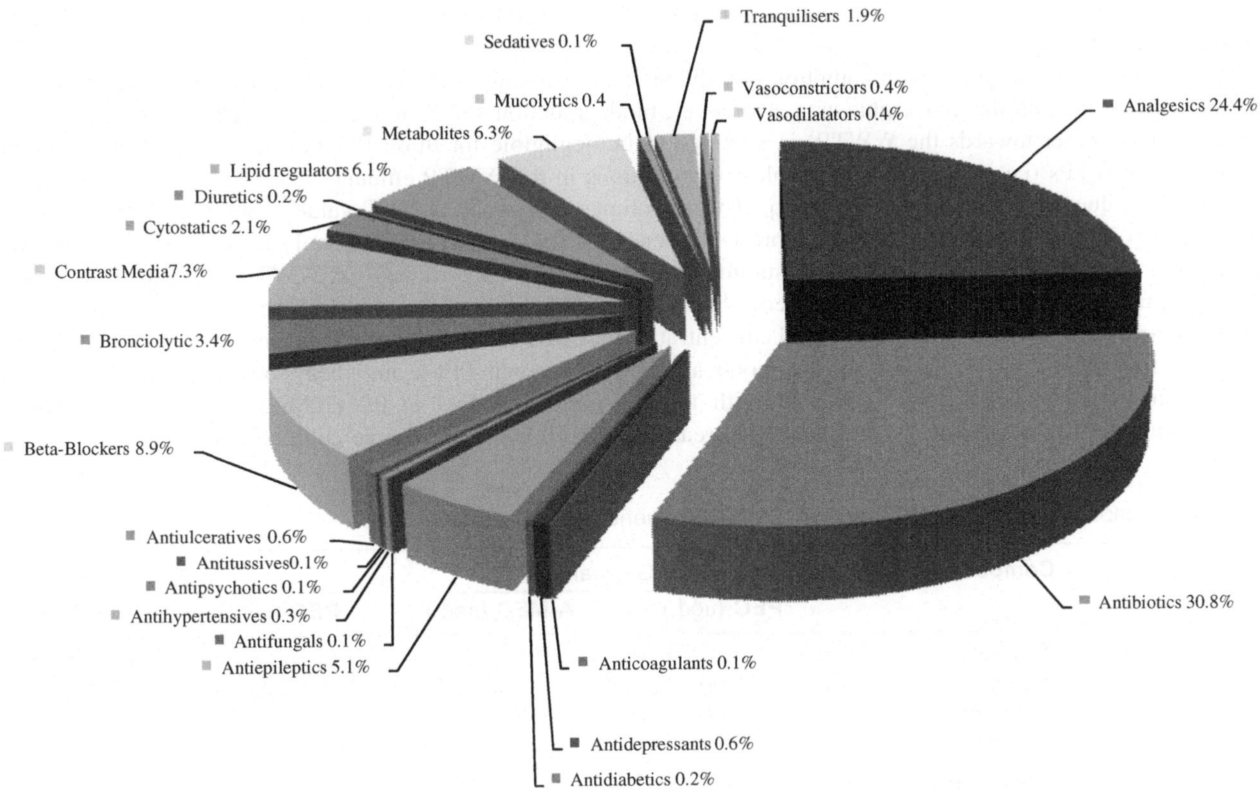

Figure 4.1 Partitioning of quantified samples based on their medical application.

4.3 RESULTS AND DISCUSSION

The current inventory list represents the state of our work on November 1st, 2007. In the following chapters, this data compilation was used for three different purposes, in view of a classification and prioritization of PPs in future strategies for environmental monitoring.

First, predicted environmental concentrations (PECs) were calculated using consumption data, pharmacokinetics, and environmental persistence.

The compilation was further used to propose and verify pharmaceutical residues as indicator substances to identify major contamination sources (e.g. wastewater from municipal WWTP discharges, run-off from agricultural fields, hospitals) in the variety of compartments of the aquatic environment.

Moreover, major gaps were identified in terms of data availability and data quality, which limit the significance of the currently available data for the establishment of an exhaustive environmental risk assessment for PPs.

4.3.1 Predicted environmental concentrations

The reliable prediction of environmental concentrations of pollutants is a pre-requisite for an environmental risk assessment (ERA), in order to understand the potential impact of pharmaceuticals in the environment (Rombke *et al.*, 2001). Such methods provide powerful tools for the prioritization of PPs according to their environmental risks.

Based on the consumption data of PPs collected in this work, predicted environmental concentrations (PEC) were calculated in raw wastewater (WWTP inf) according to the Poseidon-approach given in eq. 4.2 (Ternes and Joss, 2006), which is based on drug consumption, metabolism and excretion of the unchanged compound, and population figures.

$$PEC_{\mathrm{WWTPin}} = \frac{F_{API} \cdot E \cdot 10^9}{365 \cdot Pop \cdot AWW} \qquad [\mu g/L] \tag{4.2}$$

With F_{API}: consumption of an active pharmaceutical ingredient per year in a country [kg/a]
E: Fraction excreted in urine and faeces
AWW: amount of wastewater per capita and day [L/(cap d)], set to 250 $L \times cap^{-1} \times d^{-1}$
Pop: Population of the country [cap]

Excretion data of the unchanged compound were taken from several sources (Lienert *et al.*, 2007; Joss *et al.*, 2006; Moffat *et al.*, 2004).

The calculation is based on the assumption that the estimated country wide consumption is evenly distributed over the year and throughout the geographic area. Moreover, biodegradation of PPs in the sewer during the transport of the untreated wastewater towards the WWTPs is assumed to be negligible for those PPs not being completely degraded (<90%) in WWTPs (i.e. present in measurable concentrations in the WWTP effluent).

The PEC values calculated according to eq. II were compared with the A-MEC values obtained for the respective compounds (Chapter 4.2.2). To ensure the representativeness of the measured values, only those A-MEC values were considered, which comprise at least 20 measurements. However, A-MEC values for WWTP influents comprising a sufficient number of single measurements were obtained for Germany and Spain only. The predicted environmental concentrations (PECs) and averaged measured concentrations (A-MECs) of selected pharmaceuticals in WWTP influent are given in Table 4.4. The values show a good agreement between PECs and their corresponding A-MECs for WWTPinfluent. The results also agree well with those reported by Joss *et al.* (2006). This illustrates that the concentrations in raw wastewater can be predicted reasonably well using the simple approach of Eq. (4.2).

Table 4.4 Comparison of PEC/A-MEC values: WWTP influent.

Class	Compound	Germany		Spain	
		PEC (μg/L)	A-MEC (μg/L)	PEC (μg/L)	A-MEC (μg/L)
Analgesic	Diclofenac			0.24	0.53
	Ibuprofen	2.90	2.43	5.79	21.12
	Naproxen	5.56	3.51	0.43	0.58
	Paracetamol			25.61	30.97
Antibiotics	Clarithromycin	0.23	0.21		
	Erythromycin	0.23	0.19		
	Roxithromycin	0.47	0.16		
	Sulfamethoxazole	1.43	0.74		
	Trimethroprim	1.07	0.29		
Antiepileptic	Carbamazepine	0.54	1.60		
Beta-blocker	Atenolol	0.61	0.92		
	Metoprolol	1.20	2.05		
	Propranolol	0.19	0.81		
	Sotalol	3.29	1.26		
Tranquiliser	Diazepam	0.01	<LOQ		

Several PPs show a high average annual consumption and consequently high predicted environmental concentrations $PEC_{WWTPinf}$ in raw wastewater, although measurements in the aquatic environment in Europe have not been reported so far. Table 4.3 lists all those compounds whose $PEC_{WWTPinf}$ are higher than 1 μg/L in at least one of the four countries Germany, France, Spain, and UK. Among them, the antidiabetic drug metformin shows by far the highest $PEC_{WWTPinf}$ in all four countries. Measurements in the U.S. have revealed concentrations of this drug of 10 to 26 μg/L in WWTP influents and effluents (Benotti *et al.*, 2007), and up to 0.15 μg/L in surface water (Kolpin *et al.*, 2002). The values are in agreement with the high $PEC_{WWTPinf}$ in the four countries participating in the KNAPPE project (Table 4.5), and therefore, similar concentrations of metformin can be expected for European water bodies. Other compounds with high $PEC_{WWTPinf}$ between 18 μg/L and 82 μg/L especially in France and Germany are the antihypertensive troxerutin, the phlebotropic diosmin, the mucolytic carbocystein, the antiepileptic valproic acid, and the urostatic allopurinol.

4.3.2 Indicator pharmaceuticals

The data compilation was further used to propose and verify pharmaceutical residues as environmental indicator substances for the presence of wastewater in the different compartments of the aquatic environment. The presence of human pharmaceuticals points towards a significant proportion of wastewater in surface- and ground waters and therefore the potential contamination with a wide range of polar contaminants. In order to identify indicator substances, PPs in the data compilation were evaluated with regard to their removal in WWTPs.

Table 4.5 PPs with $PEC_{WWTPinf}$ higher than 1 μg/L, for which measurements in the aquatic environment in Europe have not been reported (PEC_a: PEC calculated assuming metabolisation as negligible (E = 1); PEC_b: PEC calculated considering only the fraction of a compound excreted non-metabolised).

Class	Compound	$PEC_{WWTPinf}$ [μg/L]							
		Germany		France		Spain		UK	
		PEC_a	PEC_b	PEC_a	PEC_b	PEC_a	PEC_b	PEC_a	PEC_b
Analgesic	Mesalazine	7.93						9.64	
Antibiotic	Sulfasalazine	5.10	0.51					9.93	0.99
Anticoagulant	Clopidogrel	1.39	0.69			1.25	0.63	1.81	0.90
Antidiabetic	Metformin	69.01	55.21	132.34	105.87			64.68	51.74
Antiepileptics	Valproic acid	7.92	1.98	20.71	5.18			5.55	1.39
Antihypertensive	Captopril	3.17	1.43			0.33	0.15	0.39	0.17
	Irbesartan	1.98				1.4		3.67	
	Lactitol	3.06						<0.001	
	Troxerutin	5.47		82.03					
	Verapamil	7.68	0.23					1.52	0.05
Contrast medium	Ioversol	5.80	5.80						
Diuretic	Torasemide	0.49				9.91		0.002	
Lipid regulator	Atorvastatin	0.63	0.010			0.69	0.01	2.32	0.02
Mucolytic	Carbocistein			42.89				2.32	
Phlebotropic	Diosmin			68.96					
Urostatic	Allopurinol	18.37	5.15	10.02	2.80			5.09	1.42

The Percentage of a compound *i* Not Removed (PNR_i) during wastewater treatment is characterized by the ratio of its measured concentrations in the WWTP effluent and influent:

$$PNR_i = \frac{concentration\ i\ in\ WWTPeffluent}{concentration\ i\ in\ WWTPinfluent} \tag{4.3}$$

This ratio was calculated from the measured concentrations of each study which reported measurements of PPs in both WWTP effluent and influent. The resulting ratios for each PP were averaged. The complete set of available average removal ratios is given by Sadeki *et al.*, 2007.

Assessment of the occurrence of pharmaceutical products in the wastewater treatment plant influents and effluents, with respect to their removal behaviour during wastewater treatment, allows to identify two different categories of potential environmental indicator substances, as described in the following.

Category I: Indicators for wastewater originated contamination

For determination of the proportion of treated wastewater in an aquatic compartment, those PPs are chosen which, upon comparison of their environmental concentrations in WWTP influents and effluents, are shown not to be removed by usual municipal wastewater treatment. Concentrations of such PPs in the WWTP effluent and influent are similar, leading to a ratio PNR_i of around 1. Most substances which are not eliminated in WWTPs are also persistent in the environment and are not easily removed by degradation or sorption to soil or sediments. Therefore, their concentrations in surface waters mainly depend on the dilution of wastewater discharged into the environment, and thus represent the wastewater proportions present in rivers and lakes.

The wastewater proportion in a surface water can be calculated by dividing the (average) concentration of a Category I indicator compound in a surface water by its (average) concentration in wastewater effluent, which is introduced into that surface water.

Table 4.6 lists all those PPs with their average ratio PNR_i above 0.7, which are eligible as indicator compounds for the proportion of treated wastewater. The calculated values of the average PNR_i are most reliable for those PPs, for which large numbers of concentration measurements from different studies have been reported. Therefore, only those PPs have been taken into consideration, for which at least 10 measurements in both influent and effluent are included into the data compilation. Moreover, their average surface water concentrations should be not less than 0.1 μg/L, in order to ensure a reliable quantification.

Substances of particular interest according to these criteria are the contrast media diatrizoate and iopamidol, the antiepileptic carbamazepine, the antibiotic erythromycin, and the beta-blocker metoprolol. Diatrizoate is especially known for its high persistence in both WWTP and the environment (Ternes, 2004). Among the compounds poorly removed in WWTP are furthermore the antibiotics clarithromycin and azithromycin, the antidepressant fluoxetine, the analgesics phenazone and propyphenazone, and the beta-blocker propanolol. However, the average concentrations of these compounds in surface water are generally below 0.03 μg/L (Table 4.10) and therefore close to the limits of quantification (LOQ) of the analytical methods.

Table 4.6 Pharmaceutical agents eligible as environmental indicator compounds for the share of treated wastewater (average $PNR_i \approx 1$).

Compound	Therapeutic class	Number of samples		Average PNR_i
		WWTP effluent	WWTP influent	
Propranolol	Beta-Blocker	182	47	1.64
Diatrizoate	Contrast Medium	10	10	1.12
Clarithromycin	Antibiotic	245	120	1.23
AMDOPH	Metabolite	12	12	1.09
N-Formyl-4-aminoantipyrine	Metabolite	12	12	1.08
Fluoxetine	Antidepressant	15	15	1.02
Phenazone	Analgesic	178	37	1.02
AMPH	Metabolite	12	12	1
Iopamidol	Contrast Medium	10	10	0.96
Carbamazepine	Antiepileptic	401	209	0.96
Erythromycin[1]	Antibiotic	215	209	0.96
CBZ-DiOH	Metabolite	15	14	0.92
Azithromycin	Antibiotic	25	19	0.82
Propyphenazone	Analgesic	73	29	0.81
4-Acetaminoantipyrine	Metabolite	12	12	0.80
Metoprolol	Beta-Blocker	208	74	0.77

Several metabolites of analgesics and carbamazepine (Table 4.6) are among the potential indicator substances. As for all metabolites only very few measurements have been reported (mostly performed within a single study), general conclusions about their applicability as environmental indicator compounds cannot be drawn from the present dataset. However, the wide variety of metabolites stable in WWTP illustrated in Table 4.6 indicates that a closer and more general survey of PP metabolites in WWTP and surface waters might be useful for future strategies of water quality assessment with regard to PPs as organic trace pollutants.

Generally, treated wastewater contaminated with some of the above-mentioned PP indicator substances is expected to contain also a large variety of other polar persistent organic compounds, besides the PPs, not being removed during wastewater treatment.

For groundwaters, the presence of the above mentioned PPs also hints an exposure to wastewater, although the wastewater proportions cannot be quantified. In particular, the iodinated X-ray contrast medium diatrizoate has repeatedly been identified in ground water with an A-MEC of 1.5 μg/L and a maximum concentration of 9.6 μg/L (Table 4.10). The antibiotic sulfamethoxazole is often identified in ground water (Table 4.10) and might therefore also serve as an indicator for wastewater exposure, although this compound is not suggested as an indicator for the proportion of treated wastewater in surface waters due to an average ratio PNR_i of 0.6.

Category II: Indicators for basic removal of polar contaminants in wastewater treatment

PPs which are completely removed during basic wastewater treatment ($PNR_i \approx 0$) can serve as indicators for the presence of raw or insufficiently treated wastewater in rivers and lakes, e. g. due to WWTP malfunction or overflow events. Table 4.7 lists all PPs eligible for this type of environmental indicator, with their average ratio PNR_i being below 0.2.

Only for 4 PPs meeting these criteria, more than 10 measurements in both WWTP influents and effluents have been published. These substances are the analgesics ibuprofen and paracetamol, the metabolite salicylic acid, and the lipid regulator bezafibrate. In particular, the analgesics paracetamol and ibuprofen are suggested as indicators for input of widely untreated wastewater. Complete elimination during wastewater treatment was also reported for ofloxacin. However, this compound was discarded as a potential indicator substance, due to its high tendency to sorb onto solid matter.

[1] Usage of the sum of erythromycin and anhydro-erythromycin is recommended.

Table 4.7 Pharmaceutical agents eligible as environmental indicator compounds for the presence of raw or insufficiently treated wastewater (average PNR_i < 0.2).

		Number of samples		
Compound	**Therapeutic class**	**WWTP effluent**	**WWTP influent**	**Average PNR_i**
Salicylic acid	Metabolite	91	88	0.22
Paracetamol	Analgesic	57	56	0.21
Ibuprofen	Analgesic	384	211	0.15
Bezafibrate	Lipid Regulator	201	58	0.14

The presence of poorly treated wastewater or even raw wastewater in surface waters might have major consequences for the water quality and the usage of these water bodies, e.g. bathing and swimming, or as drinking water resource due to contamination with pollutants and pathogens.

For both types of indicator substances, those PPs which are used during the whole year in similar quantities are most suitable as environmental indicators. For example, the antiepileptic carbamazepine is consumed by patients regularly and in constant daily doses. This is not necessarily the case for contrast media, which are applied in single high doses before X-ray diagnostics, and are excreted by patients within hours following the examination. That leads to a more irregular input of these substances into the sewage system. This effect is particularly noted in effluent discharges from small WWTPs serving a limited number of households, and tends to be attenuated in large facilities of major cities.

The presence of indicator substances of category I and II in rivers, lakes and in drinking water resources provides hints that a significant proportion of wastewater, seepage water from landfill sites or runoff water from agricultural fields is present. In addition to the PP indicator substances, these waters might contain other contaminants as well.

No PPs could be classified as indicator substances specific for the input from hospitals or veterinary medicine. Potential indicator substances for hospital input either lacked the specificity of only being used and excreted in a hospital environment or their environmental concentrations were too low for a reliable quantification.

Due to the different registration situation in the various countries, with deviating and/or overlapping use in human and veterinary medicine, no pharmaceuticals are in general fully attributable to veterinary purposes only.

4.3.3 Data gaps

Several data gaps were identified from the present data compilation.

The dataset from environmental measurements used for the assessment of the occurrence of PPs in the aquatic environment was found to be rather heterogenic in many aspects. Considerable differences exist concerning characteristics of the sampling locations and time of measurements, sampling procedures (grab samples and composite samples), analytical methods (methodology and quality assurance) and data evaluation and reporting, thus leading to a limited comparability of the collected data. These aspects are so much more critical, as for many compounds only few data is available.

Moreover, samples were frequently taken from sites with a special exposure situation, e.g. receiving waters near WWTPs, vulnerable ground waters infiltrated by wastewater or contaminated river water, and therefore these sites hardly reflect the general exposure situation. In particular, data for ground water, drinking water and marine coastal water, which are already rather limited, moreover often originate from scientific studies conducted at particularly exposed locations. Hence, these data do not show a representative view of the exposure of these compartments to contamination with PPs. The inclusion of non-public data on drinking water, such as national reports, quality surveys carried out by the drinking water suppliers, etc., would help to draw a more general picture of the exposure situation, however, access to these data was found to be rather difficult.

Concerning the consumption data of PPs used for prediction of their environmental concentrations, for several countries, they only comprise those quantities of PPs prescribed by physicians. The lack of data about the "over-the-counter" sale of non-prescription drugs leads to an underestimation of the consumption of certain PPs, in particular of analgesics.

Several PPs show a high average annual consumption and consequently high predicted environmental concentrations $PEC_{WWTPinf}$ in raw wastewater, although measurements in the aquatic environment in Europe have not been reported so far. In view of their potentially high input, a closer investigation of their occurrence and environmental fate would be desirable.

Furthermore, many pharmaceuticals are known to be extensively metabolized into phase I metabolites. Such metabolites might be of concern especially for those PPs with a high consumption and low predicted and measured environmental concentrations of the parent compound. Examples of PP with potentially relevant phase I metabolites are given in Table 4.8. However, environmental occurrence data is rare or not available at all for most known phase I metabolites. As a consequence, further work should be concentrated on the detection and quantification of metabolites and also on environmental transformation products of PP in environmental water compartments.

Table 4.8 Compounds undergoing major phase I metabolisation.

Compound	Metabolite	Proportion excreted [%]
Diclofenac	4'-hydroxydiclofenac	20
Metoprolol	4-(2-hydroxy-3-isopropylamino-proproxy)phenylacetic acid	65
Allopurinol	Oxipurinol	70
Oseltamivir	Oseltamivir hydrolyzed	75

4.4 CONCLUSIONS AND OUTLOOK

An extensive data compilation on the environmental occurrence of PPs was created, containing 58400 measurements of 178 pharmaceutical products in 22 countries in WWTPinfluent, WWTPeffluent, surface water, groundwater, bank filtrate, marine water and drinking water. Average mean environmental concentrations (A-MECs) were calculated from the compiled data for each PP, characterizing the exposure situation in the seven compartments of the aquatic environment in each country, from which measurements were available.

Additionally, consumption data of PPs for five European countries (France, Germany, Poland, Spain and the UK) was used to predict their environmental concentrations following the approach used in the EU-Project Poseidon. A comparison revealed that predicted and averaged measured environmental concentrations matched well for WWTPinfluent. The data compilation was used to establish a set of indicator substances for the determination of the wastewater share in surface waters and for determining whether a water body is influenced by poorly or raw wastewater.

Category I indicator substances for the determination of the wastewater share are:

- Diatrizoate (iodinated X-ray contrast medium)
- Iopamidol (iodinated X-ray contrast medium)
- Carbamazepine (antiepileptic)
- Erythromycin (antibiotic)[2]
- Metoprolol (betablocker)

Surface waters contaminated with these compounds are expected to contain a variety of other polar persistent organic compounds, which were also not removed during wastewater treatment.

Category II indicator substances showing the presence of non- or poorly treated wastewater are:

- Ibuprofen (Analgesic)
- Paracetamol (Analgesic)
- Salicylic acid (Analgesic)
- Bezafibrate (Lipid regulator)

The presence of these category II indicator substances which are readily (bio)degradable in wastewater treatment, in surface waters indicate the input of non- or poorly treated wastewater with potential consequences for water quality and water usage, concerning pollutants and pathogens.

Currently, the available data on the consumption and the occurrence of PPs in the aquatic environment on a European level is of very limited comparability.

Efforts should be made to achieve a common book keeping system on the usage of all pharmaceutical products within the EU, including prescribed PPs, OTC drugs, and all compounds used for diagnosis and in hospitals.

The available environmental exposure data on PPs is very limited in quantity, concerning the diversity of PPs, their metabolites and environmental transformation products, the numbers of available measurements in the different aquatic/marine compartments being measured and concerning data quality, concerning sampling, analysis and data evaluation.

Future monitoring studies should close the existing data gaps in the current knowledge and overcome the challenge of data comparability.

[2] Usage of the sum of erythromycin and anhydro-erythromycin is recommended.

Table 4.9 Mean annual consumption of selected pharmaceuticals for France, Germany, Poland, Spain and the UK (England, Wales). Data available for the period from 1999–2006.

Compound CAS-No.	Annual per capita consumption of PP (g/a)					Total annual consumption (t/a)				
	Germany	France	Spain	UK/Wales	Poland	Germany	France	Spain	UK/Wales	Poland
				ANALGESICS						
Acetylsalicylic acid 50-78-2	13.665	10.749		1.564		1124.786	638.106		81.863	
Celecoxib 184007-95-2	0.074		0.048	0.08		6.12		1.956	4.205	
Codeine 76-57-3	0.094			0.302		7.72			15.822	
Dextropropoxyphene 469-62-5		0.875		0.861			51.963		45.04	
Diclofenac 15307-86-5	0.883	0.167	0.059	0.538	0.506	72.676	9.896	2.419	28.15	19.448
Hydrocodone 125-29-1	0					0.007				
Ibuprofen 15687-27-1	3.172	3.425	2.641	2.843	5.036	261.055	203.312	108.253	148.787	193.387
Indomethacin 53-86-1	0.041			0.02	0.022	3.402			1.06	0.855
Ketoprofen 22071-15-4	0.018	0.366	0.005	0.025	0.337	1.479	21.697	0.21	1.296	12.933
Mesalazine 89-57-6	0.723			0.879		59.524			46.017	
Metamizole sodium 68-89-3	1.99					163.778				
Naproxen 22204-53-1	0.063	0.643	0.259	0.615	1.009	5.186	38.166	10.626	32.161	38.761
Paracetamol 103-90-2	4.456	47.143	3.595	15.683	4.839	366.809	2798.538	147.348	820.834	185.806
Phenazon 60-80-0	0.352					29.003				
Piroxicam 36322-90-4	0.008	0.034	0.011	0.004		0.669	2.008	0.452	0.215	
Propyphenazon 479-92-5	0.44					36.243				
Rofecoxib 162011-90-7	0.022		0.005	0		1.792		0.186	0	
Tramadol 27203-92-5	0.3	0.436		0.272		24.69	25.897		14.229	
				ANTIBIOTICS						
Amoxicillin 26787-78-0	1.198	6.496		1.535		98.57	385.612		80.321	
Ampicillin 69-53-4	0.131			0.039		10.758			2.06	
Azithromycin 83905-01-5	0.041	0.069		0.008	0.048	3.351	4.073		0.405	1.858
Chloramphenicol 56-75-7	0.004			0.002		0.32			0.104	
Chlortetracycline 57-62-5	0.003			0		0.28			0	
Ciprofloxacin 85721-33-1	0.172	0.205	0.091	0.124	0.126	14.186	12.186	3.746	6.506	4.847
Clarithromycin 81103-11-9	0.118	0.254	0.132	0.077	0.271	9.729	15.105	5.391	4.027	10.397
Clindamycin 18323-44-9	0.146			0.011	0.113	12.048			0.571	4.348
Dapsone 80-08-0	0.001			0.002		0.076			0.079	
Dicloxacillin 3116-76-5	0.001					0.104				
Doxycycline 564-25-0	0.137	0.105		0.027		11.252	6.243		1.407	
Erythromycin 114-07-8	0.257			0.536	0.166	21.124			28.031	6.367
Flucloxacillin 5250-39-5	0.028			0.558		2.345			29.213	
Flumequine 42835-25-6	0.008					0.668				
Furazolidone 67-45-8	0					0.015				
Josamycin 16846-24-5		0.216					12.802			
Metronidazole 443-48-1	0.095	0.616		0.178		7.798	36.545		9.338	
Mezlocillin 51481-65-3	0.083					6.868				

(continued)

Table 4.9 *Continued*

Compound CAS-No.	Annual per capita consumption of PP (g/a)					Total annual consumption (t/a)				
	Germany	France	Spain	UK/Wales	Poland	Germany	France	Spain	UK/Wales	Poland
Minocycline 10118-90-8	0.009					0.713				
Norfloxacin 70458-96-7	0.037			0.008	0.079	3.04			0.393	3.031
Ofloxacin 83380-47-6	0.066	0.07		0.005	0.023	5.4	4.137		0.276	0.895
Oxacillin 66-79-5	0.003					0.216				
Oxytetracycline 6153-64-6	0.025			0.476		2.032			24.904	
Penicillin G 61-33-6	0.082			0		6.742			0.007	
Penicillin V 87-08-1	1.05			0.443		86.398			23.186	
Piperacillin 13523-86-9	0.001			0		0.063			0.002	
Roxithromycin 80214-83-1	0.086	0.16		0	0.049	7.096	9.5		0	1.878
Spiramycin 8025-81-8	0.005	0.007		0		0.402	0.42		0	
Sulfadiazine 68-35-9	0.031			0		2.55			0.007	
Sulfamerazine 127-79-7	0.011					0.923				
Sulfamethoxazole 723-46-6	0.652	0.337		0.018	0.171	53.693	20.015		0.965	6.579
Sulfasalazine 599-79-1	0.465			0.906		38.275			47.426	
Tetracycline 60-54-8	0.017			0.028	0.085	1.396			1.49	3.256
Trimethoprim 738-70-5	0.147	0.056		0.14		12.13	3.346		7.337	
Vancomycin 1404-90-6	0.011	0.015		0		0.912	0.918		0.016	
ANTICOAGULANTS										
Clopidogrel 113665-84-2	0.127		0.091	0.165		10.432		3.749	8.632	
ANTIDEPRESSANTS										
Amitryptiline 50-48-6	0.081					6.676				
Citalopram 59729-33-8	0.024			0.061		1.958			3.214	
Doxepin 1668-19-5	0.065					5.385				
Fluoxetine 54910-89-3		0.063	0.034	0.06			3.74	1.38	3.115	
Paroxetine 61869-08-7	0.007		0.04	0.028		0.578		1.631	1.489	
Sertraline 79617-96-2	0.03		0.067	0.085		2.445		2.758	4.43	
Venlafaxine 93413-69-5	0.082		0.027	0.147		6.787		1.109	7.677	
ANTIDIABETICS										
Metformin 657-24-9	6.297	12.076		5.902		518.303	716.858		308.888	
ANTIEPILEPTICS										
Carbamazepine 298-46-4	0.983	0.614		0.767	0.838	80.892	36.438		40.152	32.168
Gabapentin 60142-96-3	0.202		0.046	0.75		16.592		1.868	39.237	
Phenobarbital 50-06-6	0.007					0.6				
Valproic acid 99-66-1	0.723	1.889		0.506		59.478	112.162		26.508	
ANTIFUNGALS										
Clotrimazole 23593-75-1	0.057			0.011		4.662			0.591	
Miconazole 22916-47-8	0.107			0.008		8.8			0.397	

(*continued*)

Table 4.9 *Continued*

Compound CAS-No.	Annual per capita consumption of PP (g/a)					Total annual consumption (t/a)				
	Germany	France	Spain	UK/Wales	Poland	Germany	France	Spain	UK/Wales	Poland
ANTIHYPERTENSIVES										
Amlodipine 88150-42-9	0.018		0.013	0.058		1.448		0.513	3.011	
Captopril 62571-86-2	0.29		0.024	0.035		23.833		0.992	1.85	
Enalapril 75847-73-3	0.056					4.607				
Irbesartan 138402-11-6	0.181		0.102	0.335		14.89		4.188	17.54	
Lactitol 585-86-4	0.279			0		22.967			0.001	
Nifedipine 21829-25-4	0.097					7.97				
Troxerutin 7085-55-4	0.499	7.485				41.058	444.339			
Valsartan 137862-53-4	0.101		0.127	0.15		8.283		5.187	7.83	
Verapamil 152-11-4	0.7			0.139		57.655			7.264	
ANTIPSYCHOTICS										
Risperidone 106266-06-2	0.001					0.109				
ANTITUSSIVES										
Dihydrocodeine 125-28-0	0.013			0.205		1.069			10.74	
ANTIULCERATIVES										
Cimetidine 51481-61-9	0.009					0.72				
Lansoprazole 103577-45-3	0.007		0.012	0.157		0.543		0.49	8.221	
Omeprazole 73590-58-6	0.081		0.063	0.14		6.708		2.571	7.326	
Pantoprazole 102625-70-7	0.061		0.018	0.02		5		0.757	1.068	
Ranitidine 66357-35-5	1.038	0.196	0.484	0.599		85.409	11.656	19.831	31.344	
Atenolol 29122-68-7	0.139	0.309		0.636	0.125	11.421	18.337		33.274	4.819
Betaxolol 63659-18-7	0.004			0		0.354			0.006	
Bisoprolol 66722-44-9	0.037	0.036		0.012		3.01	2.113		0.636	
Carazolol 57775-29-8	0					0.001				
Diltiazem 42399-41-7	0.129		0.136	0.427		10.656		5.561	22.345	
Metoprolol 37350-58-6	0.995	0.148		0.036	0.393	81.907	8.786		1.905	15.102
Nadolol 42200-33-9	0.003	0.016		0.001		0.213	0.938		0.042	
Pindolol 61477-96-1	0.149	0.075		0		12.275	4.476		0.004	
Propranolol 318-98-9	0.065	0.21		0.143	0.01	5.362	12.487		7.478	0.403
Sotalol 3930-20-9	0.283			0.061	0.032	23.288			3.213	1.23
BRONCHIOLYTICS										
Clenbuterol 37148-27-9	0					0.001				
Fenoterol 13392-18-2	0.003			0		0.217			0	
Salbutamol 18559-94-9	0.01			0.002	0.004	0.852			0.109	0.17
Terbutaline 23031-25-6	0.002			0.008		0.131			0.411	
Theophylline 58-55-9	1.457			0.177		119.896			9.258	

(*continued*)

Table 4.9 *Continued*

Compound CAS-No.	Annual per capita consumption of PP (g/a)					Total annual consumption (t/a)				
	Germany	France	Spain	UK/Wales	Poland	Germany	France	Spain	UK/Wales	Poland
CONTRAST MEDIA										
Diatrizoate 737-31-5	0.662					54.481				
Iohexol 66108-95-0	0.098					8.053				
Iomeprol 78649-41-9	0.603					49.616				
Iopamidol 62883-00-5	0.398					32.744				
Iopromide 73334-07-3	0.788	1.236				64.841	73.4			
Ioxithalamic acid 28179-44-4	0.108					8.895				
Ioversol 87771-40-2	0.529					43.581				
CYTOSTATICS										
Cyclophosphamid 50-18-0	0.007	0.005		0		0.538	0.282		0.017	
Ifosfamid 3778-73-2	0.002	0.002				0.187	0.121			
Tamoxifen 10540-29-1	0.009					0.78				
DIURETICS										
Furosemide 54-31-9	0.341	0.359		0.281		28.086	21.288		14.732	
Hydrochlorothiazide 58-93-5	0.201					16.514				
Torasemide 56211-40-6	0.045		0.723	0		3.7		29.654	0.008	
LIPID REGULATORS										
Atorvastatin 134523-00-5	0.058		0.051	0.212		4.753		2.074	11.1	
Bezafibrate 41859-67-0	0.382	0.462		0.146	0.007	31.454	27.426		7.66	0.273
Fenofibrate 49562-28-9	0.169	1.446		0.052	0.18	13.896	85.835		2.723	6.905
Gemfibrozil 25812-30-0	0.072			0.017	0.007	5.963			0.902	0.274
Pravastatin 81093-37-0	0.016		0.024	0.039		1.354		0.979	2.067	
Simvastatin 79902-63-9	0.065	0.117	0.023	0.257	0.028	5.318	6.943	0.925	13.458	1.06
MUCOLYTICS										
Ambroxol 18683-91-5	0.15					12.37				
Carbocistein 638-23-3		3.913		0.212			232.308		11.093	
NOOTROPICS										
Nimodipine 66085-59-4	0.001		0.033	0		0.12		1.349	0.004	
Piracetam 7491-74-9	1.249			0.022		102.784			1.174	
PHLEBOTROPICS										
Diosmin 520-27-4		6.293					373.544			
SEDATIVES										
Zopiclone 43200-80-2	0.003	0.033		0.012		0.265	1.948		0.65	
Zolpidem 82626-48-0	0.004	0.056		0.001		0.327	3.344		0.058	

(*continued*)

Table 4.9 *Continued*

Compound CAS-No.	Annual per capita consumption of PP (g/a)					Total annual consumption (t/a)				
	Germany	France	Spain	UK/Wales	Poland	Germany	France	Spain	UK/Wales	Poland
				TRANQUILLISERS						
Alprazolam 28981-97-7	0.121					10				
Bromazepam 1812-30-2	0.004	0.044				0.29	2.604			
Diazepam 439-14-5	0.011	0.008		0.012		0.888	0.463		0.649	
Lorazepam 846-49-1	0.984	0.01		0.001		81	0.585		0.04	
Oxazepam 604-75-1	0.012	0.104		0.002		0.97	6.195		0.124	
				UROSTATICS						
Allopurinol 315-30-0	1.677	0.914		0.464		138.008	54.247		24.286	
				VASOCONSTRICTORS						
Xylometazolin 526-36-3	0.007			0		0.572			0.001	
				VASODILATORS						
Pentoxifylline 6493-05-6	0.953			0.018		78.418			0.926	

Table 4.10 Structure and properties of selected pharmaceuticals; A-MEC and maximum concentrations measured in surface waters, WWTP influent, and WWTP effluent ([1]DDD given for the most relevant application [g/d]).

Name CAS-No.	Structure	Log Kow	Bioavail-ability (oral)	DDD[1] (g/d)	Renal excretion (%)		Surface water (SW)			WWTP influent			WWTP effluent		
							Samples	A-MEC (µg/L)	MAX (µg/L)	Samples	A-MEC (µg/L)	MAX (µg/L)	Samples	A-MEC (µg/L)	MAX (µg/L)
					ANALGESICS										
Acetylsalicylic acid 50-78-2	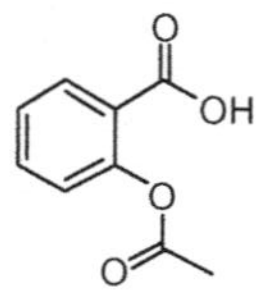	1.19	68±3	3	–	France	52	0.004	0.086				41	0.01	0.212
						Germany	69	0	0.34	2	3.2	5	88	0.18	1.51
						Romania	8	0.032	0.037						
						UK				6	0	0	6	0	0
Butalbital 77-26-9	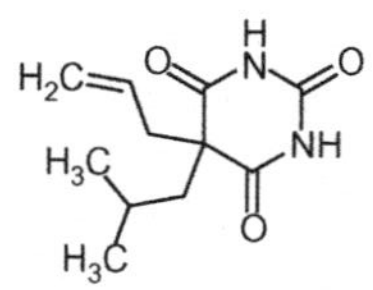	1.87	–	–	5	Germany	18	0	5.3						
Codeine 76-57-3	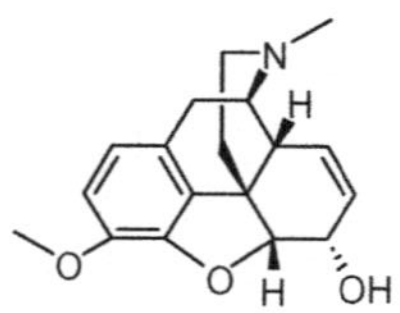	1.14	50±7	0.1	55	Germany	44	0.009	0.094	11	0.22	0.54	42	0.106	0.9
						Romania	8	0.033	0.054						
						Spain				18	3.767	11	12	3.7	8.1
						UK	12	0.01	0.034						
						USA	46	0.012	0.019	5	0.17	0.17	6	0.17	0.17
Dextropropoxyphene 469-62-5	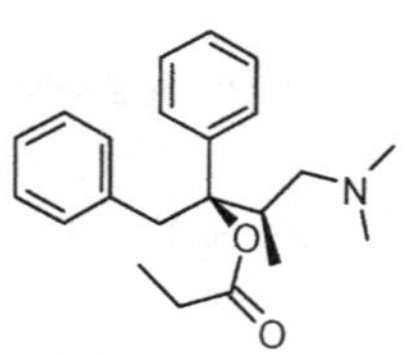	4.18	40	0.3	5	UK	47	0.004	0.682	9	0.027	0.033	14	0.113	0.585

(*continued*)

Table 4.10 *Continued*

Name CAS-No.	Structure	Log Kow	Bioavail-ability (oral)	DDD[1] (g/d)	Renal excretion (%)		Surface water (SW)			WWTP influent			WWTP effluent		
							Samples	A-MEC (µg/L)	MAX (µg/L)	Samples	A-MEC (µg/L)	MAX (µg/L)	Samples	A-MEC (µg/L)	MAX (µg/L)
Diclofenac 15307-86-5	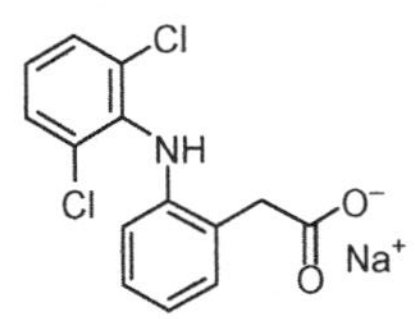	4.51	54±2	0.1	15	Austria	9	0.036	0.392	7	2.15	3.55	7	1.844	3
						Canada	211	0.015	0.194	70	0.429	2.45	112	0.118	0.895
						Croatia				5	0.25	0.54	5	0.215	0.39
						Finland	8	0.022	0.055	2	0.35	0.35	2	0.028	0.028
						France	85	0.023	0.173	5	3.309	5.218	46	0.313	0.92
						Germany	563	0.06	1.8	44	2.43	28	382	1.344	10
						Greece							6	0.07	0.89
						Italy	25	0.01	0.158				8	0.68	5.45
						Japan				11	0.251	0.251	11	0.145	0.145
						Korea	8	0.003	0.007	17	2.59	9.87	26	1.301	10.96
						Norway				1	0	0	4	0	0.03
						Spain	20	0.021	0.06	56	0.531	3.6	47	0.305	2.2
						Sweden							24	0.29	1.48
						UK	49	0.003	0.568	9	0.997	1.036	15	0.34	2.349
						USA	42	0.003	0.009	8	0.087	0.116	30	0	0.033
Dimethylami-nophenazone 58-15-1	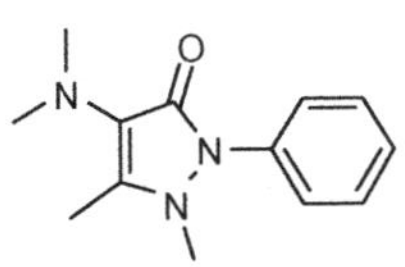	1	–	–	–	France							2	0.215	0.43
						Germany	168	0.001	0.34	14	0.137	0.96	92	0.013	1
						Italy							3	0	0
Fenoprofen 31879-05-7	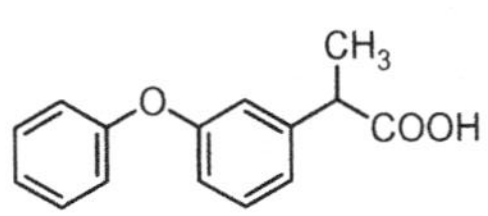	3.99	–	1.2	–	Canada	25	0	0	44	0	0	78	0	0.217
						France							2	0.235	0.28
						Germany	43	0	0				59	0	0
						Italy	15	0.004	0.022				8	0.938	5.4
Hydrocodone 125-29-1	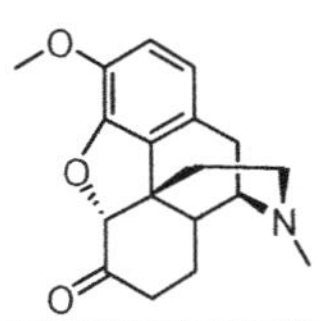	2.13	–	0.015	3	Germany	37	0.002	0.028				28	0	0.4
						Korea	8	0.002	0.002				9	0.033	0.07
						USA	8	3.113	4.15	7	0.055	0.07	9	0.012	0.068

(continued)

Table 4.10 *Continued*

Name CAS-No.	Structure	Log Kow	Bioavail-ability (oral)	DDD[1] (g/d)	Renal excretion (%)		Surface water (SW)			WWTP influent			WWTP effluent		
							Samples	A-MEC (µg/L)	MAX (µg/L)	Samples	A-MEC (µg/L)	MAX (µg/L)	Samples	A-MEC (µg/L)	MAX (µg/L)
Ibuprofen 15687-27-1	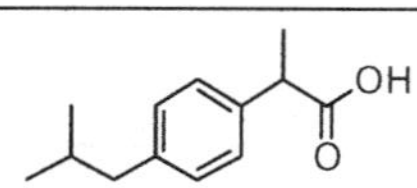	3.97	>80	1.2	16	Austria				7	1.642	3.17	7	0.012	0.035
						Canada	211	0.018	6.4	70	15.4	75.8	112	1.37	24.6
						Croatia				5	0.516	0.9	5	0.266	0.8
						Finland	8	0.023	0.069	2	13.1	13.1	2	0.026	0.026
						France	91	0.109	5.6	5	0.097	2.191	46	0.101	1.82
						Germany	407	0.011	0.53	25	3.512	14	293	0.175	3.7
						Italy	33	0.014	0.201				25	1.526	95
						Japan				11	1.966	1.966	11	0.04	0.04
						Korea	8	0.028	0.038	17	0.3	0.58	26	0.065	0.31
						Norway							4	0.235	0.68
						Romania	8	0.073	0.115						
						Spain	20	0.055	0.289	58	21.12	168	49	1.849	28
						Sweden							24	0.11	7.11
						UK	108	0.526	5.044	15	16.89	33.764	33	1.384	27.256
						USA	150	0.154	5.85	2	35.48	68.7	34	0.084	1.25
Indomethacin 53-86-1	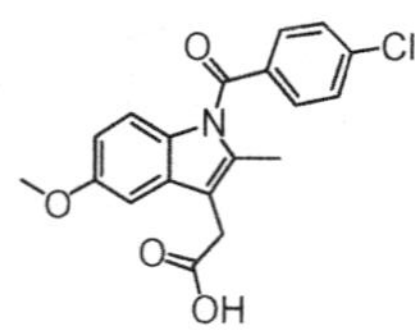	4.27 at pH 2.7	–	0.1	30	Canada	152	0	0.15	44	0.211	0.64	82	0.1	0.507
						Croatia				5	0	0	5	0	0
						France	6	0	0				3	0	0
						Germany	394	0.006	0.22	5	0	0.95	190	0.141	0.6
						Spain	10	0	0.01						
						UK	2	0.001	0.002						
						USA	23	0	0.048				20	0.012	0.029
Ketoprofen 22071-15-4	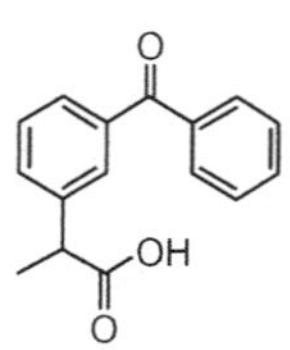	3.12	–	0.15	75	Canada	69	0.021	0.079	70	1.545	5.7	112	0.048	0.409
						Croatia				5	0.451	0.97	5	0.318	0.62
						Finland	8	0.013	0.028	2	2	2	2	0.032	0.032
						France	85	0.007	0.04	5	4.1	10.8	46	0.179	1.62
						Germany	382	0	0.12	8	0.07	0.19	218	0.077	0.38
						Italy	15	0.002	0.026				8	0.438	2.1
						Japan				11	0.979	0.979	11	0.445	0.445
						Spain	17	0	0.144	5	0.23	0.96	5	0.2	0.75
						USA	23	0	0	5	1	1	28	0.002	0.031
Meclofe-namic acid 644-62-2	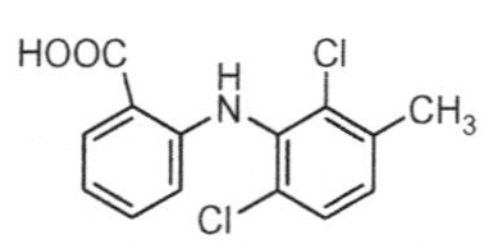	6	–	–	–	Brazil							10	0	0
						Canada	9	0.107	0.115				4	0.084	0.09
						Germany	49	0	0				10	0	0
						UK	2	0	0						

(*continued*)

Table 4.10 *Continued*

Name CAS-No.	Structure	Log Kow	Bioavail-ability (oral)	DDD[1] (g/d)	Renal excretion (%)		Surface water (SW)			WWTP influent			WWTP effluent		
							Samples	A-MEC (μg/L)	MAX (μg/L)	Samples	A-MEC (μg/L)	MAX (μg/L)	Samples	A-MEC (μg/L)	MAX (μg/L)
Mefenamic acid 61-68-7	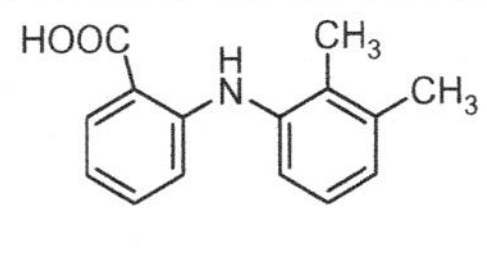	5.1	–	1	–	Croatia				5	0.005	0.005	5	0.007	0.01
						Japan				11	0.221	0.221	11	0.062	0.062
						Spain	10	0.002	0.003						
						UK	78	0.03	0.242	15	0.172	0.363	27	0.186	0.396
Metamizole sodium 68-89-3	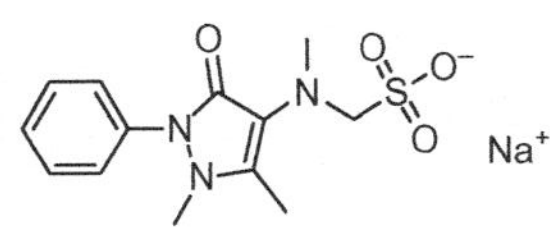	0.71	–	3	–	Spain				12	14	24	12	4.9	7.5
Naproxen 22204-53-1	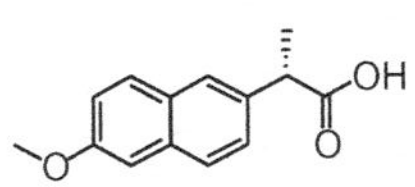	3	99	0.5	70	Austria	5	0.013	0.382						
						Canada	299	0.019	4.5	70	13.52	611	112	2.616	33.9
						Croatia				5	0.099	0.19	5	0.108	0.16
						Finland	8	0.015	0.032	2	4.9	4.9	2	0.043	0.043
						France	85	0.018	0.275	5	0.207	9.4	46	0.127	2.667
						Germany	201	0.008	0.39	14	0.577	1.3	194	0.092	0.94
						Italy	15	0	0.022				8	1.154	5.22
						Japan				11	0.276	0.276	11	0.099	0.099
						Korea	8	0.011	0.018				9	0.105	0.483
						Spain	27	0.012	0.247	39	0.582	8.62	39	0.091	0.45
						Sweden							24	0.41	2.15
						USA	64	0.022	0.195	8	19.58	22.5	40	0.637	24.6
Paracetamol 103-90-2	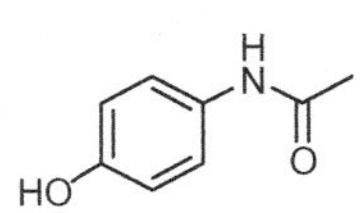	0.46	70–90	3	52	Canada	16	0	3.6				16	0	9
						Croatia				5	10.19	26.09	5	2.102	5.99
						Czech Rep.	9	0.015	0.106						
						France	34	0.01	0.463				2	5.704	11.3
						Germany	77	0.002	0.066				49	0	6
						Korea	8	0.033	0.073	16	0.07	0.26	25	0.041	0.16
						Spain	13	0.037	0.25	56	30.97	246	47	0.28	6.2
						UK	92	0.118	1.388	15	4.199	69.57	30	0.055	0.218
						USA	8	0.001	0.002	7	48.96	61	9	0.574	0.86
Phenazon 60-80-0	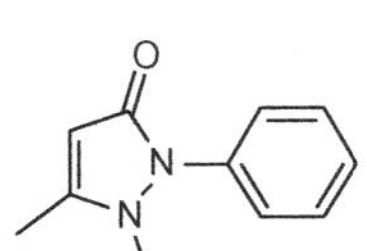	0.23	–	3	–	Canada	16	0	0	18	0	0	34	0	0
						Czech Rep.	9	0	0.035						
						France							2	0	0
						Germany	493	0.031	2.5	32	0.171	0.45	264	0.088	0.9
						Italy							3	0	0.37

(continued)

Table 4.10 *Continued*

Name CAS-No.	Structure	Log Kow	Bioavail-ability (oral)	DDD[1] (g/d)	Renal excretion (%)		Surface water (SW)			WWTP influent			WWTP effluent		
							Samples	A-MEC (μg/L)	MAX (μg/L)	Samples	A-MEC (μg/L)	MAX (μg/L)	Samples	A-MEC (μg/L)	MAX (μg/L)
Piroxicam 36322-90-4	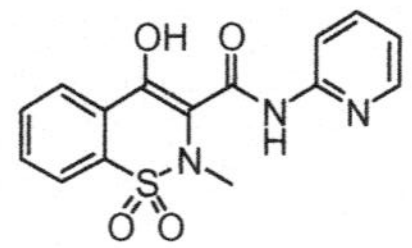	3.06	–	0.02	10	Germany	40	0	0				32	0	0
Propyphena-zon 479-92-5	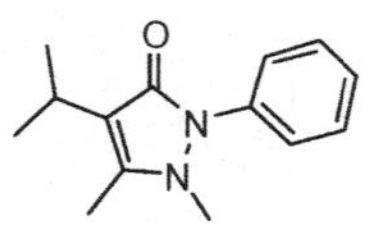	1.94	–	3	–	Croatia				5	0	0	5	0	0
						Czech Rep.	9	0.008	0.069						
						France	6	0	0				3	0	0
						Germany	313	0.042	0.88	24	0.135	0.25	95	0.074	0.99
						Spain	17	0	0.059	5	0.5	0.85	5	0.2	0.45
Tramadol 27203-92-5	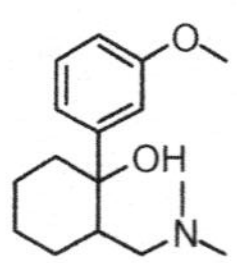	2.51	70–75	0.3	10–30	Germany	4	0.025	0.052						
						UK	12	0.358	2.108						
					ANTIBIOTICS										
Amoxicillin 26787-78-0	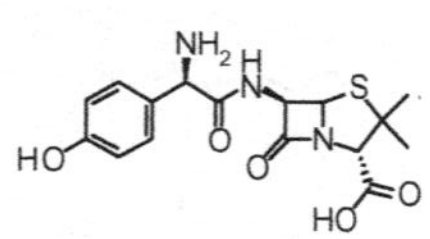	0.87	93±10	1	65	China	24	0	0						
						France				19	0	0	19	0	0
						Germany	50	0	0.1				67	0	0.147
						Italy	15	0	0				9	0.005	0.005
						UK	12	0.035	0.245						
						USA	21	0	0						
Ampicillin 69-53-4	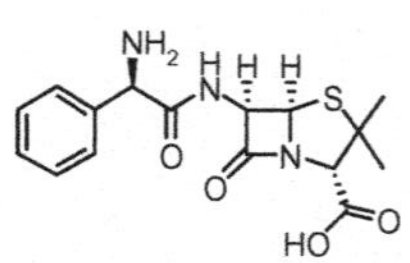	0.1	–	2	30	France	6	0	0				3	0	0
						Germany	36	0	0				67	0	0
						USA	21	0	0						

(*continued*)

Table 4.10 *Continued*

Name CAS-No.	Structure	Log Kow	Bioavail-ability (oral)	DDD[1] (g/d)	Renal excretion (%)		Surface water (SW)			WWTP influent			WWTP effluent		
							Samples	A-MEC (µg/L)	MAX (µg/L)	Samples	A-MEC (µg/L)	MAX (µg/L)	Samples	A-MEC (µg/L)	MAX (µg/L)
Azithromycin 83905-01-5	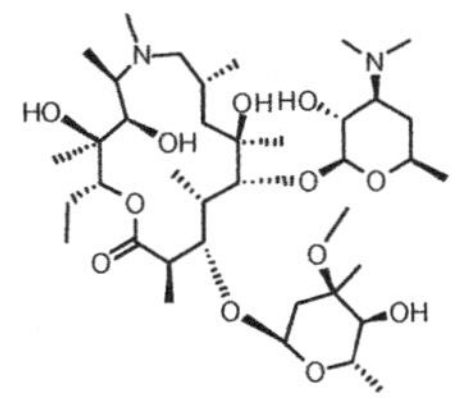	4.02	34±19	0.5	12	Croatia				5	0.152	0.3	5	0.096	0.21
						France	6	0.005	0.005				3	0.101	0.101
						Germany	29	0	0.14				67	0	0.135
						Spain	17	0.012	0.068	5	0.16	0.45	5	0.14	0.3
						Switzerland				9	0.17	0.38	15	0.16	0.4
Chloramphe-nicol 56-75-7	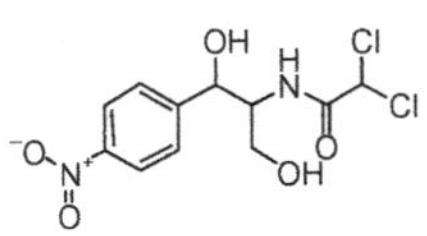	1.14	–	3	90	Canada	136	0	0						
						China	24	0.084	0.266						
						France				19	0	0	19	0	0
						Germany	215	0	1.3	8	0	0.19	71	0	0.56
						UK	12	0	0						
Chlortetracy-cline 57-62-5	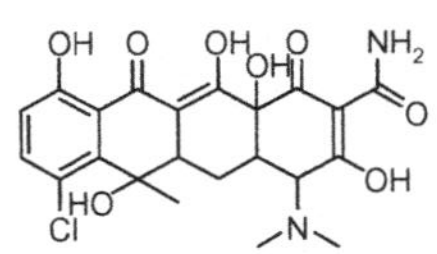	−0.62	–	1	15	Canada	136	0.176	0.192						
						France	6	0	0				3	0.056	0.056
						Germany	43	0	0				72	0	0
						Korea				2	85.98	171	2	45.47	90.9
						USA	235	0.15	0.69	3	0.31	0.31	3	0.42	0.42
Ciprofloxacin 85721-33-1	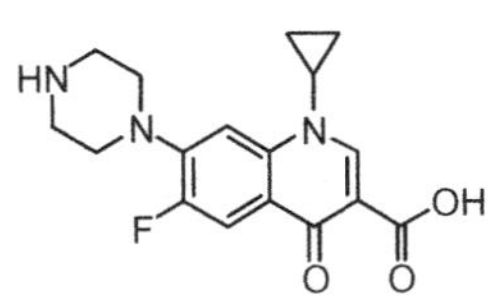	0.28	60±12	1	53	Canada							16	0.132	0.4
						Finland	11	0.008	0.036	24	0.397	4.23	24	0.066	0.13
						France	6	0.006	0.006	19	0	0.166	24	0.035	0.101
						Germany	55	0	0.028				67	0	0.144
						Italy	15	0	0.026				20	0.208	0.514
						Sweden				13	0.166	0.3	13	0.023	0.06
						USA	128	0.024	0.36	17	0.53	1.4	29	0.262	5.6
Clarithromy-cin 81103-11-9	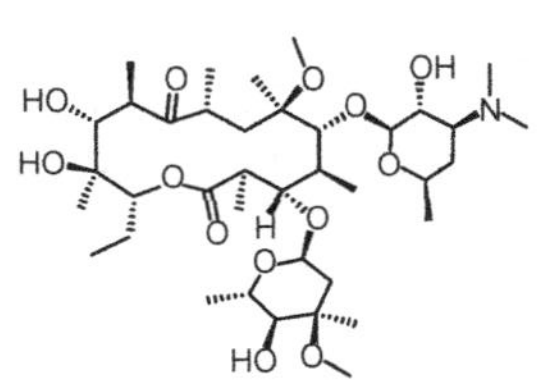	2.6	55±8	1	18	Canada							8	0.087	0.536
						France	6	0.008	0.075				3	0.117	0.117
						Germany	272	0	0.95	114	0.214	1.47	287	0.109	4.5
						Italy	15	0.002	0.02				17	0.017	0.073
						Switzerland				9	0.38	0.6	18	0.225	0.35

(continued)

Table 4.10 *Continued*

Name CAS-No.	Structure	Log Kow	Bioavail-ability (oral)	DDD[1] (g/d)	Renal excretion (%)		Surface water (SW)			WWTP influent			WWTP effluent		
							Samples	A-MEC (µg/L)	MAX (µg/L)	Samples	A-MEC (µg/L)	MAX (µg/L)	Samples	A-MEC (µg/L)	MAX (µg/L)
Clindamycin 18323-44-9	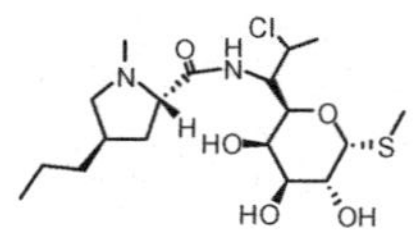	2.16	87	1.8	13	France	6	0.001	0.001	19	0	0	22	0.001	0.01
						Germany	35	0.005	0.03	3	0	0	79	0.011	0.13
						USA	10	0	0.14				8	0	1
Cloxacillin 61-72-3	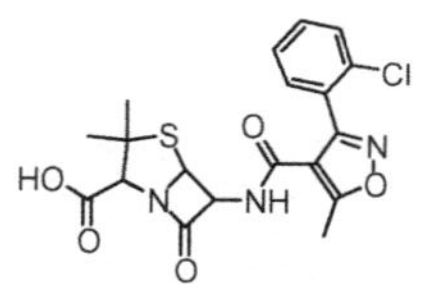	2.44	–	2	35	France	6	0	0				3	0.01	0.01
						Germany	14	0	0				4	0	0
						USA	21	0	0	2	0.02	0.02			
Dapsone 80-08-0	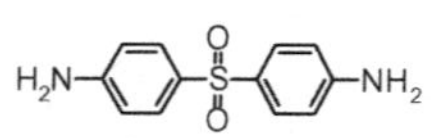	0.97	93±8	0.05	60	Germany	14	0	0						
Demeclocy-cline 127-33-3	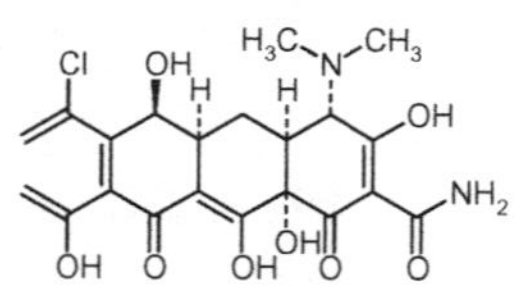	−1.14	–	0.6	–	Korea				2	1.71	3.15	2	1.365	2.7
						USA	36	0	0						
Dicloxacillin 3116-76-5	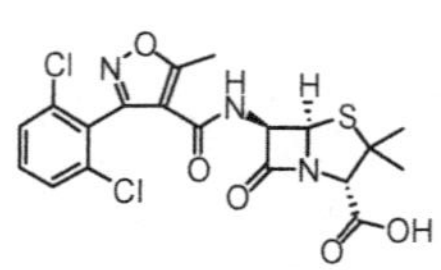	2.91	50–85	2	60	France	6	0	0				3	0	0
						Germany	57	0	0				63	0	0
Doxycycline 564-25-0	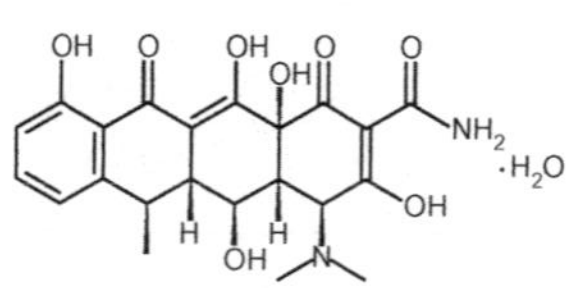	−0.02	93	0.1	40	Canada	136	0.007	0.073				8	0.038	0.046
						France	6	0	0	19	0	0	22	0.01	0.073
						Germany	35	0	0				56	0	0
						Korea				2	0.11	0.22	2	0.015	0.03
						Sweden				10	0.083	2.48	10	0.227	0.915
						USA	151	0.001	0.02	3	10	10	3	10.9	10.9

(*continued*)

Table 4.10 *Continued*

Name CAS-No.	Structure	Log Kow	Bioavail-ability (oral)	DDD[1] (g/d)	Renal excretion (%)		Surface water (SW)			WWTP influent			WWTP effluent		
							Samples	A-MEC (µg/L)	MAX (µg/L)	Samples	A-MEC (µg/L)	MAX (µg/L)	Samples	A-MEC (µg/L)	MAX (µg/L)
Erythromycin 114-07-8	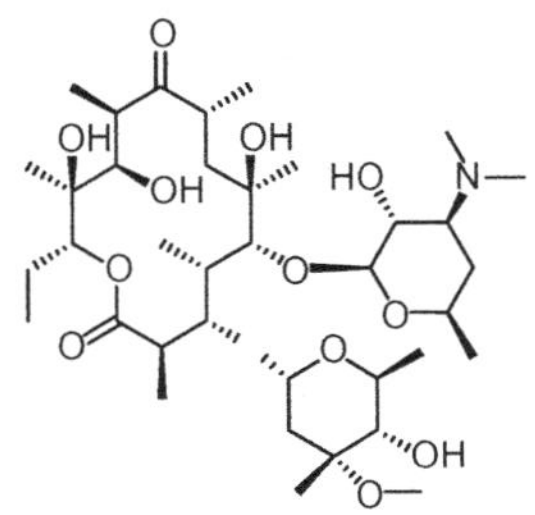	2.54	35±25	2	8	Canada	224	0.003	0.051				8	0.08	0.838
						China	24	0.245	0.636						
						Croatia				5	0	0	5	0	0
						France	6	0.025	0.075				3	0.35	0.35
						Germany	301	0.026	1.7	98	0.189	1.51	278	0.252	6
						Italy	15	0.003	0.016				17	0.043	0.353
						Korea	8	0.003	0.005				9	0.12	0.294
						Spain	21	0.033	0.2	11	0.069	0.25	5	0.15	0.28
						Switzerland				9	0.07	0.19	18	0.068	0.199
						UK	37	0.032	1.022	9	0.113	0.141	15	0.157	1.842
						USA	148	0.101	1209	101	0.332	3.9	102	0.111	1.1
Flucloxacillin 5250-39-5	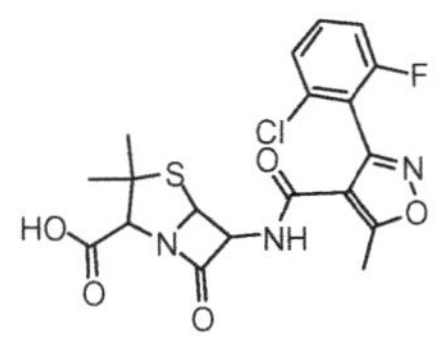	2.58	–	2	–	France	6	0	0				3	0	0
						Germany							51	0	0.023
Flumequine 42835-25-6	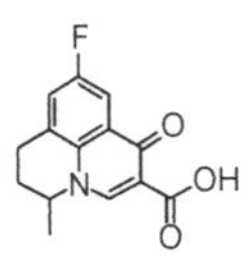	1.6	–	1.2	–	France				19	0.006	0.025	19	0.017	0.035
						USA	21	0	0						
Furazolidone 67-45-8	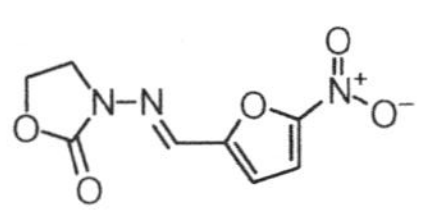	−0.04	–	–	–	Germany	14	0	0						
Fusidine 6990-06-3	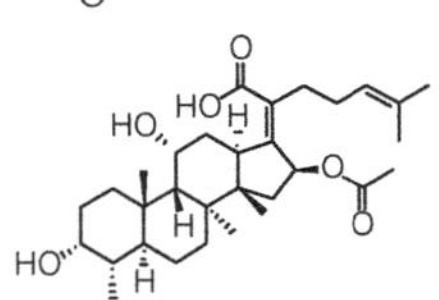	4.97	–	–	–	France				19	0	0.07	19	0	0
Josamycin 16846-24-5	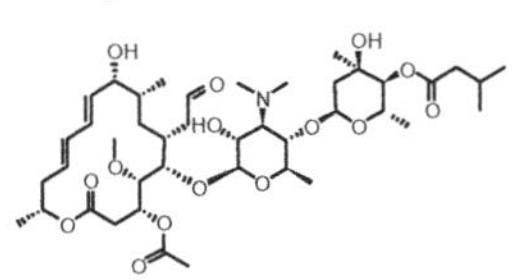	2.39	–	2	–	France				19	0	0	19	0.019	0.27

(*continued*)

Table 4.10 *Continued*

Name CAS-No.	Structure	Log Kow	Bioavail-ability (oral)	DDD[1] (g/d)	Renal excretion (%)		Surface water (SW)			WWTP influent			WWTP effluent		
							Samples	A-MEC (µg/L)	MAX (µg/L)	Samples	A-MEC (µg/L)	MAX (µg/L)	Samples	A-MEC (µg/L)	MAX (µg/L)
Lincomycin 154-21-2	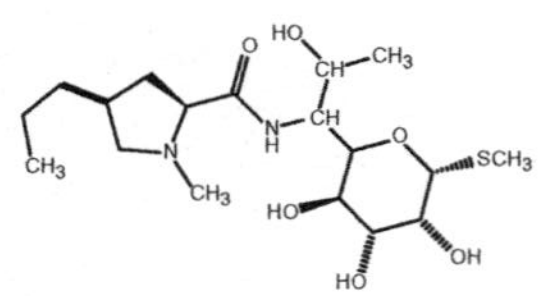	0.6	–	–	–	Canada	224	0.017	0.355						
						Italy	15	0.03	0.249				9	0.03	0.03
						USA	24	0	0	6	0	0			
Meclocycline 2013-58-3	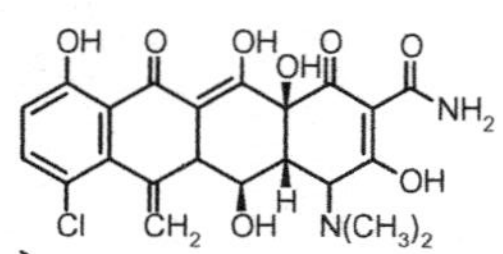	–	–	–	–	Canada	136	0	0						
						Korea				2	0.785	1.07	2	0.275	0.37
						USA	15	0.03	0.11						
Methicillin 61-32-5	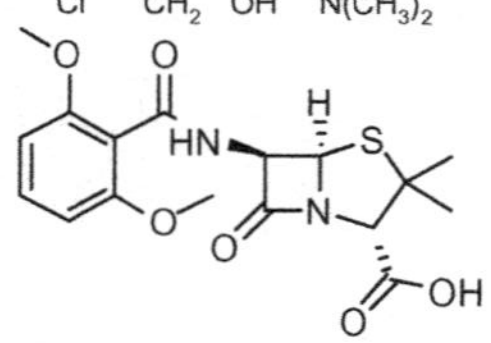	1.22	–	–	–	France	6	0	0				3	0	0
						Germany	14	0	0				4	0	0
Metronida-zole 443-48-1	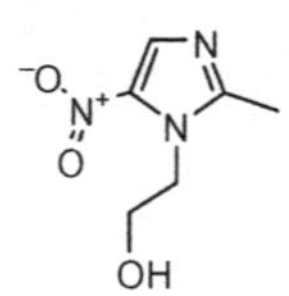	−0.02	99±8	2	20	France				19	0	0	19	0.011	0.039
						Germany	16	0.003	0.044						
						Spain				6	5.9	9.4			
						UK	12	0	0						
Mezlocillin 51481-65-3	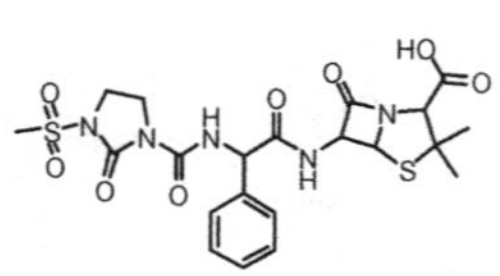	2	–	6	–	France	6	0	0				3	0	0
						Germany	21	0	0.012				51	0	0.02
Minocycline 10118-90-8	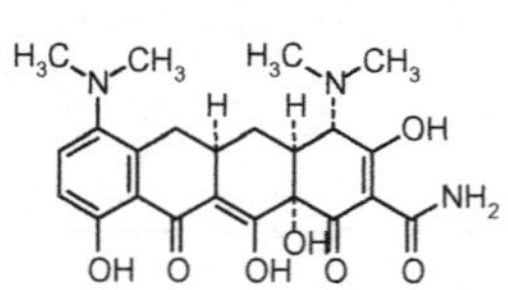	0.1	–	0.2	–	Korea				2	4.64	8.9	2	0	0
						USA	36	0	0						

(*continued*)

Table 4.10 *Continued*

Name CAS-No.	Structure	Log Kow	Bioavail-ability (oral)	DDD[1] (g/d)	Renal excretion (%)		Surface water (SW)			WWTP influent			WWTP effluent		
							Samples	A-MEC (µg/L)	MAX (µg/L)	Samples	A-MEC (µg/L)	MAX (µg/L)	Samples	A-MEC (µg/L)	MAX (µg/L)
Nafcillin 147-52-4	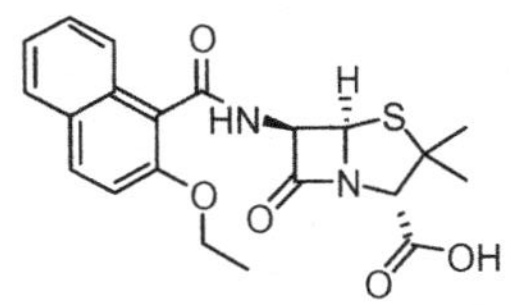	3.21	–	–	–	France	6	0	0				3	0	0
						Germany	14	0	0				4	0	0
Norfloxacin 70458-96-7	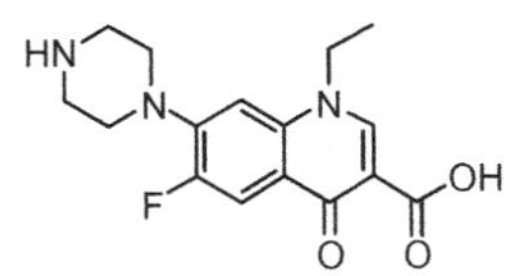	–1	–	0.8	28	Canada							16	0.042	0.112
						China	24	0.081	0.251						
						Finland	11	0	0	24	0.081	0.96	24	0	0.11
						France				19	0	0.059	21	0.037	0.24
						Italy							3	0.06	0.07
						Sweden				13	0.135	0.319	13	0.024	0.071
						USA	151	0.091	0.12	15	0.054	0.25	9	0.113	0.33
Ofloxacin 83380-47-6	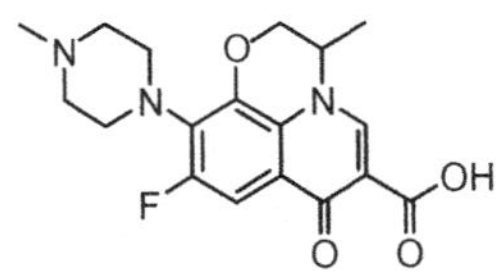	0.35	100	0.4	80	Canada							16	0.136	0.548
						China	24	0.044	0.108						
						Croatia				5	0	0	5	0	0
						Finland	11	0	0.005	24	0.072	0.35	24	0.014	0.03
						France	6	0.005	0.005	19	0	0	24	0.094	0.51
						Germany	21	0	0.06	3	0	0.22	54	0.01	0.19
						Italy	7	0.033	0.306				20	0.584	1.081
						Spain	17	0.033	0.146						
						Sweden				10	0.153	0.287	10	0.045	0.12
						USA	24	0.088	0.109	6	0.435	1	8	0.255	0.35
Oxacillin 66-79-5	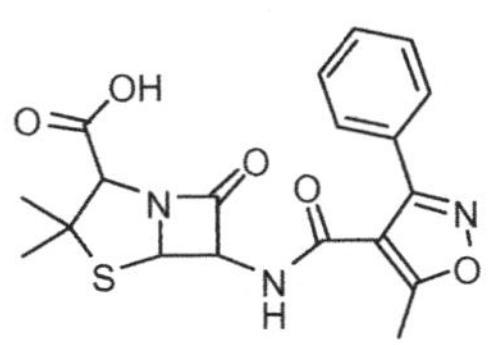	2.38	–	2	–	France	6	0	0				3	0	0
						Germany	57	0	0				71	0	0.03
						USA	21	0	0						
Oxytetracy-cline 6153-64-6	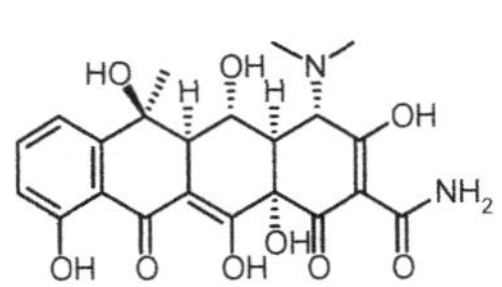	–1.6	–	1	–	Canada	136	0	0						
						France	6	0	0	19	0	0	22	0.009	0.063
						Germany	35	0	0				56	0	0.02
						Italy	8	0	0.019				8	0	0
						Korea				2	11.93	23.62	2	3.265	6.53
						USA	238	0.12	0.34	9	15.67	47	3	4.2	4.2

(*continued*)

Table 4.10 *Continued*

Name CAS-No.	Structure	Log Kow	Bioavail-ability (oral)	DDD[1] (g/d)	Renal excretion (%)		Surface water (SW)			WWTP influent			WWTP effluent		
							Samples	A-MEC (μg/L)	MAX (μg/L)	Samples	A-MEC (μg/L)	MAX (μg/L)	Samples	A-MEC (μg/L)	MAX (μg/L)
Penicillin G 61-33-6	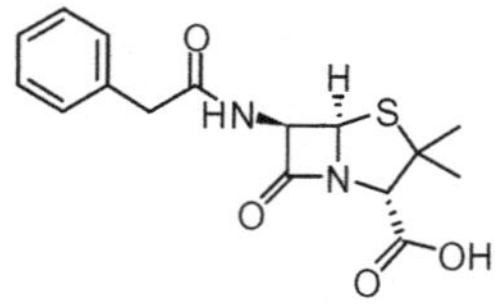	1.83	30	–	79	France	6	0	0				3	0	0
						Germany	57	0	0				71	0	0
						USA	24	0	0	6	0	0			
Penicillin V 87-08-1	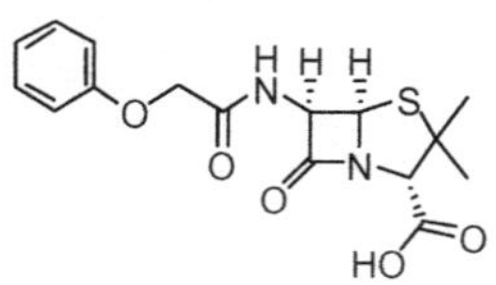	2.09	–	–	–	France	6	0	0				3	0	0
						Germany	49	0	0.02				55	0	0.013
						USA	24	0	0	6	0	0			
Piperacillin 13523-86-9	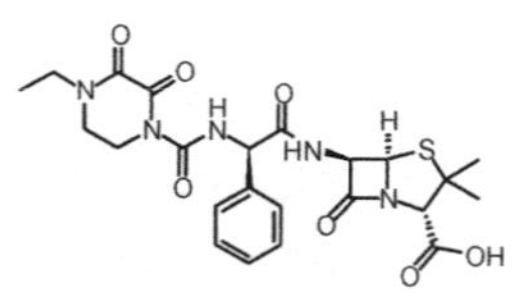	1.75	–	0.015	–	France	6	0	0				3	0	0
						Germany	21	0	0				51	0	0.04
Rifampicin 13292-46-1	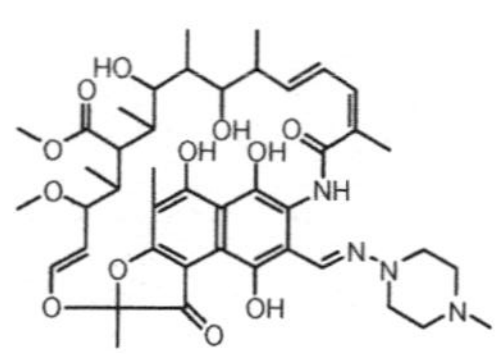	4.24	–	–	7±3	France				19	0	0	19	0	0.009
Roxithromycin 80214-83-1	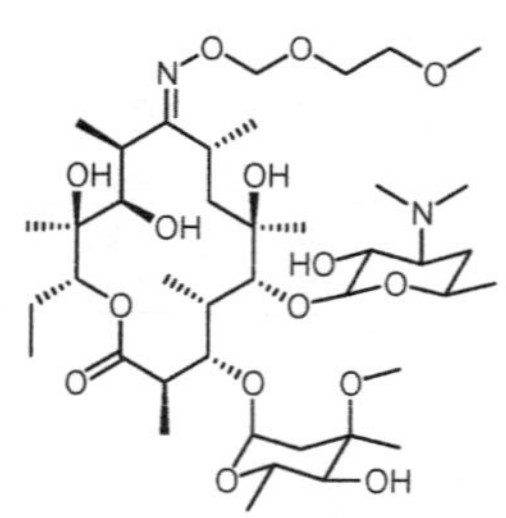	2.5	50	0.3	–	Austria				7	0.187	0.5	7	0.098	0.23
						Canada	139	0.001	0.002				11	0.006	0.018
						China	24	0.041	0.169						
						France	11	0.013	0.395	19	0	0	22	0.041	0.14
						Germany	303	0	0.56	117	0.163	1	292	0.101	1.7
						Switzerland				9	0.02	0.04	18	0.011	0.031
						USA	140	0.037	0.18	3	1.5	1.5	3	0.87	0.87

(continued)

Table 4.10 *Continued*

Name CAS-No.	Structure	Log Kow	Bioavail-ability (oral)	DDD[1] (g/d)	Renal excretion (%)		Surface water (SW)			WWTP influent			WWTP effluent		
							Samples	A-MEC (μg/L)	MAX (μg/L)	Samples	A-MEC (μg/L)	MAX (μg/L)	Samples	A-MEC (μg/L)	MAX (μg/L)
Spiramycin 8025-81-8	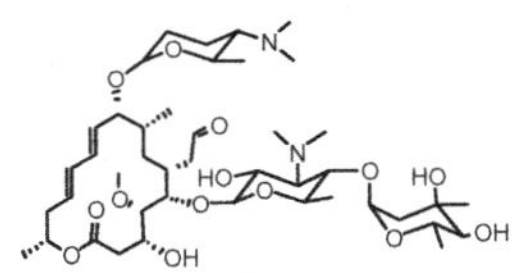	2.49	–	3	–	France	6	0.004	0.004	19	0	0.158	22	0.318	4
						Germany	43	0	0				67	0	0.04
						Italy	15	0.005	0.074				17	0.043	0.161
Sulfachlor-pyridazine 80-32-0	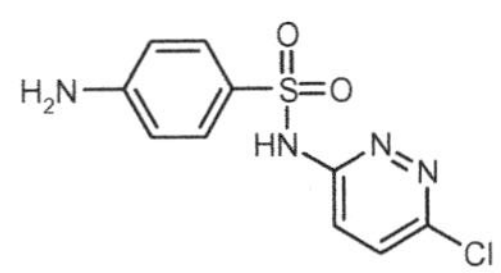	0.3	–	–	–	Canada	224	0.003	0.02						
						Korea				2	4.045	6.53	2	0.03	0.06
						USA	36	0	0						
Sulfadiazine 68-35-9	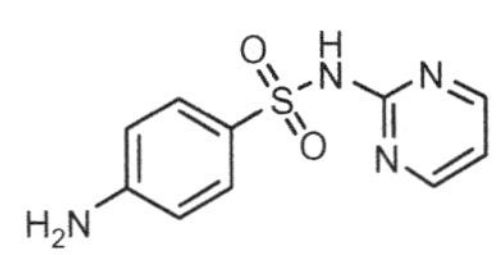	0.12	–	0.6	50	Canada							8	0.019	0.019
						China	30	0.099	0.336	3	0	0	3	0	0
						Germany	14	0	0						
						Italy	3	0	0.236						
						USA	21	0	0						
Sulfadime-thoxine 122-11-2	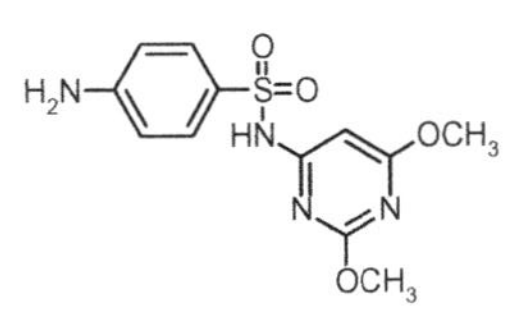	1.6	–	0.5	50	Canada	136	0.001	0.056						
						Italy	3	0.028	0.074						
						Korea				2	6.345	12.23	2	0	0
						USA	15	0	0.04						
Sulfamera-zine 127-79-7	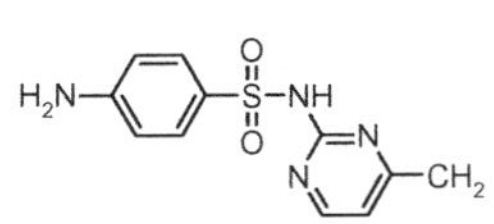	−0.1	–	3	60	Canada	136	0	0						
						Greece							24	0	0
						Korea				2	2.49	3.45	2	0	0
						USA	36	0	0.06						

(continued)

Table 4.10 *Continued*

Name CAS-No.	Structure	Log Kow	Bioavail-ability (oral)	DDD[1] (g/d)	Renal excretion (%)		Surface water (SW)			WWTP influent			WWTP effluent		
							Samples	A-MEC (μg/L)	MAX (μg/L)	Samples	A-MEC (μg/L)	MAX (μg/L)	Samples	A-MEC (μg/L)	MAX (μg/L)
Sulfametha-zine 57-68-1	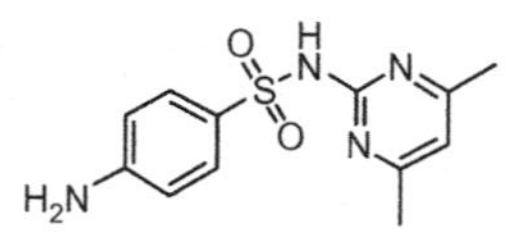	0.14	–	–	50	Canada	224	0.002	0.408				8	0.363	0.363
						China	30	0.134	0.34	3	0.58	0.58	3	0	0
						France	6	0.002	0.002	19	0	0.002	22	0	0.004
						Germany	170	0	0.07				59	0	0.2
						Greece							24	0	0
						Korea				2	50.62	97.23	2	0	0
						USA	224	0.096	0.22	8	0.112	0.3	3	0.36	0.36
Sulfamethox-azole 723-46-6	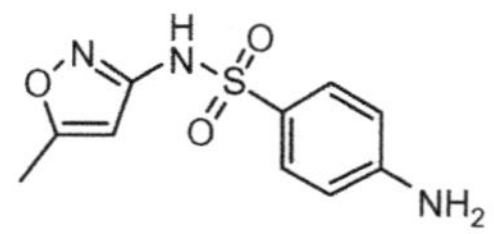	0.89	100	2	98	Canada	136	0.003	0.009				8	0.243	0.871
						China	30	0.101	0.33	3	0.64	0.64	3	0.56	0.56
						Croatia				5	0.59	0.87	5	0.39	0.82
						France	38	0.01	0.24	19	0.225	5.4	24	0.206	0.34
						Germany	308	0.03	0.48	25	0.742	9	103	0.307	4.7
						Greece							6	0.194	0.4
						Italy	21	0.003	0.402				20	0.123	0.317
						Korea	8	0.02	0.036	2	0.225	0.45	11	0.118	0.407
						Spain	17	0.015	0.169	5	0.45	0.96	5	0.4	0.8
						Sweden				10	0.302	0.674	10	0.248	0.304
						Switzerland				9	0.09	0.15	15	0.09	0.35
						UK	36	0.003	0.06	9	0	0	14	0	0.132
						USA	257	0.135	1.9	30	0.782	2.8	44	0.389	6
Sulfapyridine 144-83-2	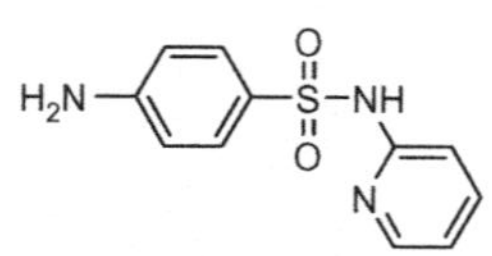	0.4	–	1	45	China	6	0	0	3	0	0	3	0	0
						Greece							24	0	0
						Italy	3	0.066	0.121						
						UK	12	0.009	0.039						
Sulfathiazole 72-14-0	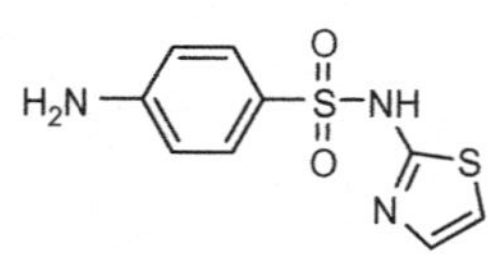	–0.4	–	–	–	Canada	136	0.001	0.016						
						China	6	0.26	0.52	3	0.35	0.35	3	0	0
						Greece							24	0	0
						Korea				2	584.62	1158.68	2	2.225	4.27
						USA	36	0	0.01						
Tetracycline 60-54-8	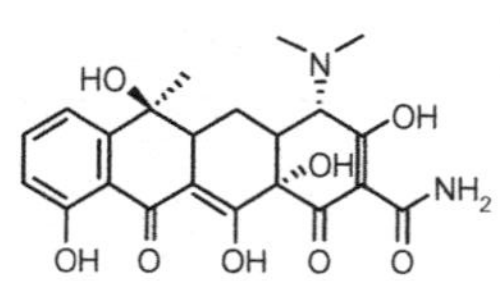	1.4	77	1	58±8	Canada	136	0	0				8	0.151	0.977
						France	6	0	0	19	0	0	22	0.009	0.066
						Germany	35	0	0				56	0	0
						Korea				2	0.545	0.98	2	0.705	1.41
						USA	245	0.038	0.11	11	13.46	48	19	0.646	3.6

(*continued*)

Table 4.10 *Continued*

Name CAS-No.	Structure	Log Kow	Bioavail-ability (oral)	DDD[1] (g/d)	Renal excretion (%)		Surface water (SW)			WWTP influent			WWTP effluent		
							Samples	A-MEC (µg/L)	MAX (µg/L)	Samples	A-MEC (µg/L)	MAX (µg/L)	Samples	A-MEC (µg/L)	MAX (µg/L)
Thiamphenicol 15318-45-3	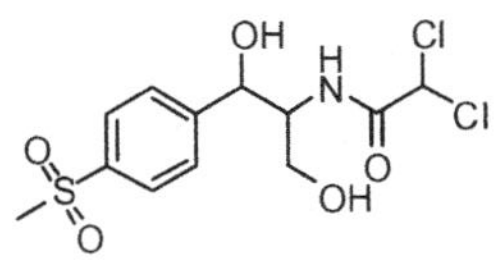	−0.27	–	1.5	–	France				19	0	0	19	0	0
Trimethoprim 738-70-5	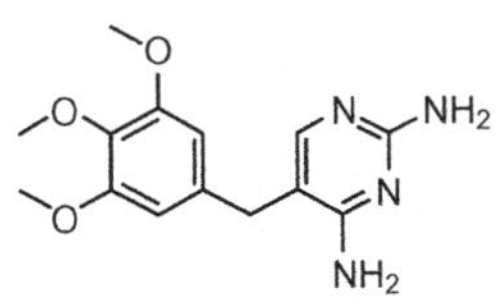	0.91	>63	0.4	63±10	Canada	224	0.002	0.015				19	0.205	0.344
						Croatia				5	1.172	4.22	5	0.29	0.31
						France	33	0.001	0.014	19	0.026	1.05	24	0.019	0.17
						Germany	299	0	0.2	23	0.291	1.1	203	0.07	1.5
						Greece							6	0.12	0.3
						Italy							3	0.04	0.13
						Korea	8	0.004	0.005				9	0.053	0.188
						Spain	17	0.02	0.069	11	0.186	0.65	5	0.11	0.23
						Sweden				13	0.52	1.86	37	0.284	1.88
						Switzerland				9	0.29	0.44	15	0.07	0.31
						UK	54	0.006	0.569	9	0.263	0.3	12	0.238	0.322
						USA	242	0.076	0.8	30	1.652	7.9	41	0.238	2.4
Tylosin 1401-69-0	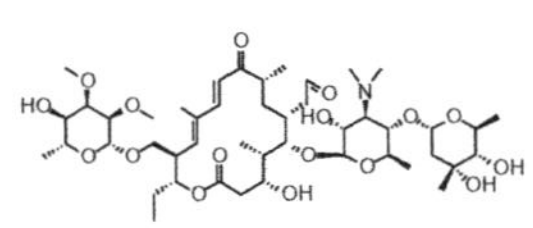	1.67	–	–	–	Canada	136	0	0						
						France	6	0	0	19	0	0	22	0	0
						Germany	167	0	0.02				143	0	0.09
						Italy	8	0	0.003				8	0.001	0.001
						USA	133	0.031	0.28	105	0.153	1.5	99	0.046	0.72
Vancomycin 1404-90-6	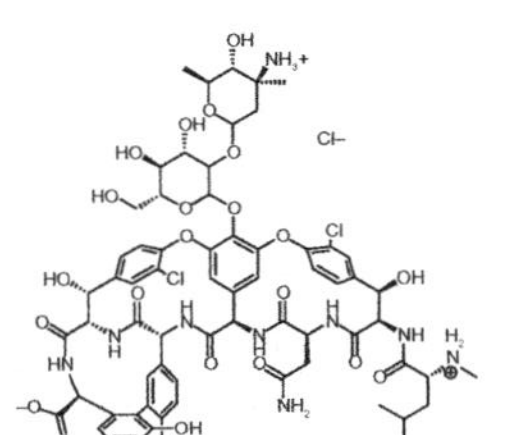	1.18	–	2	79±11	France	6	0	0				3	0	0
						Germany	21	0	0				51	0	0

(*continued*)

Table 4.10 *Continued*

Name CAS-No.	Structure	Log Kow	Bioavail-ability (oral)	DDD[1] (g/d)	Renal excretion (%)		Surface water (SW)			WWTP influent			WWTP effluent		
							Samples	A-MEC (µg/L)	MAX (µg/L)	Samples	A-MEC (µg/L)	MAX (µg/L)	Samples	A-MEC (µg/L)	MAX (µg/L)
					ANTICOAGULANTS										
Warfarin 81-81-2	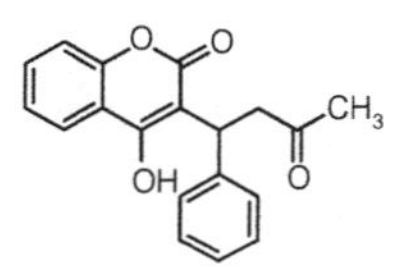	2.6	–	0.0075	–	USA				5	0.17	0.17	6	0	0
Antidepressants															
Amitriptyline 50-48-6	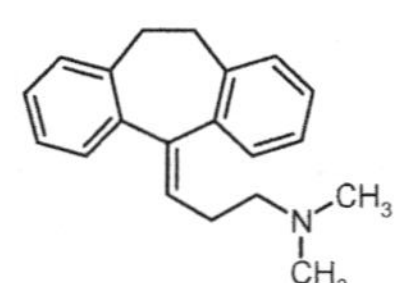	4.92	–	0.075	2	France	7	0	0				2	0.003	0.006
						UK	12	0	0						
Citalopram 59729-33-8	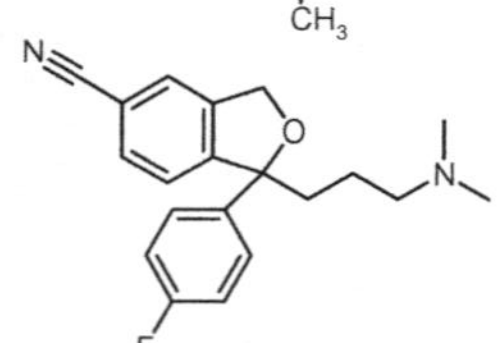	3.74	–	0.02	19	Austria							24	0.054	0.322
						Norway				4	0.089	0.612	3	0.062	0.382
Doxepin 1668-19-5	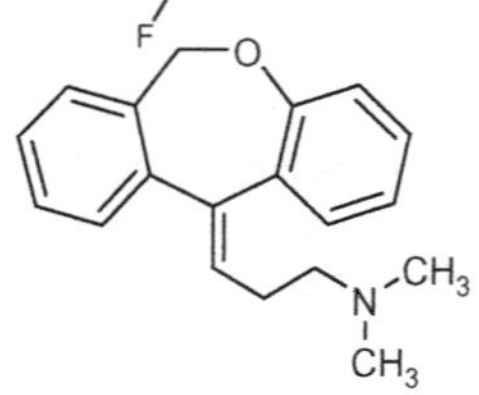	4.29	–	0.1	0	France	7	0	0				2	0	0
Fluoxetine 54910-89-3	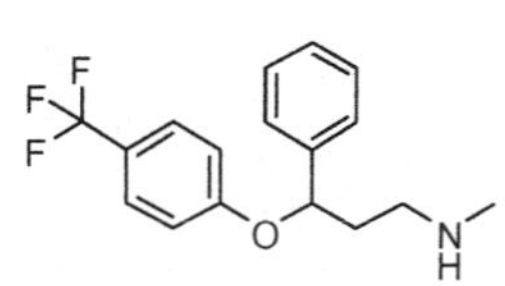	3.82	–	0.02	28	Canada							19	0.011	0.099
						Croatia				5	0	0	5	0	0
						France	27	0	0.004	5	0.209	1.236			
						Korea	8	0	0				9	0.002	0.002
						Norway				4	0.002	0.002	3	0.001	0.001
						Spain	10	0	0						
						UK	6	0.02	0.034						
						USA	27	0.001	0.002	13	0.239	0.6	17	0.207	0.56

(continued)

Table 4.10 *Continued*

Name CAS-No.	Structure	Log Kow	Bioavail-ability (oral)	DDD[1] (g/d)	Renal excretion (%)		Surface water (SW)			WWTP influent			WWTP effluent		
							Samples	A-MEC (µg/L)	MAX (µg/L)	Samples	A-MEC (µg/L)	MAX (µg/L)	Samples	A-MEC (µg/L)	MAX (µg/L)
Fluvoxamine 54739-18-3	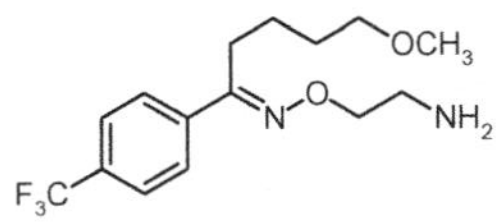	2.89	–	0.1	4	Norway				4	0.002	0.004	3	0	0.001
Imipramine 50-49-7	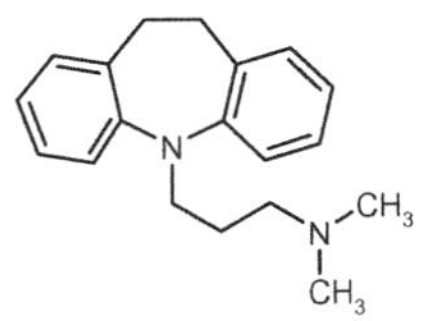	2.5	–	0.1	–	France	7	0	0				2	0	0
Paroxetine 61869-08-7	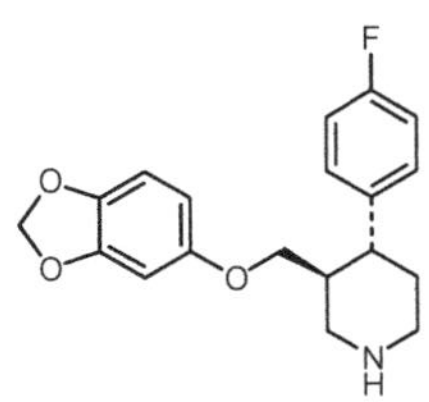	3.95	–	0.02	–	Croatia				5	0	0	5	0	0
						Norway				4	0.002	0.012	3	0.001	0.002
						Spain	10	0	0						
Sertraline 79617-96-2	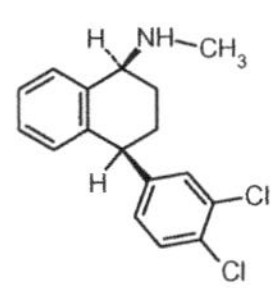	5.29	–	0.05	–	Norway				4	0.002	0.016	3	0.002	0.002
Venlafaxine 93413-69-5	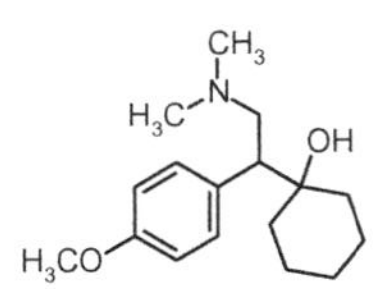	0.43	–	0.1	–	Austria							24	0	0.206
ANTIDIABETICS															
Metformin 657-24-9	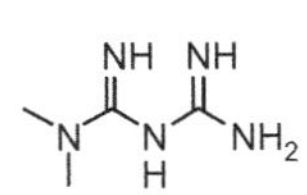	–1.8	52±5	2	50	USA	84	0.11	0.15	5	26	26	6	11	11

(*continued*)

Table 4.10 *Continued*

Name CAS-No.	Structure	Log Kow	Bioavail-ability (oral)	DDD[1] (g/d)	Renal excretion (%)		Surface water (SW)			WWTP influent			WWTP effluent		
							Samples	A-MEC (µg/L)	MAX (µg/L)	Samples	A-MEC (µg/L)	MAX (µg/L)	Samples	A-MEC (µg/L)	MAX (µg/L)
					ANTIEPILEPTICS										
Carbamazepine 298-46-4	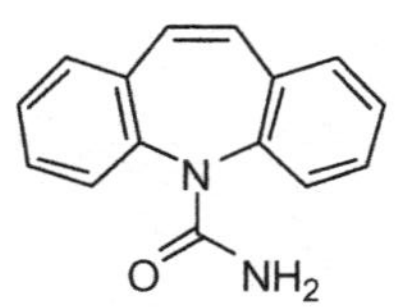	2.25	>70	1	5	Austria	4	0.047	0.133	7	1.258	2.41	7	1.132	1.67
						Canada	293	0.017	0.65	29	0.5	1.9	64	0.307	2.3
						Croatia				5	0.42	0.95	5	0.41	0.63
						Finland	11	0.037	0.08	24	0.288	0.82	24	0.489	2.44
						France	89	0.024	1.15	5	0.694	1.333	46	0.394	2.519
						Germany	619	0.227	2.5	68	1.6	4	351	1.187	22
						Italy	23	0.009	0.345				23	0.286	1.318
						Korea	8	0.025	0.061	16	1.84	9.42	25	0.171	0.97
						Norway				1	0	0	4	0	0
						Romania	8	0.068	0.075						
						Spain	22	0.059	0.279	58	0.132	0.95	49	0.168	0.6
						Sweden							24	0.44	1.18
						UK	14	0.17	0.794						
						USA	20	0.075	0.104	13	0.173	0.274	15	0.119	0.187
Phenytoin 57-41-0	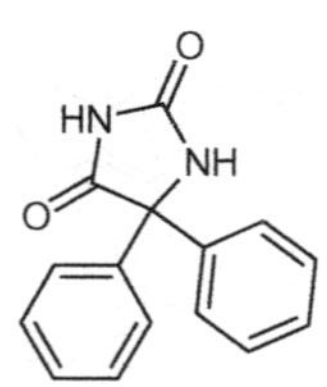	2.47	–	0.3	3	Korea	8	0.004	0.009				9	0.036	0.181
						USA	14	0.076	0.088	8	0.329	0.402	6	0.287	0.287
Gabapentin 60142-96-3	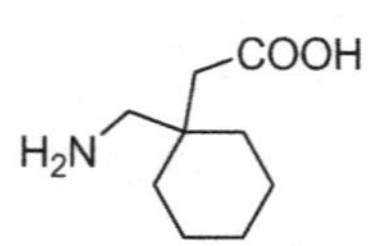	−1.10	–	1.8	–	UK	12	0.046	0.098						
Pentobarbital 76-74-4	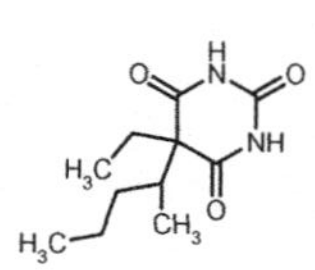	1.9	–	0.1	–	Germany	18	0.445	5.4						
Phenobarbital 50-06-6	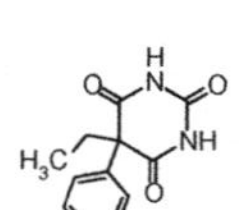	1.5	–	0.1	43	Germany	18	0	1.5						

(*continued*)

Table 4.10 *Continued*

Name CAS-No.	Structure	Log Kow	Bioavail-ability (oral)	DDD[1] (g/d)	Renal excretion (%)		Surface water (SW)			WWTP influent			WWTP effluent		
							Samples	A-MEC (μg/L)	MAX (μg/L)	Samples	A-MEC (μg/L)	MAX (μg/L)	Samples	A-MEC (μg/L)	MAX (μg/L)
ANTIFUNGALS															
Clotrimazole 23593-75-1	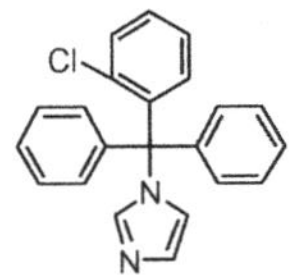	5.48	–	0.1	–	France	27	0	0						
						Germany	1	0.005	0.005						
						UK	18	0.022	0.034	9	0.031	0.033	9	0.014	0.027
Miconazole 22916-47-8	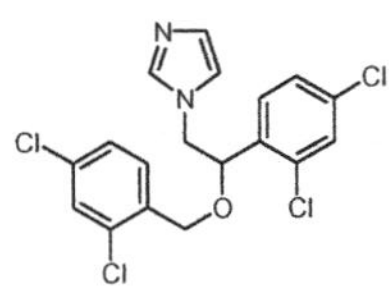	6.25	–	1	1	UK	6	0.008	0.009						
ANTIHYPERTENSIVES															
Enalapril 75847-73-3	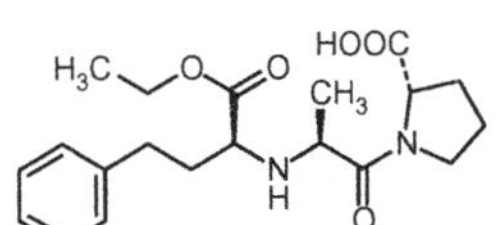	0.07	–	0.01	20	Italy	15	0	0.001				9	0	0
						USA	12	0	0	6	0.035	0.035	6	0.001	0.001
Labetalol 36894-69-6	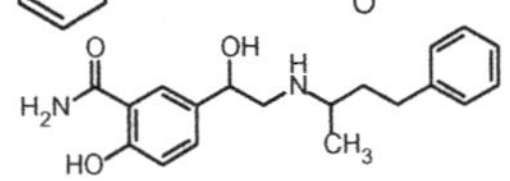	3.1	–	0.6	–	Canada							7	0.081	0.175
Nifedipine 21829-25-4	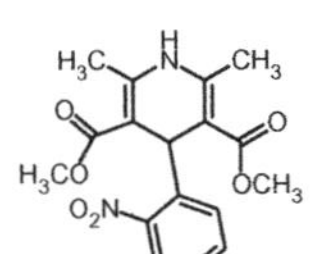	2.2	–	0.03	–	Canada	6	0	0						
						Germany	11	0	0	5	0	0	19	0	0.089
						USA				5	0	0	6	0	0
Valsartan 137862-53-4	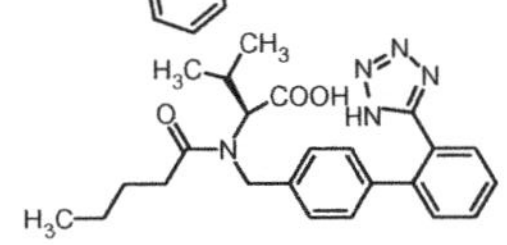	–	–	0.08	–	UK	12	0.012	0.133						
ANTIPSYCHOTICS															
Risperidone 106266-06-2	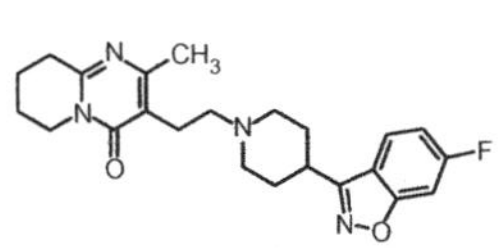	3.49	–	0.005	–	USA	12	0	0	6	0	0	6	0	0

(continued)

Table 4.10 *Continued*

Name CAS-No.	Structure	Log Kow	Bioavail-ability (oral)	DDD[1] (g/d)	Renal excretion (%)		Surface water (SW)			WWTP influent			WWTP effluent		
							Samples	A-MEC (μg/L)	MAX (μg/L)	Samples	A-MEC (μg/L)	MAX (μg/L)	Samples	A-MEC (μg/L)	MAX (μg/L)
					ANTITUSSIVES										
Dihydroco-deine 125-28-0	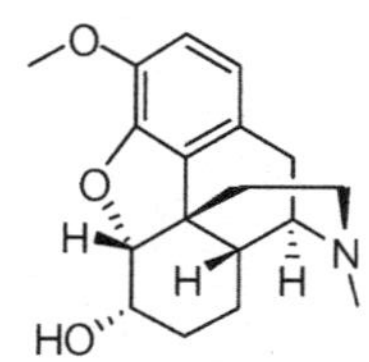	1.58	20	0.15	–	Germany	36	0.008	0.079	6	0.1	0.33	38	0.026	0.3
					ANTIULCERATIVES										
Cimetidine 51481-61-9	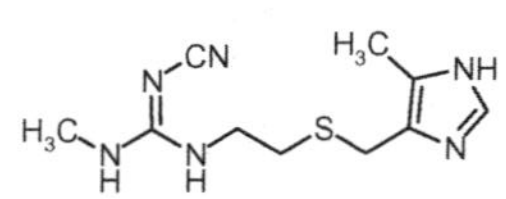	0.40	–	0.8	75	UK	12	0	0.011						
						USA				5	0.014	0.014	6	0.012	0.012
Lansoprazole 103577-45-3	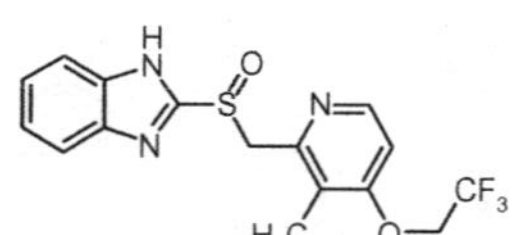	1.77	–	0.03	–	Croatia				5	0	0	5	0	0
						Spain	10	0	0						
Omeprazole 73590-58-6	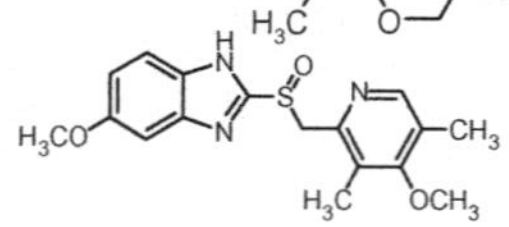	2.23	–	0.02	–	Germany	11	0	0	5	0	0	19	0	0
						Italy	16	0	0				17	0	0
Ranitidine 66357-35-5	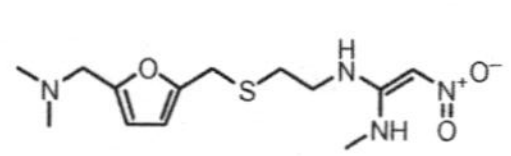	0.27	52±11	0.3	38	Croatia				5	0.188	0.29	5	0.135	0.2
						Italy	15	0.001	0.038				17	0.282	0.61
						Spain	17	0.019	0.142	6	0.98	1.7			
						UK	12	0	0						
						USA	84	0.01	0.01	5	0.33	0.33	6	0.062	0.062
					Beta-BLOCKERS										
Acebutolol 37517-30-9	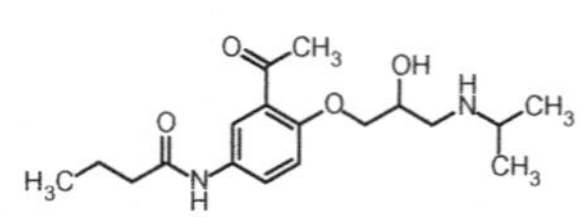	1.71	–	0.4	–	Canada							7	0.308	0.662
						Finland	11	0.006	0.014	24	0.3	1.04	24	0.146	0.255
						France							2	0.105	0.13
						Italy							3	0.04	0.11

(*continued*)

Table 4.10 *Continued*

Name CAS-No.	Structure	Log Kow	Bioavail-ability (oral)	DDD[1] (g/d)	Renal excretion (%)		Surface water (SW)			WWTP influent			WWTP effluent		
							Samples	A-MEC (μg/L)	MAX (μg/L)	Samples	A-MEC (μg/L)	MAX (μg/L)	Samples	A-MEC (μg/L)	MAX (μg/L)
Atenolol 29122-68-7	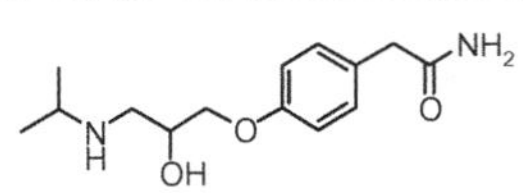	0.57	58±16	0.075	40	Canada							7	0.987	1.68
						Croatia				5	0.395	0.74	5	0.4	1.15
						Finland	11	0.022	0.055	24	0.728	1.71	24	0.264	1.18
						France	33	0.007	0.1				3	0.57	0.57
						Germany	225	0	0.07	22	0.923	2.3	132	0.277	1.8
						Italy	15	0.023	0.241				17	0.418	1.168
						Spain	17	0.142	0.465	11	1.959	122	5	0.28	1.2
						Sweden							24	0.19	0.47
						UK	12	0.012	0.06						
						USA	12	0.432	0.432	6	3.06	3.06	6	0.879	0.879
Betaxolol 63659-18-7	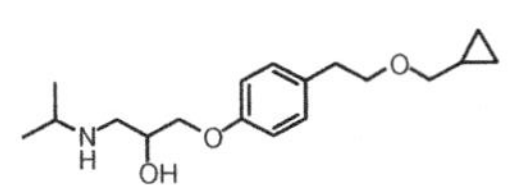	2.81	80 -90	0.02	15	France							26	0	0
						Germany	87	0	0.028	2	0.87	0.87	84	0.024	0.19
						Italy							3	0	0
Bisoprolol 66722-44-9	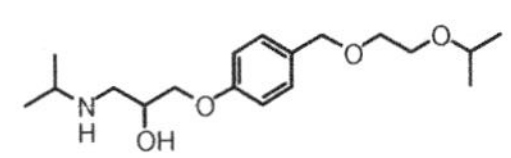	1.87	90	0.01	55	Canada							7	0.024	0.071
						France	6	0.01	0.01				27	0.703	2.838
						Germany	317	0	2.9	17	0.074	0.5	200	0.073	2
Carazolol 57775-29-8	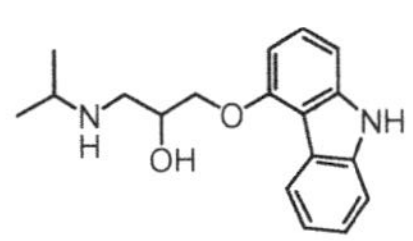	3.59	–	–	–	Germany	76	0	0.11	2	0.19	0.19	85	0.002	0.12
Diltiazem 42399-41-7	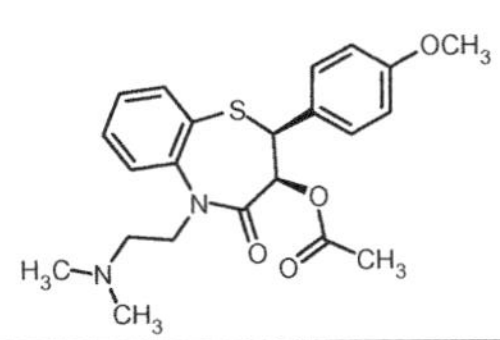	2.79	–	0.24	4	Canada	6	0	0						
						UK	12	0	0.001						
						USA				5	0.057	0.057	6	0.053	0.053

(*continued*)

Table 4.10 *Continued*

Name CAS-No.	Structure	Log Kow	Bioavail-ability (oral)	DDD[1] (g/d)	Renal excretion (%)		Surface water (SW)			WWTP influent			WWTP effluent		
							Samples	A-MEC (µg/L)	MAX (µg/L)	Samples	A-MEC (µg/L)	MAX (µg/L)	Samples	A-MEC (µg/L)	MAX (µg/L)
Metoprolol 37350-58-6	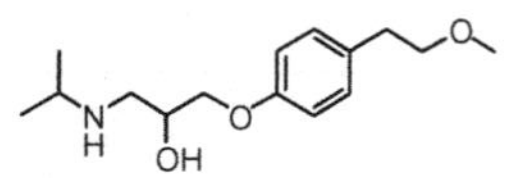	1.88	38±14	0.15	7	Canada							7	0.244	0.402
						Croatia				5	0	0	5	0	0
						Finland	11	0.049	0.116	24	1.094	1.46	24	0.802	1.6
						France	33	0.002	0.07				29	0.437	1.774
						Germany	335	0.025	2.2	22	2.05	7.2	232	0.842	9.12
						Italy							3	0.01	0.1
						Spain	13	0	0.014	30	0.02	0.09	30	0.01	0.14
						Sweden							24	0.08	0.39
						UK	12	0.03	0.155						
						USA	23	0.131	0.571				20	0.282	2.269
Nadolol 42200-33-9	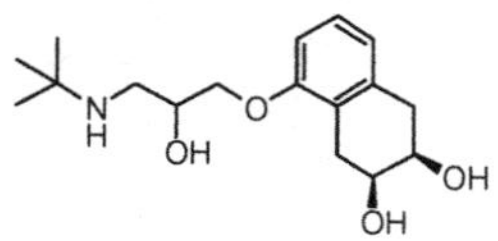	0.71	–	0.16	19	Canada							7	0.056	0.076
						Germany	76	0	0				83	0.009	0.14
Pindolol 61477-96-1	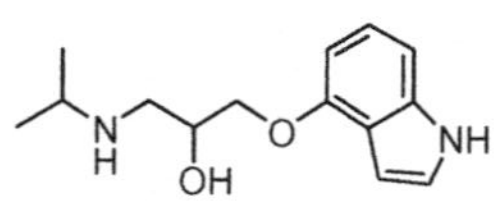	1.8	–	14	25	France	6	0.01	0.01				3	0	0
						Germany	14	0	0						
Propranolol 318-98-9	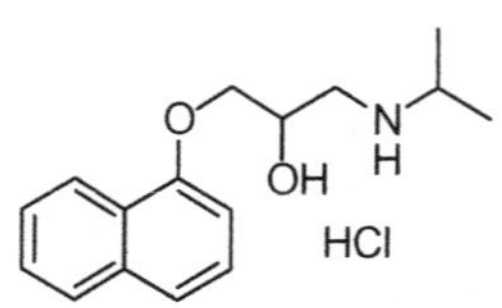	−0.45 at pH 2.0	26±10	0.16	23	Canada	6	0	0				7	0.03	0.05
						Croatia				5	0.168	0.29	5	0.29	0.47
						France	33	0.002	0.04				29	0.366	1.111
						Germany	257	0.002	0.59	28	0.814	10	223	0.072	0.8
						Italy							3	0.01	0.09
						Spain	20	0	0.063	41	0.212	6.5	35	0.033	0.52
						UK	62	0.02	0.215	9	0.07	0.119	15	0.23	0.373
						USA	22	0	0.074				19	0.009	0.2
Sotalol 3930-20-9	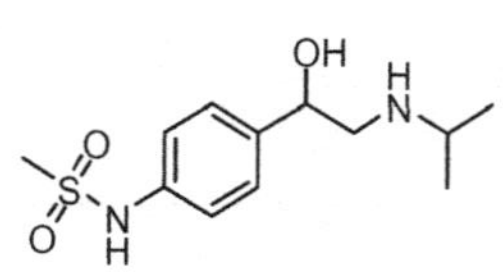	0.24	90	0.16	88	Canada							7	0.264	0.429
						Croatia				5	0.167	0.2	5	0.185	0.21
						Finland	11	0.041	0.086	24	0.706	3.28	24	0.26	1.12
						France	6	0.063	0.063				3	0.87	0.87
						Germany	225	0.044	0.95	22	1.255	3.526	132	1.109	6.5
						Spain	10	0	0.07						

(*continued*)

Table 4.10 *Continued*

Name CAS-No.	Structure	Log Kow	Bioavail-ability (oral)	DDD[1] (g/d)	Renal excretion (%)		Surface water (SW)			WWTP influent			WWTP effluent		
							Samples	A-MEC (µg/L)	MAX (µg/L)	Samples	A-MEC (µg/L)	MAX (µg/L)	Samples	A-MEC (µg/L)	MAX (µg/L)
Timolol 26839-75-8	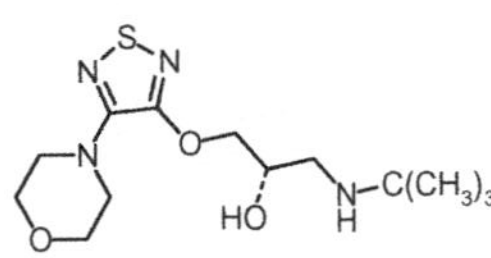	1.8	–	0.02	–	Canada							7	0	0
						Germany	68	0.002	0.068				29	0	0.07
					BRONCHIOLYTICS										
Clenbuterol 37148-27-9	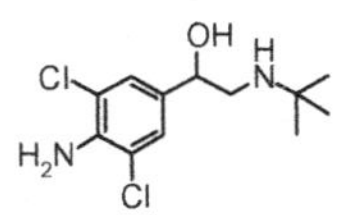	2.94	–	–	–	France	7	0	0				2	0.003	0.006
						Germany	148	0	0.11	14	0	0	206	0	0.1
Fenoterol 13392-18-2	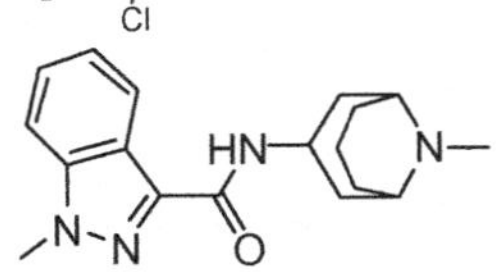	1.22	–	0.01	37	Germany	76	0	0.061				83	0	0.06
Salbutamol 18559-94-9	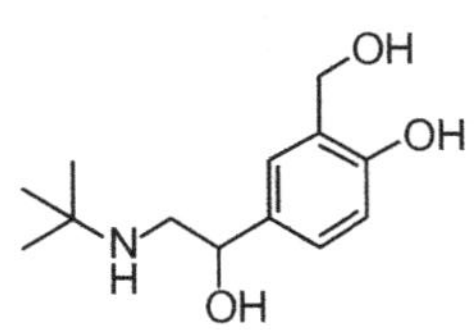	0.64	50	0.012	75	Canada							7	0.007	0.01
						France	7	0	0				2	0	0
						Germany	323	0	0.19	16	0.026	0.21	208	0	0.17
						Italy	15	0.001	0.002				17	0.008	0.018
						UK	72	0.115	0.471	6	0.113	0.141	18	0.109	0.17
						USA				5	13	13	6	8.1	8.1
Terbutaline 23031-25-6	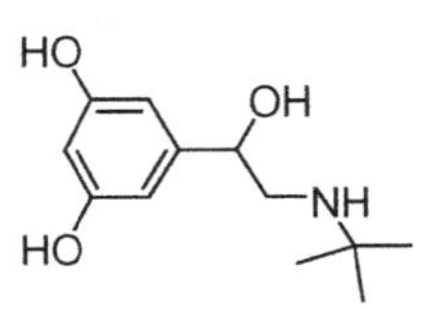	0.9	–	0.015	10	France	7	0	0				2	0	0
						Germany	176	0	0.03	14	0.053	0.37	142	0.002	0.6
						Germany	594	0.15	100	17	1.941	3.3	106	0.425	15.8
					CONTRAST MEDIA										
Diatrizoate 737-31-5	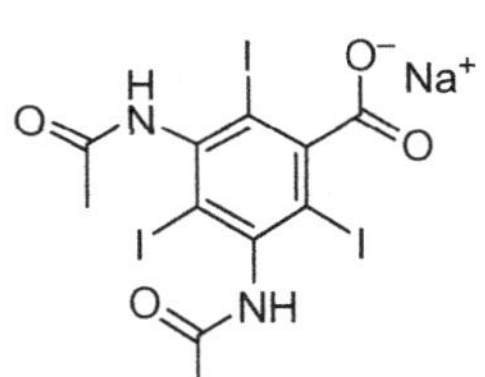	1.37	–	–	100	France	6	0.129	0.129				3	2.37	2.37

(continued)

Table 4.10 *Continued*

Name CAS-No.	Structure	Log Kow	Bioavail-ability (oral)	DDD[1] (g/d)	Renal excretion (%)		Surface water (SW)			WWTP influent			WWTP effluent		
							Samples	A-MEC (µg/L)	MAX (µg/L)	Samples	A-MEC (µg/L)	MAX (µg/L)	Samples	A-MEC (µg/L)	MAX (µg/L)
Iohexol 66108-95-0	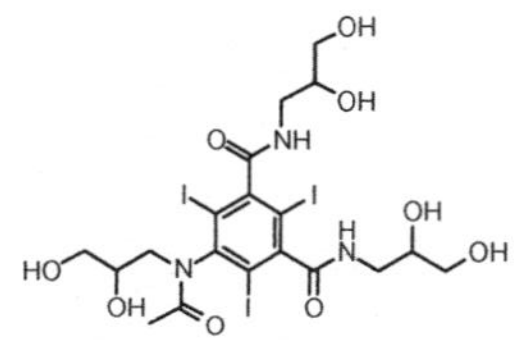	–3.05	–	–	100	France	6	0.115	0.115				3	1.64	1.64
						Germany	498	0.074	0.5	4	9	9	4	1	7
Iomeprol 78649-41-9	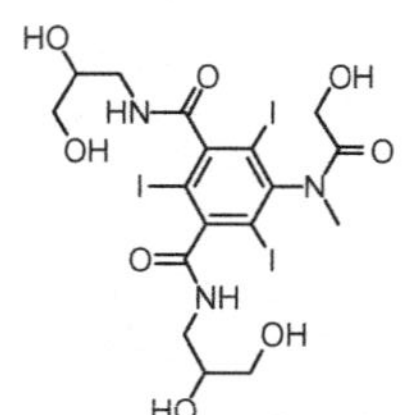	–2.52	–	–	100	France	6	0.14	0.411				3	1.59	1.59
						Germany	583	0.132	0.89	17	3.095	10	72	0.356	10
Iopamidol 62883-00-5	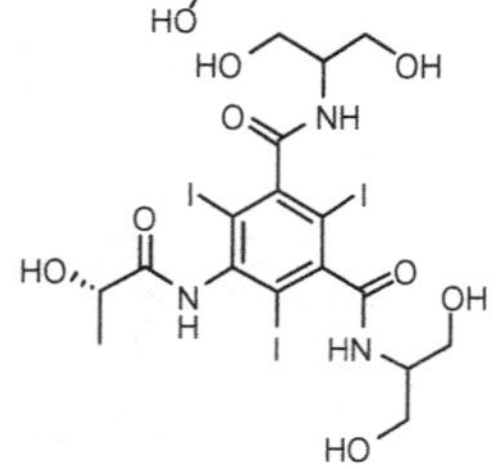	–2.42	–	–	100	France	6	0.456	0.456				3	8.03	8.03
						Germany	596	0.22	2.8	17	2.059	9	85	0.615	15
Iopanoic acid 96-83-3	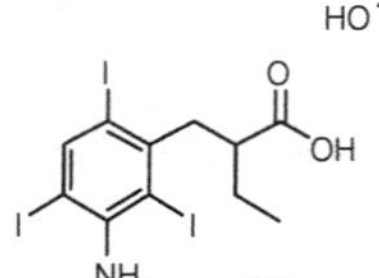	5.78	–	–	100	France	6	0	0				3	0	0
Iopromide 73334-07-3	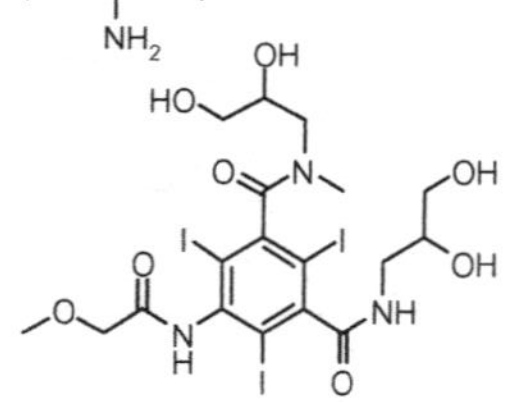	–2.05	–	–	100	Austria				7	0.036	0.18	7	0	0
						France	10	0.004	0.017				3	0.25	0.25
						Germany	601	0.112	4	19	8.816	22	86	1.051	20
						Korea	8	0.134	0.361				9	2.359	4.03
						USA	8	0.004	0.009	2	0.008	0.017	3	0.012	0.012

(continued)

Table 4.10 *Continued*

Name CAS-No.	Structure	Log Kow	Bioavail-ability (oral)	DDD[1] (g/d)	Renal excretion (%)		Surface water (SW)			WWTP influent			WWTP effluent		
							Samples	A-MEC (μg/L)	MAX (μg/L)	Samples	A-MEC (μg/L)	MAX (μg/L)	Samples	A-MEC (μg/L)	MAX (μg/L)
Ioxaglic acid 59017-64-0	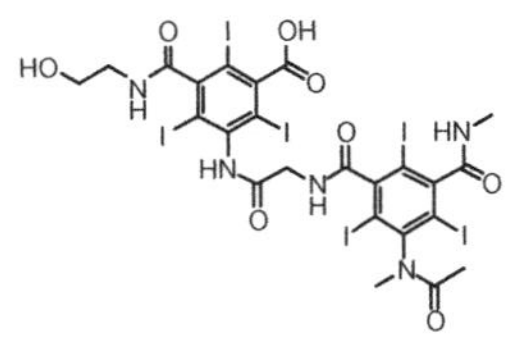	2.51	–	–	100	France	6	0.039	0.039				3	0.58	0.58
Ioxithalamic acid 28179-44-4	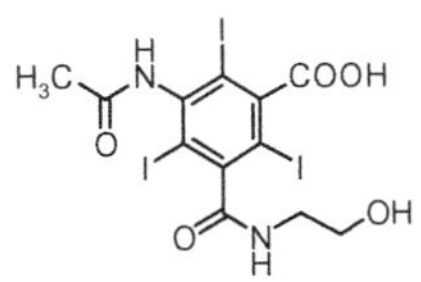	1.36	–	–	100	France	6	0.036	0.438				3	0.44	0.44
						Germany	50	0.01	0.08	6	0.17	0.17	34	0.041	0.21
					CYTOSTATICS										
Cyclopho-sphamid 50-18-0	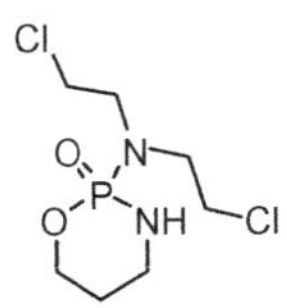	0.63	74±22	–	25	Canada	16	0	0	18	0	0	53	0.001	0.008
						Germany	239	0	0.1	14	0	0	130	0	0.15
						Italy	16	0	0				17	0.003	0.009
						Romania	8	0.023	0.065						
						Switzerland	5	0	0	5	0.005	0.011	4	0.003	0.01
						UK				6	0	0	6	0	0
Ifosfamid 3778-73-2	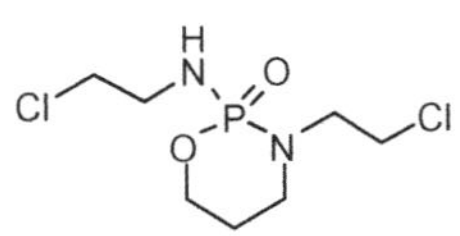	0.86	–	–	–	Canada				18	0	0	18	0	0
						Germany	224	0	0.18	12	0	0.04	122	0	2.9
						Switzerland	5	0	0	5	0.005	5	4	0	0.006
Tamoxifen 10540-29-1	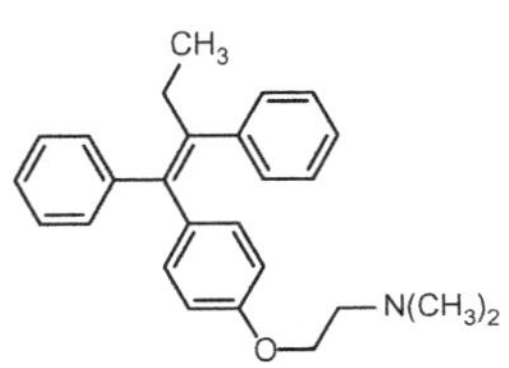	6.3	–	0.02	–	UK	20	0.053	0.212	9	0.153	0.215	9	0.199	0.369

(*continued*)

Table 4.10 *Continued*

Name CAS-No.	Structure	Log Kow	Bioavail-ability (oral)	DDD[1] (g/d)	Renal excretion (%)		Surface water (SW)			WWTP influent			WWTP effluent		
							Samples	A-MEC (µg/L)	MAX (µg/L)	Samples	A-MEC (µg/L)	MAX (µg/L)	Samples	A-MEC (µg/L)	MAX (µg/L)
DIURETICS															
Furosemide 54-31-9	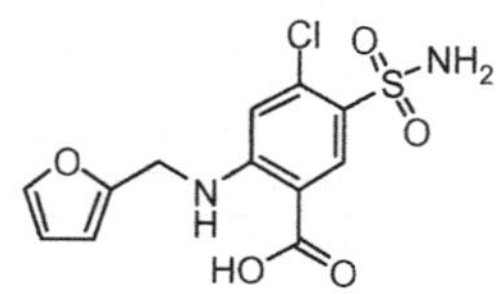	2.03	71±35	0.04	90	France	27	0	0.039						
						Italy	15	0.004	0.255				17	0.589	2.102
Hydrochloro-thiazide 58-93-5	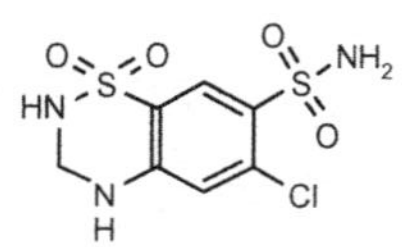	–0.07	71±18	0.025	82	Italy	15	0.006	0.256				17	0.488	1.253
LIPID REGULATORS															
Atorvastatin 134523-00-5	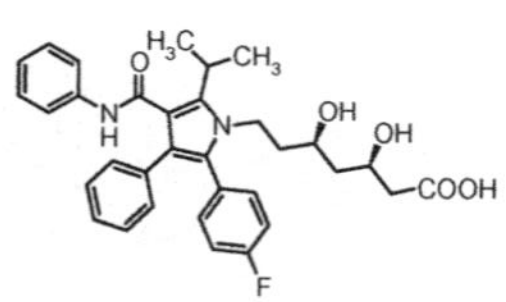	6.36	–	0.01	1	Canada	6	0	0.001	3	0.076	0.076	6	0.03	0.037
						USA	12	0	0	6	0.201	0.201	6	0	0
Bezafibrate 41859-67-0	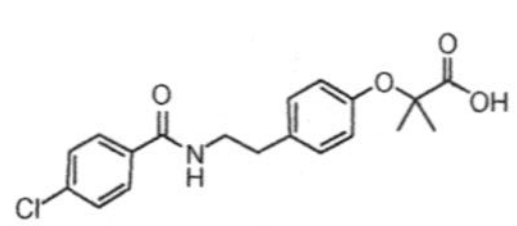	4.2	–	0.6	68	Austria	9	0.008	0.02	7	3.526	9.43	7	0.05	0.13
						Canada	196	0.003	0.47	18	0.6	4.7	53	0.202	0.81
						Croatia				5	0.023	0.05	5	0	0.01
						Finland	8	0.008	0.02	2	0.42	0.42	2	0.011	0.011
						France	34	0.031	0.78				5	0.718	1.07
						Germany	283	0.082	3.1	17	4.135	7.5	258	1.16	4.8
						Italy	31	0.013	0.076				20	0.039	0.91
						Spain	20	0.018	0.363	30	0	0	30	0.18	0.34
Etofibrate 31637-97-5	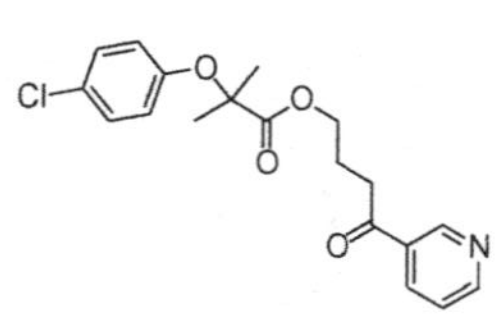	3.43	–	–	–	Germany	63	0	0				20	0	0
Fenofibrate 49562-28-9	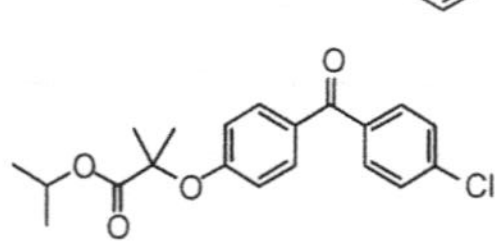	4.3	–	0.2	2	Canada				36	0	0	39	0	0
						France	33	0.005	0.025				5	0.214	0.31
						Germany	121	0	0				20	0	0.03
						Italy							3	0.16	0.16
						USA				5	0	0	6	0	0

(continued)

Table 4.10 *Continued*

Name CAS-No.	Structure	Log Kow	Bioavail-ability (oral)	DDD[1] (g/d)	Renal excretion (%)		Surface water (SW)			WWTP influent			WWTP effluent		
							Samples	A-MEC (µg/L)	MAX (µg/L)	Samples	A-MEC (µg/L)	MAX (µg/L)	Samples	A-MEC (µg/L)	MAX (µg/L)
Gemfibrozil 25812-30-0	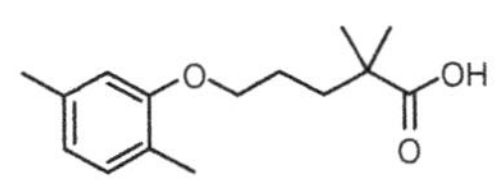	3.9	98±1	1.2	30	Canada	299	0.004	58	62	0.479	36.53	104	0.436	2.09
						Croatia				5	0.155	0.36	5	0.12	0.32
						France	85	0.007	0.086				46	0.06	1.34
						Germany	206	0.012	0.51				105	0.318	1.5
						Italy	8	0.001	0.044				3	0.84	4.76
						Korea	8	0.007	0.009	16	0	0	25	0.003	0.017
						Spain	17	0.035	0.497						
						Sweden							24	0.56	2.07
						USA	142	0.045	0.79	8	3.73	4.77	29	0.027	0.234
Mevastatin 73573-88-3	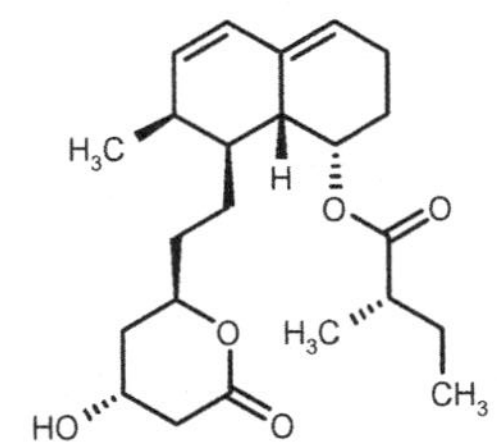	3.95	–	–	–	Croatia				5	0	1.17	5	0.383	0.8
						Spain	10	0	0	5	0.55	1.25	5	0.35	0.85
Pravastatin 81093-37-0	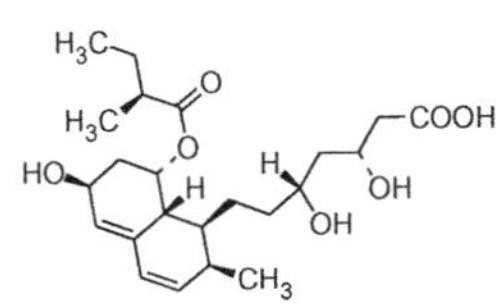	–0.23	–	0.02	–	Canada	3	0	0	3	0.117	0.117	3	0.059	0.059
						Croatia				5	0	0	5	0	0
						Spain	10	0	0						
Simvastatin 79902-63-9	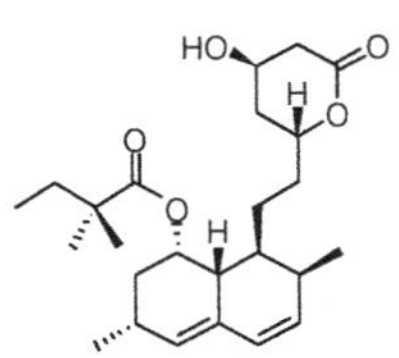	4.68	< 5	0.015	–	Canada	3	0	0	3	0.004	0.004	3	0.001	0.001
						France	3	0	0				3	0	0
						Germany	48	0	0				7	0	0
						UK	12	0	0						
						USA	12	0.001	0.001	6	5	5	6	0	0
					METABOLITES										
1-Hydroxyi-buprofen -	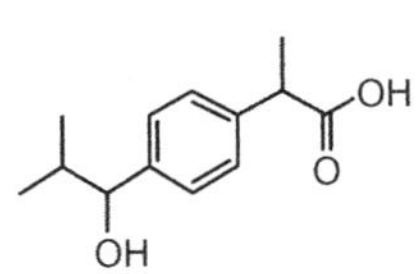	–	–	–	–	France	27	0	0						

(*continued*)

Table 4.10 *Continued*

Name CAS-No.	Structure	Log Kow	Bioavail-ability (oral)	DDD[1] (g/d)	Renal excretion (%)		Surface water (SW)			WWTP influent			WWTP effluent		
							Samples	A-MEC (µg/L)	MAX (µg/L)	Samples	A-MEC (µg/L)	MAX (µg/L)	Samples	A-MEC (µg/L)	MAX (µg/L)
2-Hydroxyi-buprofen -	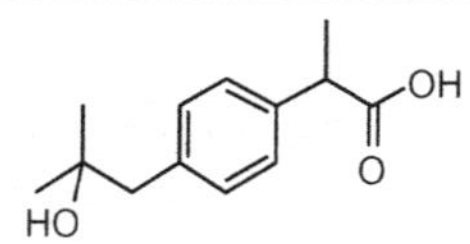	2.41	–	–	–	Canada				8	1.081	1.957	8	0.883	1.563
						France	27	0	0.114						
						Germany	7	0.032	0.101				10	0.92	5.96
						Norway							4	0.485	1.13
4-Chloroben-zoic acid 83-15-8	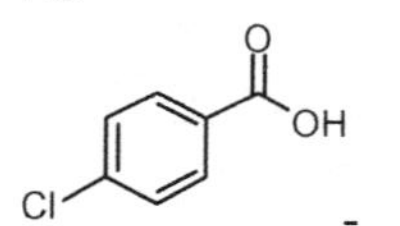	0.4	–	–	–	France	27	0.002	0.011						
6-o-des-methyl naproxen 60756-73-2	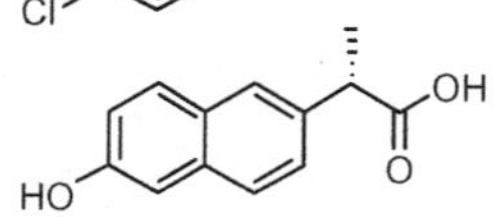	3.18	–	–	–	France	27	0	0.032						
AMDOPH -	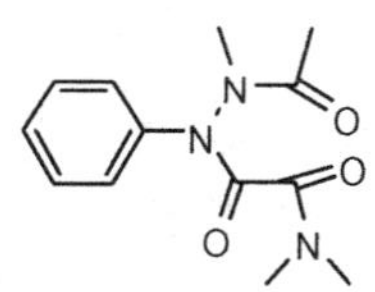	3.73	–	–	–	Germany	22	0.213	0.35	12	0.55	0.71	25	0.428	0.73
AMPH 38604-70-5	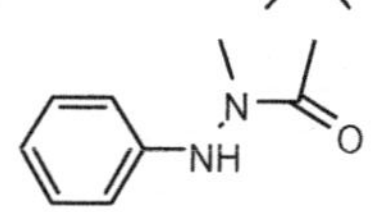	1.24	–	–	–	Germany	22	0.038	0.16	12	0.11	0.15	25	0.074	0.19
CBZ-10OH 29331-92-8	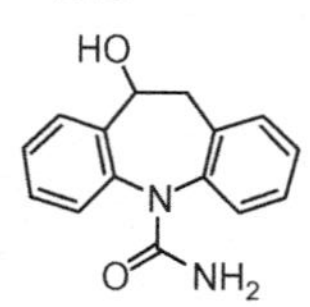	1,26	–	–	–	Canada	3	0	0	3	0.008	0.008	3	0.009	0.009
CBZ-2OH 68011-66-5	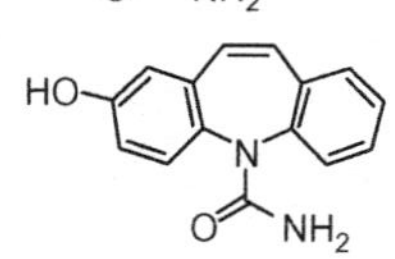	2.21	–	–	–	Canada	3	0	0	3	0.121	0.121	3	0.132	0.132
CBZ-3OH -	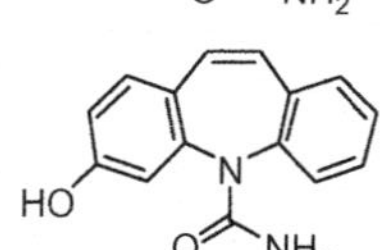	2.25	–	–	–	Canada	3	0	0	3	0.095	0.095	3	0.102	0.102

(continued)

Table 4.10 *Continued*

Name CAS-No.	Structure	Log Kow	Bioavailability (oral)	DDD[1] (g/d)	Renal excretion (%)		Surface water (SW)			WWTP influent			WWTP effluent		
							Samples	A-MEC (μg/L)	MAX (μg/L)	Samples	A-MEC (μg/L)	MAX (μg/L)	Samples	A-MEC (μg/L)	MAX (μg/L)
CBZ-DiH 3564-73-6	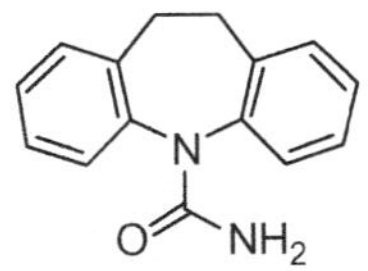	2.36	–	–	–	Germany	3	0.002	0.004						
CBZ-DiH 3564-73-6	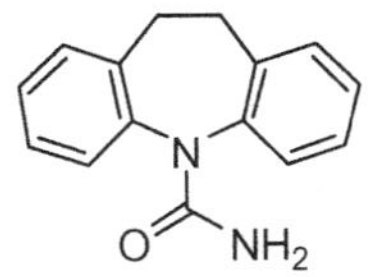	2.36	–	–	–	Germany	3	0.002	0.004						
CBZ-DiOH 35079-97-1	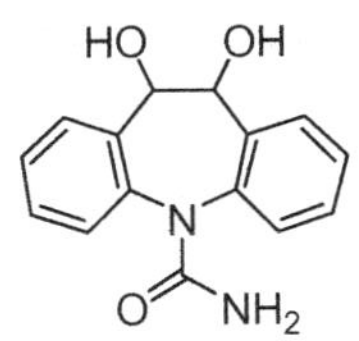	0.76	–	–	–	Canada	3	0.002	0.002	3	1.572	1.572	3	1.325	1.325
						Germany	3	0.058	0.14	11	2	3.7	12	2	3.6
CBZ-EP	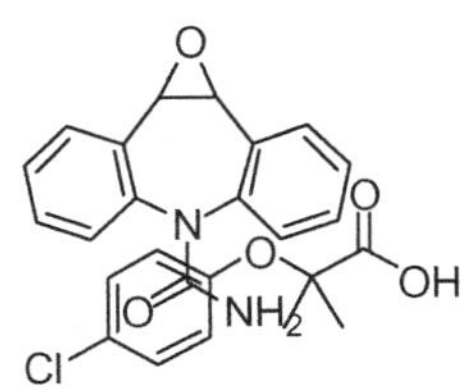	1.58	–	–	–	Canada	3	0	0	3	0.047	0.047	3	0.052	0.052
						Spain				12	0.35	0.5	12	0.16	0.3
Clofibric acid 882-09-7		3.1	–	–	76	Austria	4	0.025	0.044						
						Canada	196	0.011	0.175	62	0.011	0.22	100	0.03	0.283
						Croatia				5	0.072	0.11	5	0.028	0.03
						France	27	0	0.014				2	0	0
						Germany	291	0.026	0.55	34	0.434	1.2	297	0.235	3.29
						Italy	8	0.001	0.006				11	0.064	0.68
						Japan				11	0.028	0.028	11	0.014	0.014
						Korea				17	1.51	4.38	17	0.31	0.74
						Norway				1	0	0	4	0	0
						Spain	20	0.007	0.02	30	0.04	2.02	30	0.025	0.12
						UK	42	0.01	0.111	15	0.203	0.651	18	0.01	0.044
						USA	20	0.038	0.63				2	0	0

(*continued*)

Table 4.10 *Continued*

Name CAS-No.	Structure	Log Kow	Bioavail-ability (oral)	DDD[1] (g/d)	Renal excretion (%)		Surface water (SW)			WWTP influent			WWTP effluent		
							Samples	A-MEC (µg/L)	MAX (µg/L)	Samples	A-MEC (µg/L)	MAX (µg/L)	Samples	A-MEC (µg/L)	MAX (µg/L)
Fenofibric acid 42017-89-0	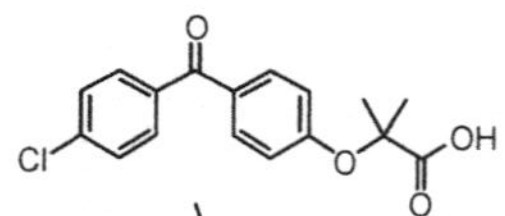	4.3	–	–	–	Brazil	2	0.06	0.54				2	0.07	0.4
						France	27	0	0.21						
						Germany	145	0.014	0.28	2	1.1	1.1	166	0.223	1.2
N-Formyl-4-aminoantipyr-ine 1672-5-8-8	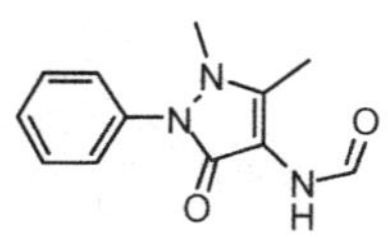	0.04	–	–	–	Czech Rep.	9	0	0.8						
						Germany	33	0.201	6.5	12	1.3	1.9	25	0.838	2
Norfluoxetine 56161-73-0	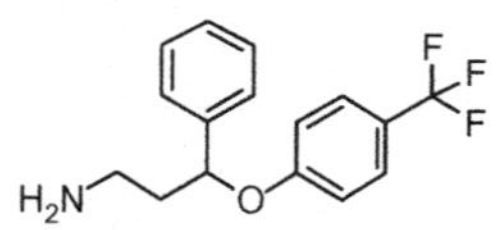	3.8	–	–	–	Canada							11	0	0
o-Hydroxy Atorvastatin -	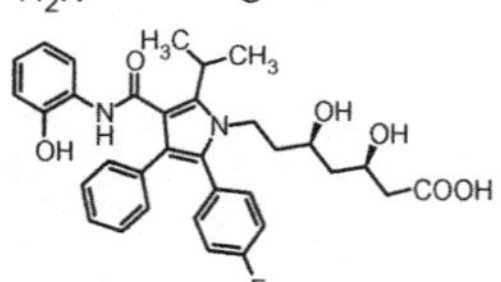	–	–	–	–	USA	12	0.004	0.004	6	0.196	0.196	6	0	0
p-Hydroxy Atorvastatin -	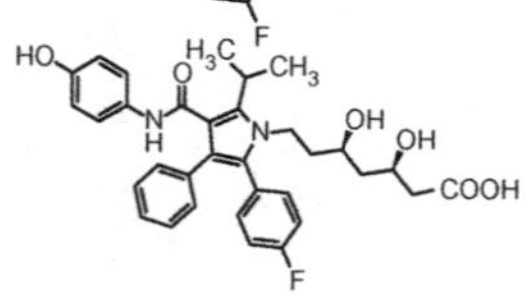	–	–	–	–	USA	12	0.005	0.005	6	0.28	0.28	6	0	0
Salicylic acid 69-72-7	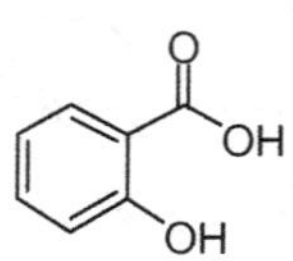	2.26	–	–	–	Canada	31	0.079	17	52	11.1	27.8	75	0.385	35
						France	27	0.008	0.613						
						Germany	35	0.025	4.1				36	0	0.14
						Italy	15	0.036	0.205				5	6.2	12
						Korea				16	23.92	88.99	16	2.43	6.73
						Spain	3	0.013	0.019	30	0	0	30	0.045	0.19
Simvastatin hydroxy acid 121009-77-6	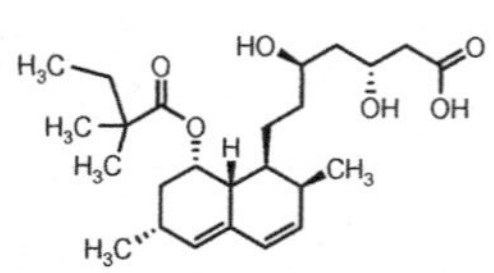	–	–	–	–	USA	12	0	0	6	0.01	0.01	6	0	0

(*continued*)

Table 4.10 *Continued*

Name CAS-No.	Structure	Log Kow	Bioavail-ability (oral)	DDD[1] (g/d)	Renal excretion (%)		Surface water (SW)			WWTP influent			WWTP effluent		
							Samples	A-MEC (µg/L)	MAX (µg/L)	Samples	A-MEC (µg/L)	MAX (µg/L)	Samples	A-MEC (µg/L)	MAX (µg/L)
					MUCOLYTICS										
Ambroxol 18683-91-5	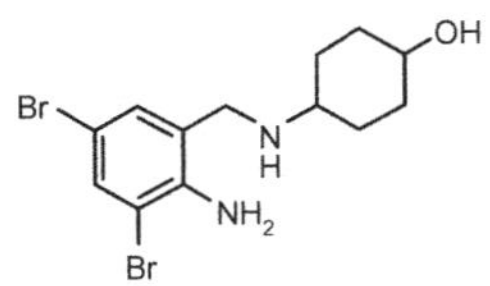	−0.65	–	0.12	80	Czech Rep. Germany	9 33	0 0.002	0 0.13				3	0	0
					SEDATIVES										
Secobarbital 76-73-3	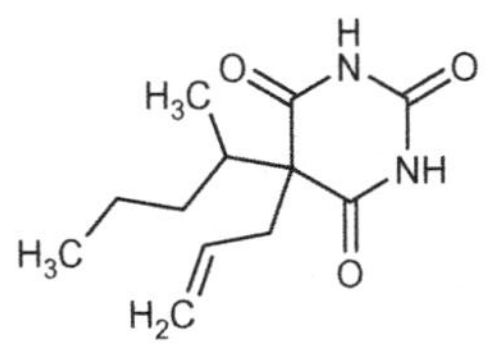	2.0	–	0.1	–	Germany	18	0	0.1						
Zolpidem 82626-48-0	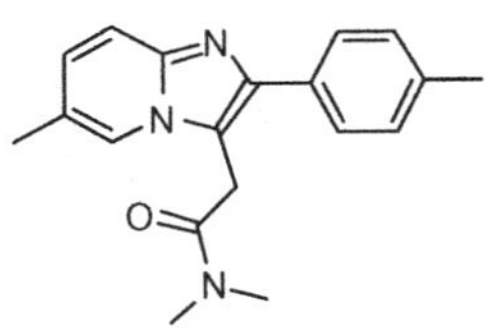	3.15	72±7	0.01	< 1	France	27	0	0.004	5	0.06	1.119			
					TRANQUILLISERS										
Alprazolam 28981-97-7	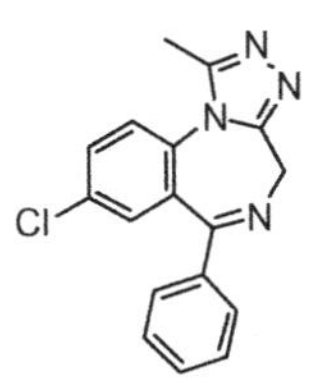	2.12	88±16	1	20	France	27	0	0.002						
Bromazepam 1812-30-2	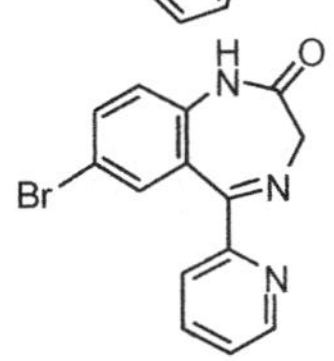	2.05	–	0.01	2	France	27	0	0.134						

(continued)

Table 4.10 *Continued*

Name CAS-No.	Structure	Log Kow	Bioavail-ability (oral)	DDD[1] (g/d)	Renal excretion (%)		Surface water (SW)			WWTP influent			WWTP effluent		
							Samples	A-MEC (μg/L)	MAX (μg/L)	Samples	A-MEC (μg/L)	MAX (μg/L)	Samples	A-MEC (μg/L)	MAX (μg/L)
Diazepam 439-14-5	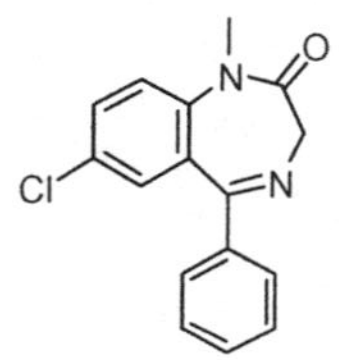	2.99	100±14	0.01	6	Belgium				3	0.59	1.18	3	0	0.66
						France	43	0	0				5	0	0
						Germany	314	0	0.033	32	0	0.1	227	0	0.1
						Italy	16	0	0.002				17	0	0
						Korea							2	0	0
						Romania	8	0.029	0.034						
						USA	20	0.001	0.001	8	3.75	5	9	0.003	0.004
Lorazepam 846-49-1	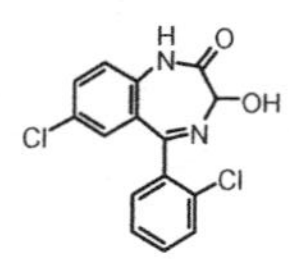	2.39	93±10	2.5	75	France	27	0	0.015	5	0.133	0.221			
Meproba-mate 57-53-4	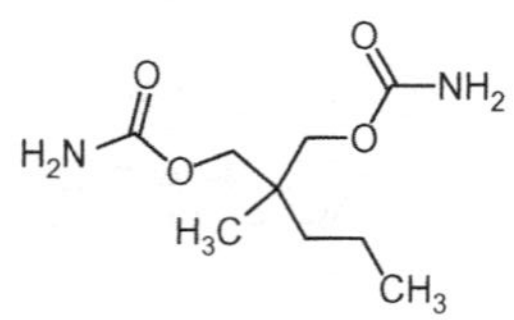	0.7	–	1.2	–	Korea	8	0	0				9	0.005	0.006
						USA	14	0.263	0.306	8	1.166	1.44	6	1.27	1.27
Nordazepam 1088-11-5	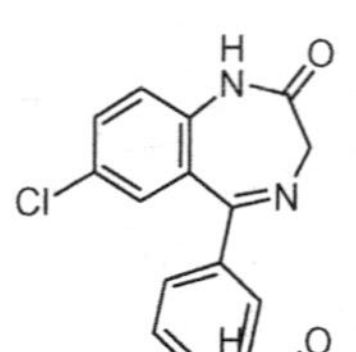	2.9	–	0.015	–	France	7	0	0.002				2	0.004	0.008
Oxazepam 604-75-1	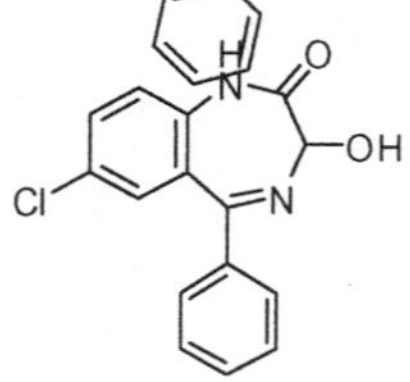	2.24 (at pH 7.4)	–	0.05	–	France	27	0.038	0.814	5	0.674	1.574			
						Germany	11	0.09	0.4	11	0.48	0.86	12	0.32	0.63

(continued)

Table 4.10 *Continued*

Name CAS-No.	Structure	Log Kow	Bioavail-ability (oral)	DDD[1] (g/d)	Renal excretion (%)		Surface water (SW)			WWTP influent			WWTP effluent		
							Samples	A-MEC (µg/L)	MAX (µg/L)	Samples	A-MEC (µg/L)	MAX (µg/L)	Samples	A-MEC (µg/L)	MAX (µg/L)
VASOCONSTRICTORS															
Xylometazo-lin 526-36-3	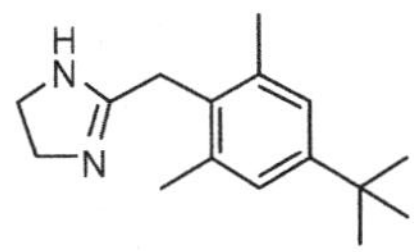	4.68	–	0.8	–	Czech Rep.	9	0	0						
						Germany	32	0.001	0.009						
VASODILATORS															
Pentoxifylline 6493-05-6	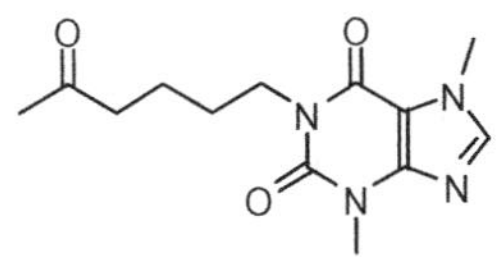	0.56	–	1	1	Canada	16	0	0				35	0.011	0.045
						Germany	99	0	0.26						
						Korea	8	0.002	0.002				9	0.003	0.004
						Romania	8	0.209	0.299						
						USA	8	0	0	2	0	0	3	0.002	0.002

Table 4.11 A-MEC and maximum concentrations of selected pharmaceuticals measured in ground water, bank filtrate, drinking water and marine water.

Name CAS-No.		Ground water (GW)			Bank Filtrate (BF)			Drinking Water (DW)			Marine Water (MW)		
		samples	A-MEC (μg/L)	MAX (μg/L)	samples	A-MEC (μg/L)	samples	A-MEC (μg/L)	MAX (μg/L)	MAX (μg/L)	samples	A-MEC (μg/L)	MAX (μg/L)
ANALGESICS													
Acetylsalicylic acid	France							7	0	0.001			
50-78-2	Germany							55	0	0			
Codeine 76-57-3	Germany	1	0	0									
	USA										44	0.005	0.105
Diclofenac 15307-86-5	Austria	7	0	0									
	Canada							3	0	0			
	France	2	0	0				7	0.002	0.002	8	0.032	0.101
	Germany	143	0	0.93	57	0.003	0.43	58	0	0.006			
	Italy	2	0	0				6	0	0			
	USA							8	0	0			
Dimethylaminophenazone 58-15-1	Germany	33	0.024	0.4				4	0	0			
Fenoprofen 31879-05-7	Germany	7	0	0									
Hydrocodone 125-29-1	Germany	1	0	0									
	USA							2	0	0	38	0.002	0.025
Ibuprofen 15687-27-1	Austria	7	0	0									
	Canada							3	0.003	0.003			
	France	2	0	0	3	0	0	8	0	0.001	8	0	0.021
	Germany	101	0	0	30	0	0.2	58	0	0.003	2	0	0
	Italy	2	0	0				6	0	0			
	Norway										12	0	0.001
	USA							22	0.082	1.35			
Indomethacin 53-86-1	Germany	88	0	0	49	0	0.005	55	0	0			
Ketoprofen 22071-15-4	France	2	0	0				7	0	0.003	8	0	0
	Germany	94	0	0	30	0	0	25	0	0			
	USA										38	0	0
Naproxen 22204-53-1	Canada							6	0	0			
	France	2	0	0				7	0	0	8	0.011	0.214
	Germany	77	0	0.14	30	0	0						
	USA							10	0	0			
Paracetamol 103-90-2	France	2	0	0				7	0.01	0.211	8	0.013	0.162
	Germany	42	0	0	84	0	0						
	USA							2	0	0	44	0.018	0.188
Phenazon 60-80-0	Germany	159	0.234	3.95	116	0.159	1.25	22	0.123	0.4			
	USA										38	0	0
Piroxicam 36322-90-4	Germany	49	0	0									
Propyphenazon 479-92-5	Germany	75	0.133	1.23	131	0.041	1.465	8	0.1	0.12			

(continued)

Table 4.11 *Continued*

Name CAS-No.		Ground water (GW)			Bank Filtrate (BF)			Drinking Water (DW)			Marine Water (MW)		
		samples	A-MEC (μg/L)	MAX (μg/L)	samples	A-MEC (μg/L)	samples	A-MEC (μg/L)	MAX (μg/L)	MAX (μg/L)	samples	A-MEC (μg/L)	MAX (μg/L)
ANTIBIOTICS													
Amoxicillin 26787-78-0	China										20	0	0
	Germany	16	0	0.1									
Ampicillin 69-53-4	Germany	10	0	0									
Azithromycin 83905-01-5	Germany	10	0	0									
Chloramphenicol 56-75-7	China										20	0	0
	Germany	98	0	0	10	0	0						
Chlortetracycline 57-62-5	Germany	58	0	0									
	USA							6	0	0			
Ciprofloxacin 85721-33-1	Germany	10	0	0									
	USA							6	0	0			
Clarithromycin 81103-11-9	Germany	152	0	0	12	0	0						
Cloxacillin 61-72-3	Germany	37	0	0									
Dapsone 80-08-0	Germany	6	0	0									
Demeclocycline 127-33-3	USA							6	0	0			
Dicloxacillin 3116-76-5	Germany	53	0	0									
Doxycycline 564-25-0	Germany	37	0	0									
	USA							6	0	0			
Erythromycin 114-07-8	China										20	0.003	0.005
	Germany	152	0	0	12	0	0						
	USA							8	0	0.005			
Flumequine 42835-25-6	USA							6	0.002	0.002			
Furazolidone 67-45-8	Germany	6	0	0									
Lincomycin 154-21-2	USA							6	0	0			
Methicillin 61-32-5	Germany	37	0	0									
Metronidazole 443-48-1	Germany	6	0	0									
Minocycline 10118-90-8	USA							6	0	0			
Nafcillin 147-52-4	Germany	37	0	0									
Norfloxacin 70458-96-7	China										20	0.011	0.028
	USA							12	0	0			
Ofloxacin 83380-47-6	China										20	0.008	0.016
Oxacillin 66-79-5	Germany	53	0	0									
Oxolinic acid 14698-29-4	USA							6	0.003	0.004			
Oxytetracycline 6153-64-6	Germany	48	0	0									
	USA							6	0	0			
Penicillin G 61-33-6	Germany	48	0	0									

(continued)

Table 4.11 *Continued*

Name CAS-No.		Ground water (GW)			Bank Filtrate (BF)			Drinking Water (DW)			Marine Water (MW)		
		samples	A-MEC (μg/L)	MAX (μg/L)	samples	A-MEC (μg/L)	samples	A-MEC (μg/L)	MAX (μg/L)	MAX (μg/L)	samples	A-MEC (μg/L)	MAX (μg/L)
Penicillin V 87-08-1	Germany	43	0	0									
Roxithromycin 80214-83-1	Austria	7	0.004	0.01									
	China										20	0.006	0.031
	France				4	0	0	2	0	0			
	Germany	152	0	0	12	0	0						
Spiramycin 8025-81-8	Germany	11	0	0									
Sulfachlorpyridazine 80-32-0	USA							6	0	0			
Sulfadiazine 68-35-9	China							3	0	0	20	0	0
	Germany	17	0	0.05									
	Italy							4	0	0			
Sulfadimethoxine 122-11-2	Italy							4	0	0.011			
	USA	18	0.05	0.068				6	0	0			
Sulfamerazine 127-79-7	USA							6	0	0			
Sulfamethazine 57-68-1	China							3	0	0	20	0	0
	Germany	124	0	0.24									
	USA	18	0.078	0.215				6	0	0			
Sulfamethoxazole 723-46-6	China							3	0	0	20	0	0
	France	2	0	0	4	0.011	0.058	2	0	0	8	0.028	0.24
	Germany	162	0.006	0.47	20	0.032	0.079						
	Italy	2	0	0				10	0.003	0.08			
	USA							14	0.001	0.003	44	0.006	0.081
Sulfapyridine 144-83-2	China							3	0	0			
	Italy							4	0	0			
Sulfathiazole 72-14-0	China							3	0	0			
	Germany	11	0	0.02									
Tetracycline 60-54-8	Germany	48	0	0									
	USA							6	0	0			
Trimethoprim 738-70-5	France	2	0	0							8	0	0
	Germany	162	0	0.04	20	0	0						
	USA							14	0	0	44	0.005	0.125
Tylosin 1401-69-0	Germany	59	0	0									
	USA							6	0	0.004			
ANTICOAGULANTS													
Warfarin 81-81-2	USA										44	0	0.005

(*continued*)

Table 4.11 *Continued*

Name CAS-No.		Ground water (GW)			Bank Filtrate (BF)			Drinking Water (DW)			Marine Water (MW)		
		samples	A-MEC (μg/L)	MAX (μg/L)	samples	A-MEC (μg/L)	samples	A-MEC (μg/L)	MAX (μg/L)	MAX (μg/L)	samples	A-MEC (μg/L)	MAX (μg/L)
ANTIDEPRESSANTS													
Amitryptiline 50-48-6	France							7	0	0			
Doxepin 1668-19-5	France							7	0	0			
Fluoxetine 54910-89-3	France	2	0	0							8	0	0
	USA							13	0	0	17	0	0.007
Imipramine 50-49-7	France							7	0	0			
ANTIDIABETICS													
Metformin 657-24-9	USA										44	0	0.092
ANTIEPILEPTICS													
Carbamazepine 298-46-4	Austria	7	0	0									
	Canada							3	0	0			
	France	2	0.012	0.023	3	0.053	0.071	8	0	0.043	8	0.329	0.997
	Germany	215	0.015	1.1	70	0.082	1	12	0	0.03			
	Italy	2	0	0				6	0.005	0.005			
	USA							8	0	0	44	0.005	0.078
Phenytoin 57-41-0	USA							8	0.001	0.001			
ANTIFUNGALS													
Clotrimazole 23593-75-1	France	2	0	0							8	0	0
ANTIHYPERTENSIVES													
Enalapril 75847-73-3	USA							6	0	0			
Nifedipine 21829-25-4	Canada							3	0	0			
	USA										41	0	0.004
ANTIPSYCHOTICS													
Risperidone 106266-06-2	USA							6	0	0			
ANTITUSSIVES													
Dihydrocodeine 125-28-0	Germany	1	0	0									
ANTIULCERATIVES													
Cimetidine 51481-61-9	USA										44	0	0.067
Ranitidine 66357-35-5	USA										44	0	0
Beta-BLOCKERS													
Atenolol 29122-68-7	France	2	0	0							8	0.016	0.245
	Germany	103	0	0	32	0	0						
	USA							6	0	0			
Betaxolol 63659-18-7	Germany	29	0	0									

(continued)

Table 4.11 *Continued*

Name CAS-No.		Ground water (GW)			Bank Filtrate (BF)			Drinking Water (DW)			Marine Water (MW)		
		samples	A-MEC (μg/L)	MAX (μg/L)	samples	A-MEC (μg/L)	samples	A-MEC (μg/L)	MAX (μg/L)	MAX (μg/L)	samples	A-MEC (μg/L)	MAX (μg/L)
Bisoprolol 66722-44-9	Germany	117	0	0.074	32	0	0.11						
Carazolol 57775-29-8	Germany	23	0	0									
Diltiazem 42399-41-7	Canada							3	0	0			
	USA										44	0.001	0.013
Metoprolol 37350-58-6	France	2	0	0							8	0.005	0.118
	Germany	126	0	0.065	36	0	0.03						
Nadolol 42200-33-9	Germany	21	0	0									
Pindolol 61477-96-1	Germany	6	0	0									
Propranolol 318-98-9	Canada							3	0	0			
	France	2	0	0							8	0.021	0.224
	Germany	141	0	0.096	32	0	0						
Sotalol 3930-20-9	Germany	102	0	0.56	40	0	0.37						
Timolol 26839-75-8	Germany	20	0	0									
BRONCHIOLYTICS													
Clenbuterol 37148-27-9	France							7	0	0			
	Germany	87	0	0	32	0	0						
Fenoterol 13392-18-2	Germany	3	0	0									
Salbutamol 18559-94-9	France							7	0	0			
	Germany	93	0	0	32	0	0						
	USA										17	0	0
Terbutaline 23031-25-6	France							7	0	0			
	Germany	76	0	0	16	0	0						
CONTRAST MEDIA													
Diatrizoate 737-31-5	Germany	89	1.468	9.6	30	0.011	4	10	0.021	0.085			
Iodipamide 606-17-7	Germany	8	0.01	0.01									
Iohexol 66108-95-0	Germany	32	0.002	0.01				2	0.039	0.04			
Iomeprol 78649-41-9	Germany	59	0.001	0.01	30	0	0.16	2	0.086	0.092			
Iopamidol 62883-00-5	Germany	96	0.064	2.4	30	0	1.4	15	0.011	0.098			
Iopanoic acid 96-83-3	Germany	8	0.01	0.01									
Iopromide 73334-07-3	Austria	7	0.002	0.01									
	France				4	0	0	2	0	0			
	Germany	96	0.001	0.21	30	0	0.04	15	0.01	0.086			
	USA							2	0.002	0.005			
Ioxaglic acid 59017-64-0	Germany	8	0.01	0.01									
Ioxithalamic acid 28179-44-4	Germany	42	0.002	0.049				10	0	0			

(*continued*)

Table 4.11 *Continued*

Name CAS-No.		Ground water (GW)			Bank Filtrate (BF)			Drinking Water (DW)			Marine Water (MW)		
		samples	A-MEC (μg/L)	MAX (μg/L)	samples	A-MEC (μg/L)	samples	A-MEC (μg/L)	MAX (μg/L)	MAX (μg/L)	samples	A-MEC (μg/L)	MAX (μg/L)
CYTOSTATICS													
Cyclophosphamid 50-18-0	Germany	85	0	0	32	0	0	12	0	0			
Ifosfamid 3778-73-2	Germany	85	0	0	32	0	0	12	0	0			
DIURETICS													
Furosemide 54-31-9	France	2	0	0							8	0.01	0.118
LIPID REGULATORS													
Atorvastatin 134523-00-5	USA							6	0	0			
Bezafibrate 41859-67-0	Austria	7	0.004	0.023									
	France	2	0	0							8	0.003	0.026
	Germany	66	0	0.19	49	0	0.11	55	0	0.027			
	Italy	2	0	0				6	0.001	0.002			
Etofibrate 31637-97-5	Germany	7	0	0				12	0	0			
Fenofibrate 49562-28-9	France	2	0	0							8	0	0.027
	Germany	33	0	0				12	0	0			
	USA										38	0	0
Gemfibrozil 25812-30-0	Canada							3	0	0			
	France	2	0	0				7	0	0	8	0	0.018
	Germany	22	0	0				55	0	0			
	Italy	2	0	0				6	0	0.001			
	USA							8	0	0.002			
Simvastatin 79902-63-9	Germany	6	0	0									
	USA							6	0	0			
METABOLITES													
1-Hydroxyibuprofen -	France	2	0	0							8	0	0
2-Hydroxyibuprofen -	France	2	0	0							8	0	0.563
	Norway										12	0	0.002
4-Chlorobenzoic acid 83-15-8	France	2	0	0							8	0	0.035
4-Hydroxyantipyrine 1672-63-5	Germany	42	0	0	84	0	0						
6-o-desmethyl naproxen 60756-73-2	France	2	0	0							8	0	0.161
AMDOPH -	Germany	2	1.2	1.2				2	0.9	0.9			
AMPH 38604-70-5	Germany							2	0.03	0.03			
Clofibric acid 882-09-7	France	2	0	0							8	0	0
	Germany	179	0	11	49	0.007	7.3	151	0.02	0.17	15	0	0.019
	USA							20	0	0			

(continued)

Table 4.11 *Continued*

Name CAS-No.		Ground water (GW)			Bank Filtrate (BF)			Drinking Water (DW)			Marine Water (MW)		
		samples	A-MEC (µg/L)	MAX (µg/L)	samples	A-MEC (µg/L)	samples	A-MEC (µg/L)	MAX (µg/L)	MAX (µg/L)	samples	A-MEC (µg/L)	MAX (µg/L)
Fenofibric acid 42017-89-0	France	2	0	0							8	0	0
	Germany	31	0	0.598				57	0	0.042			
o-Hydroxy Atorvastatin -	USA							6	0	0			
p-Hydroxy Atorvastatin -	USA							6	0	0			
Salicylic acid 69-72-7	Canada							3	0.004	0.004			
	France	2	0.037	0.043									
Simvastatin hydroxy acid 121009-77-6	USA							6	0	0			
					MUCOLYTICS								
Ambroxol 18683-91-5	Germany	42	0	0	84	0	0						
					SEDATIVES								
Zolpidem 82626-48-0	France	2	0	0							8	0	0.005
					TRANQUILLISERS								
Alprazolam 28981-97-7	France	2	0	0							8	0	0
Bromazepam 1812-30-2	France	2	0.117	0.161							8	0	0.356
Diazepam 439-14-5	France	2	0	0				7	0	0	8	0	0.004
	Germany	95	0	0	32	0	0	12	0	0			
	USA							8	0	0			
Lorazepam 846-49-1	France	2	0	0							8	0.013	0.013
Meprobamate 57-53-4	USA							8	0.006	0.008			
Nordazepam 1088-11-5	France							7	0	0			
Oxazepam 604-75-1	France	2	0.002	0.003							8	0.519	2.183
					VASOCONSTRICTORS								
Xylometazolin 526-36-3	Germany	42	0	0	84	0	0						
					VASODILATORS								
Pentoxifylline 6493-05-6	Germany	13	0	0									
	USA							2	0	0			

REFERENCES

Benotti, M.J. and Brownawell, B.J. (2007) Distribution of pharmaceuticals in an urban estuary during both Dry- and Wet weather conditions. *Environmental Science and Technology* **41**, 5795–5802.

Besse, J.-P. and Garric, J. (2008) Human pharmaceuticals in surface waters Implementation of a prioritization methodology and application to the French situation. *Toxicology Letters* **176**, 104–123.

Besse, J.P. and Garric, J. (2007) Médicaments à usage humain: risque d'exposition et effets sur les milieux récepteurs, rapport AE RM&C. pp. 238.

Bruchet, A., Hochereau, C., Picard, C., Decottignies, V., Rodrigues, J.M. and Janex-Habibi, M.L. (2005) Analysis of drugs and personal care products in French source and drinking waters: the analytical challenge and examples of application. *Water Science And Technology* **52**(8), 53–61.

Buerge, I.J., Buser, H.-R., Poiger, T. and Müller, M.D. (2006) Occurrence and fate of the cytostatic drugs cyclophosphamide and ifosfamide in wastewater and surface waters *Environ. Sci. Techno.* **40**(23), 7242–7250.

Bund/Länderausschuss für Chemikaliensicherheit (BLAC), Arzneimittel in der Umwelt – Auswertung der Untersuchungsergebnisse. 2003: Hamburg.

Buser, H.-R., Müller, M.D. and Theobald, N. (1998) Occurrence of the pharmaceutical drug clofibric acid and the herbicide mecoprop in various swiss lakes and in the North Sea. *Environ. Sci. Technol.* **32**(1), 188–192.

Gros, M., Petrovic, M. and Barcelo, D. (2006) Multi-residue analytical methods using LC-tandem MS for the determination of pharmaceuticals in environmental and wastewater samples: A review. *Analytical and Bioanalytical Chemistry* **386**(4), 941–952.

Heberer, T., Mechlinski, A., Fanck, B., and Knappe, A. (2004) Field studies on the fate and transport of pharmaceutical residues in bank filtration. *Ground Water Monitoring & Remediation* **24**(2), 70–77.

Joss, A., Carballa, M., Kreuzinger, N., Siegrist, H. and Zabczynski, S. (2006) Wastewater Treatment. In *Human Pharmaceuticals, Hormones and Fragrances: The challenge of micropollutants in urban water management*, (ed. T.A. Ternes, A. Joss), IWA Publishing, London.

Kasprzyk-Hordern, B., Dinsdalea, R.M. and Guwya, A.J. (2008) The effect of signal suppression and mobile phase composition on the simultaneous analysis of multiple classes of acidic/neutral pharmaceuticals and personal care products in surface water by solid-phase extraction and ultra performance liquid chromatography–negative electrospray tandem mass spectrometry. *Talanta* **74**, 1299–312.

Kolpin, D.W., Furlong, E.T., Meyer, M.T., Thurman, E.M., Zaugg, S.D., Barber, L.B., and Buxton, H.T. (2002) Pharmaceuticals, hormones, and other organic wastewater contaminants in U.S. streams, 1999–2000: A national reconnaissance. *Environ. Sci. Technol.* **36**, 1202–1211.

Lienert, J., Güdel, K., Escher, B. (2007) Screening method for ecotoxicological hazard assessment of 42 pharmaceuticals considering human metabolism and excretory routes. *Environment Science & Technology* **41**, 4471–4478.

Metcalfe, C.D., Koenig, B.G., Bennie, D.T., Servos, M., Ternes, T.A. and Hirsch, R. (2003) Occurrence of neutral and acidic drugs in the effluents of Canadian sewage treatment plants. *Environmental Toxicology and Chemistry* **22**(12), 2872–2880.

Moffat, A.C., Osselton, M.D., Widdop, B., eds. Clarkes Analysis of Drugs and Poisons. 2004, Pharmaceutical Press; 3 edition (May 30, 2004).

Paxeus, N. (2004) Removal of selected non-steroidal anti-inflammatory drugs (NSAIDs), gemfibrozil, carbamazepine, beta-blockers, trimethoprim and triclosan in conventional wastewater treatment plants in five EU countries and their discharge to the aquatic environment. *Water Science and Technology* **50**(5), 253–260.

Petrovic, M., Gonzalez, S. and Barcelo, D. (2003) Analysis and removal of emerging contaminants in wastewater and drinking water. *Trends in analytical chemistry* **22**(10), 685–756.

Reddersen, K., Heberer, T. and Duennbier, U. (2002) Identification and significance of phenazone drugs and their metabolites in ground- and drinking water. *Chemosphere* **49**, 539–544.

Roberts, P.H. and Thomas, K.V. (2006) The occurence of selected pharmaceuticals in wastewater effluent and surface waters of the lower tyne catchment. *Science of the Total Environment* **356**, 143–153.

Römbke, J., Knacker, T. and Teichmann, H. (2001) Environmental Risk Assessment of Pharmaceuticals: A Proposal with Special Emphasis on European Aspects. In *Pharmaceuticals and Personal Care Products in the Environment: Scientific and Regulatory Issues*, (eds. C.G. Daughton, T. Jones-Lepp), pp. 304–319, American Chemical Society, Washington, D.C.

Sadezky, A., Löffler, D. and Ternes, T. (2008) Proposal of an environmental indicator and classification system of pharmaceutical product residues for environmental management – Deliverable D1.2 of the EU Project KNAPPE., German Federal Institute of Hydrology: Koblenz.

Schlusener, M.P. and Bester, K. (2005) Determination of steroid hormones, hormone conjugates and macrolide antibiotics in influents and effluents of sewage treatment plants utilising high-performance liquid chromatography/tandem mass spectrometry with electrospray and atmospheric pressure chemical ionisation. *Mass Spectrometry* **19**(22), 3269–3278.

Sebastine, I.M. and Wakeman, R.J. (2003) Consumption and environmental hazards of pharmaceutical substances in the UK. *Process Safety and Environmental Protection* **81**(B4), 229–235.

Seitz, W., Weber, W.H., Jiang, J.-Q., Lloyd, B.J., Maier, M., Maier, D. and Schulz, W. (2006) Monitoring of iodinated X-ray contrast media in surface water. *Chemosphere* **64**, 1318–1324.

Tamtam, F., Mercier, F., Le Bot, B., Eurin, J., Dinh, Q.T., Clément, M. and Chevreuil, M. (2008) Occurrence and fate of antibiotics in the Seine River in various hydrological conditions. *Sci. Total Environ.* **393,** 84–95.

Ternes T. (2004) EU-Project Poseidon, Final report of the EU-Project Poseidon, Ternes Editor.

Ternes, T.A., Joss, A., eds. (2006) Human Pharmaceuticals, Hormones and Fragrances: The challenge of micropollutants in urban water management, IWA Publishing, London.

Ternes, T.A., Meisenheimer, M., McDowell, D., Sacher, F., Brauch, H.J., Gulde, B.H., Preuss, G., Wilme, U. and Seibert, N.Z. (2002) Removal of pharmaceuticals during drinking water treatment. *Environmental Science & Technology* **36**(17), 3855–3863.

Ternes, T.A. (1998) Occurrence of Drugs in German Sewage Treatment Plants and Rivers. *Water Research* **32**(11), 3245–3260.

The McGraw-Hill Companies, McGraw-Hill AccessMedicine. 2007: http://www.accessmedicine.com/

U.S. National Library of Medicine, Hazardous Substances Data Bank (HSDB) of the Toxicology data network (Toxnet) (2007) http://www.toxnet.nlm.nih.gov/cgi-bin/sis/htmlgen?HSDB

von Gunten, U., Janex-Habibi, M.-L., Ternes, T.A. and Weber, L. (2006) Removal of PPCP during drinking water treatment. In *Human Pharmaceuticals, Hormones and Fragrances: The challenge of micropollutants in urban water management*, (eds. T.A. Ternes, A. Joss), IWA Publishing, London.

Webb, S.F. (2004) ERA of human pharmaceuticals II – aquatic risk characterization. In *Pharmaceuticals in the environment*, (ed. K. Kümmerer), Springer, Berlin.

Weigel, S., Kuhlmann, J. and Hühnerfuss, H. (2002) Drugs and personal care products as ubiquitous pollutants: occurence and distribution of clofibric acid, caffeine and DEET in the North Sea. *The Science of the Total Environment* **295**, 131–141.

Wiegel, S., Aulinger, A., Brockmeyer, R., Harms, H., Löffler, J., Reincke, H., Schmidt, R., Stachel, B., von Tümplingen, W. and Wanke, A. (2004) Pharmaceuticals in the river elbe and its tributaries. *Chemosphere* **57**, 107–126.

World Health Organization, WHO Collaborating Centre for Drug Statistics Methodology. 2006: http://www.whocc.no/atcddd/

Zuehlke, S., Duennbier, U. and Heberer, T. (2004) Determination of polar drug residues in sewage and surface water applying liquid chromatography-tandem mass spectrometry. *Analytical Chemistry* **76**(22), 6548–6554.

Zwiener, C. (2007) Occurrence and analysis of pharmaceuticals and their transformation products in drinking water treatment. *Anal. Bioanal. Chem.* **387**, 1159–1162.

Chapter 5

Performance of conventional treatment processes of the most resistant PPs

Sebastian Żabczyński, Dagmara Buntner, Korneliusz Miksch and Benoit Roig

5.1 INTRODUCTION

For several years the presence of PP's has been investigated in various aqueous matrices, such as wastewater, sewage treatment plant effluents, surface water, groundwater and drinking water. Sewage treatment plants may be considered as the main point of collection and subsequent release of PPs into the environment. Recent data from various research projects in Europe indicate that the normal operation of sewage-treatment plants (STP's) results in only a limited reduction in the overall load of PPs. Hence, as much as 80% of the total load of pharmaceuticals entering sewage-treatment plants may be discharged into surface waters.

However, since the load of pharmaceutical products increases, it is crucial to improve the elimination efficiency of these compounds. Identifying so-called flaws of the existing technologies gives the opportunity for the future development.

The main objectives of this chapter is to

(1) Identify groups of human pharmaceuticals according to their removal rates by current biological sewage treatments and trying to establish links with their physico-chemical properties;
(2) Comparison of the technologies in regard to PPs removal;
(3) Identification of different PPs removal rates in relation to the regional specificities such as climate or most drugs consumption;
(4) Delineation the factors that determine PPs susceptibility to conventional biological treatment.

In order to fulfil above-mentioned points, the national and international publications were scanned and the most important data (for example occurrence of the PPs in the primary and secondary sludge, WWTP configuration, the sludge concentration, sludge treatment technologies) were gathered. Till now, several databases with scientific journals were evaluated (for example, ACS Publications, Directory of Open Access Journals (DOAJ), EBSCOhost research databases, ProQuest databeses, ISI Emerging Market, ScienceDirectOnSite, SpringerLink database).

Wastewater and drinking water treatment technologies in regard to PPs removal have been investigated. As a first step, the data concerning pharmaceutical compounds removal was collected and set in following order:

- Sewage treatment plant (STP) characteristics – including reactor configuration, sludge retention time (SRT), hydraulic retention time (HRT), temperature (or sampling season), red-ox conditions, influent and effluent concentration of particular PPs;
- Removal – usually described as the difference between influent and effluent concentration, however the specification concerning biological transformation and sorption onto sludge was given, if the data was available.

 Pharmaceuticals in the Environment: Current knowledge and need assessment to reduce presence and impact. Edited by Benoit Roig. ISBN: 9781843393146. Published by IWA Publishing, London, UK.

In order to highlight the differences of the crucial parameters impact on the elimination of PPs in the second step literature data was divided according to the following factors:

1 ***Sludge retention time*** (***SRT***): Sludge retention time is defined as the average retention time of sludge flocs before they are removed as excess sludge or lost through the effluent. That is the time microorganisms have for growth: the lower the sludge retention time, the less time they have to double in order to prevent washout of the population. SRT is a crucial parameter in wastewater treatment, which determines the biodiversity of the sludge, the amount of predominant bacteria in the active biomass and their enzymatic activity. According to the literature data three ranges of SRT were chosen:

- SRT < 10 days – carbon removal;
- 10 < SRT < 20 days – nutrient removal;
- SRT > 20 days – nutrient removal plus sludge stabilization (typical for small STPs).

2 ***Hydraulic retention time*** (***HRT***): Hydraulic retention time can be defined as a measure of the average length of time that a soluble compound remains in a constructed bioreactor (Wikipedia). Following IUPAC Compendium of Chemical Terminology **residence time** (**hydraulic retention time**) is the average time a particle or volume element of the culture resides in a bioreactor (or other device) through which a liquid medium continuously flows. Residence time of the liquid is the reciprocal of dilution rate.

To compare the impact of this parameter three most common ranges were chosen:

- HRT < 12 hours,
- 12 < HRT < 25 hours,
- HRT > 25 hours.

3 ***Reactor configuration***: In order to compare the removal efficiencies of reported compounds regarding to the WWTP configuration, the following treatments were taken into account:

- CASP – Conventional Activated Sludge Processes (N, C, P removal),
- CASP-C – Conventional Activated Sludge Processes with carbon removal only,
- CASP-P – Conventional Activated Sludge Processes focused on phosphorus removal,
- CASP-TF – Conventional Activated Sludge Processes combined with Trickling Filter,
- CASP-OD – Conventional Activated Sludge Processes as Oxidation Ditch,

In the following these 5 configurations will be mentioned under CASP acronym

- SBR – Sequencing Batch Reactor,
- MBR – Membrane Biological Reactor,
- FBR – Fixed Bed Reactor,
- Lagoon.

4 ***Sorption into sludge***: One of the significant processes in PPs removal mechanisms and directly link with the treatment of the sewage sludge is sorption. Sorption is regarded as being composed by two distinct reactions: sorption from liquid to the solids and desorption. Sorption equilibrium is reached, when the rate of both reactions is equal. No data is currently available to estimate the rate of sorption and desorption (Ternes and Joss, 2007).

A sorption or distribution coefficient (K_d) is commonly used to describe this process. The K_d value is ratio of the sorbed phase concentration to the solution phase concentration at equilibrium (Ternes and Joss, 2007):

$$K_d = \frac{X_{part}}{S} = \frac{X}{X_{SS} \cdot S} \tag{1}$$

X concentration sorbed onto sludge, per unit reactor volume, μg/L,
X_{part} concentration sorbed, per amount of sludge dry matter, μg/gSS,
K_d solid-water distribution coefficient, L/gSS,
X_{SS} suspended solids concentration in raw wastewater or production of suspended solids in primary and/or secondary treatment per L of wastewater, gSS/L,
S dissolved concentration, μg/L.

The K_d is usually distinguished for primary sludge and secondary sludge. It is caused by different properties of the sludges – the composition and pH of the sludge seems to be crucial.

5 ***Advanced technologies***: Among advanced treatment, we have considered:

- physicochemical treatment, which could be defined as a treatment process in which -biological and physicochemical processes are intermixed,
- combined biological-physical-chemical treatment, which differs from tertiary treatment because of the mixed construction of the all processes in the flow scheme.

– Advanced oxidation technologies such as ozonation, which is rather widely used in wastewater and drinking water treatment for oxidation and disinfection, O_3/H_2O_2, UV/H_2O_2, Fe^{2+}/H_2O_2 used for highly reactive $^{\cdot}OH$ radicals,
– combination of microfiltration with nanofiltration or reverse osmosis, which allowing micropollutant retention by molecular sieving,

All pharmaceutical compounds found in the literature were divided into therapeutic groups: antibiotics, anticonvulsants, antiinflammatories (analgesic), β-blockers, hormones, tranquilizers, X-ray contrast media and lipid regulators.

In the following, for each pharmacological family, the influence of treatment conditions such as SRT, HRT, reactor configuration and the adsorption into sludges (when known) have been reported.

SRT graph report the % of the published data describing the removal (%) of a PP for different age of sludge (<10, between 10 and 20 or >20 days).

HRT graph report the% of the published data describing the removal (%) of a PP for different time in a bioreactor (<12, between 12 and 25 or >25 hours)

Reactor configuration reports the number of the published data describing the removal (%) of a PP by using a particular reactor.

5.2 ANTIBIOTICS

5.2.1 Sludge Retention Time (SRT)

SRT seems not to have significant influence on Sulfamethoxazole and Roxithromycin removal (Figure 5.1), although slight increase tendency can be observed with longer SRT. Overall removal ranges from 0 to 80% and 0–90%, respectively. In contradiction, sludge retention time above 50 days decreases the removal of the other antibiotic, Trimethoprim to less than 10%. However, with SRT below 20 days the removal of this compound is around 90%. The same situation occurs in case of Clarythromycin; with SRT $<$ 20 days the removal rate varies from 75 to 90%, while with SRT $>$ 50 days the removal drops to 10%. Ciprofloxacin and Ofloxacin are rather highly removed within wastewater treatment, although there is only data concerning SRT $<$ 20 days available. The removal of both compounds is above 75%.

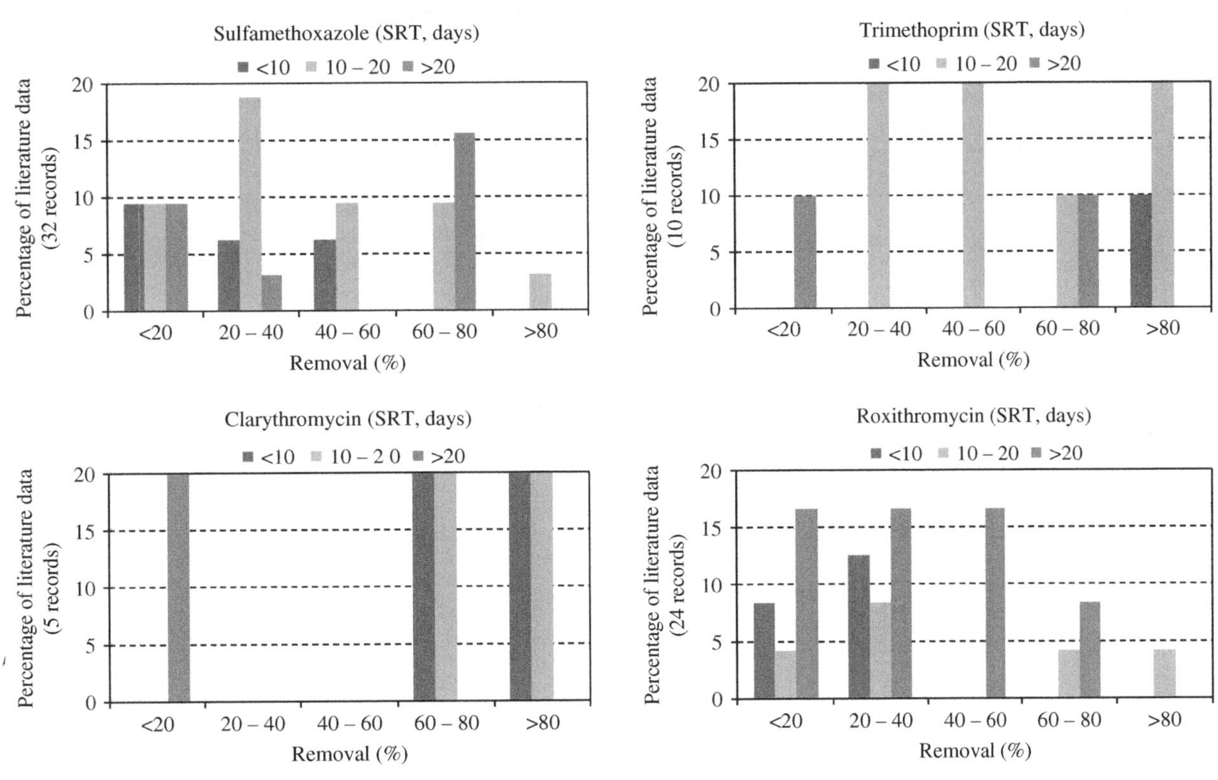

Figure 5.1 SRT influence on the removal of antibiotics (Radjenovic *et al.*, 2007; 2009, Ghosh *et al.*, 2009; Lin *et al.*, 2009a; Yu *et al.*, 2009; Carballa *et al.*, 2007a; Batt *et al.*, 2007; Vieno *et al.*, 2007; Batt *et al.*, 2006; Ternes and Joss, 2006; Kim *et al.*, 2005). *Continued over.*

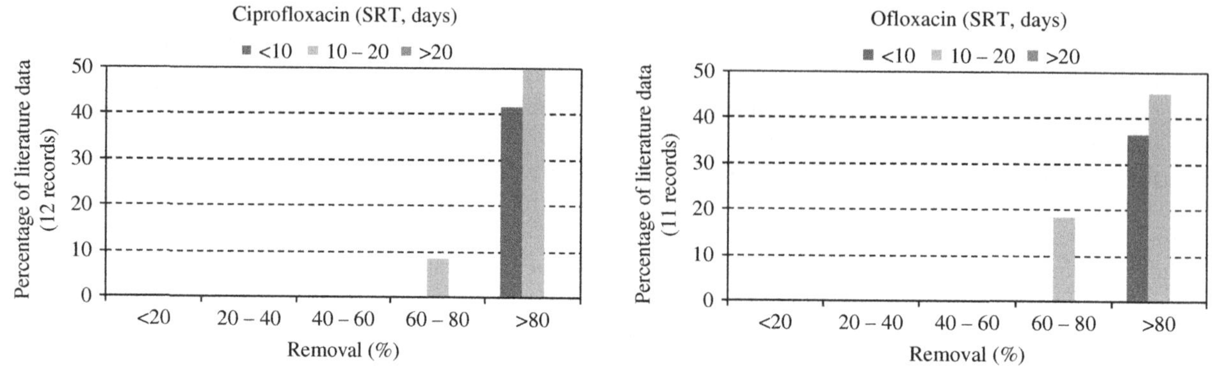

Figure 5.1 *Continued.* SRT influence on the removal of antibiotics (Radjenovic *et al.*, 2007; 2009, Ghosh *et al.*, 2009; Lin *et al.*, 2009a; Yu *et al.*, 2009; Carballa *et al.*, 2007a; Batt *et al.*, 2007; Vieno *et al.*, 2007; Batt *et al.*, 2006; Ternes and Joss, 2006; Kim *et al.*, 2005).

(Anhydro-)erythromycine shows no dependency on the sludge retention time with high (>90%) removal rate for SRT < 20 and > 50 days.

There is a significant lack of data concerning ***Norfloxacin*** and ***Sulfapyridine***. Ternes and Joss (2006), for the sludge retention time typical for nutrient elimination, presented removal below 20% and above 30%, respectively.

The ***Tetracycline*** removal efficiency was shown dependant of the reduction of the SRT.) reached the removal of 85.1% of tetracycline with 10 days of SRT (SBR with concentrations of tetracyclines ranging from 0.1 to 0.6 μg/L) whereas, 78.4% are only obtained with 3days of SRT.

No records of ***Clindamycin*** with SRT data was found. However, these antibiotics were found in the STPs' effluents at concentrations up to 1 μg/L.

5.2.2 Hydraulic Retention Time (HRT)

No clear dependency on the HRT impact on the removal of Sulfamethoxazole can be stated. However, it seems that the range of 12–25 h characterizes the highest removal rate of this compound (Figure 5.2). HRT > 25 hours represents the elimination either below 20% or above 60%. Additionally, reported no removal of Sulfamethoxazole for hydraulic retention time equal to 237 hours.

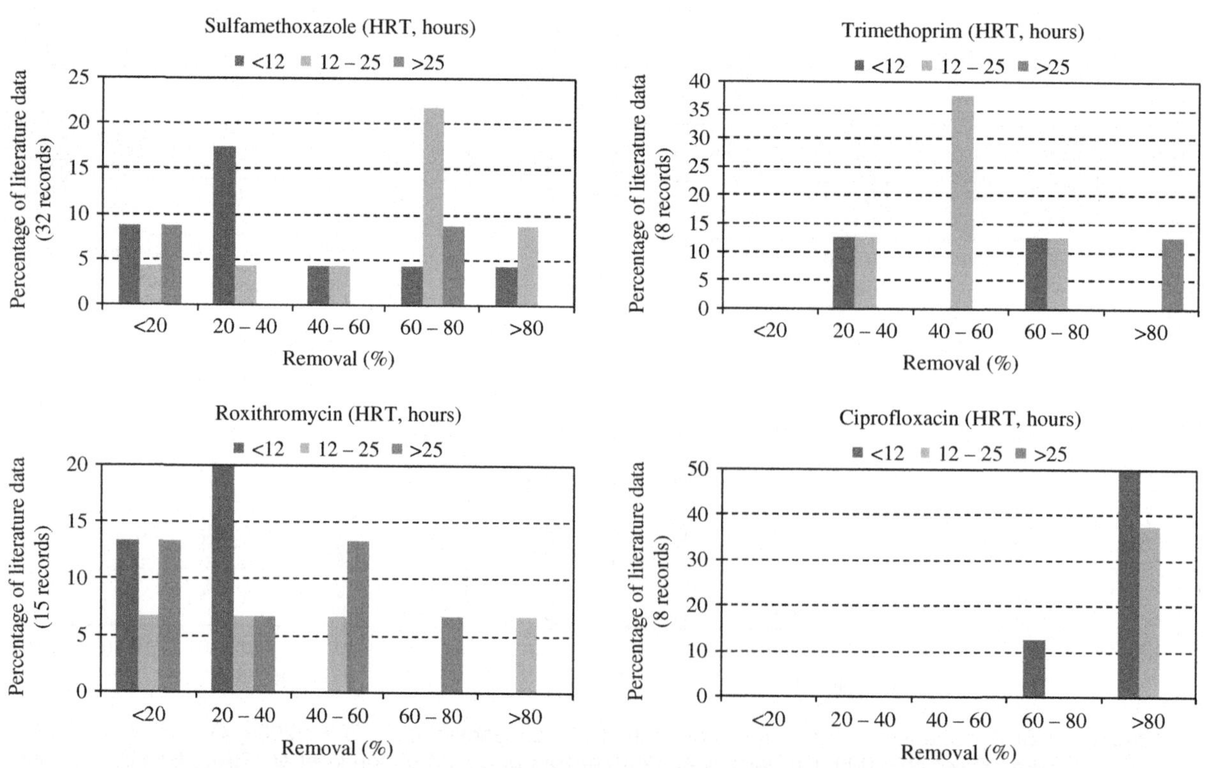

Figure 5.2 HRT influence on the removal of antibiotics (Clara *et al.*, 2005b; Bendz *et al.*, 2005; Kim *et al.*, 2005; Vieno *et al.*, 2007; and other). *Continued over.*

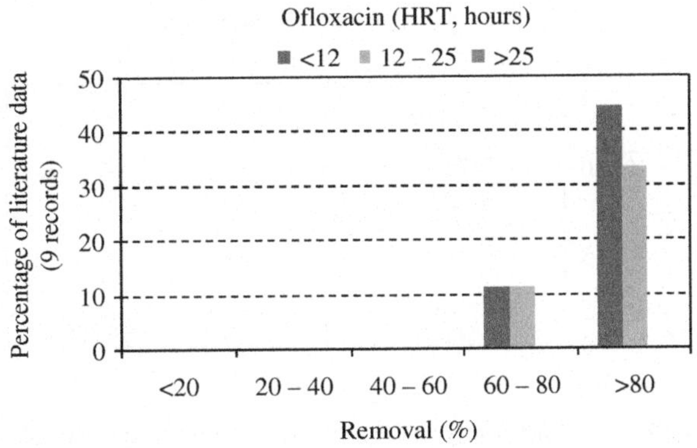

Figure 5.2 *Continued.* HRT influence on the removal of antibiotics (Clara *et al.*, 2005b; Bendz *et al.*, 2005; Kim *et al.*, 2005; Vieno *et al.*, 2007; and other).

Bendz *et al.* (2005) presented data concerning removal of ***Trimethoprim*** at HRT of 15–16 hours. The elimination of the studied compound was approximately 50%. However, that is the only record regarding this compound, therefore the establishment of the HRT impact on the removal rate is impossible.

At HRT above 12 hours ***Roxithromycin*** proves a variety of removal efficiencies, ranging from no removal to 80%. Additionally, HRT below 12 hours results in poor removal, within the range of 0–30%.

Ciprofloxacin and ***Ofloxacin*** are highly removed at HRT range up to 25 hours (Figure 2.1). No data concerning hydraulic retention time above 25 hours is available. Moreover, obtained information originate from one article by Vieno *et al.* (2007).

The removal efficiency of ***Tetracycline*** at HRT 7h (SBR) was not significantly different than that observed at HRT 24h, while maintaining a constant SRT (Kim *et al.*, 2005).

No data concerning HRT influence on the removal of ***Azithromycin***, ***Clarithromycin***, ***Norfloxacin***, ***Sulfapyridine***, and ***Clindamycin*** could be found.

5.2.3 Reactor configuration

Poor or no removal of Sulfamethoxazole was reported for CASP (Figure 5.3). SBR, MBR and FBR seem to be much more effective in the elimination of this compound. Similarly, Trimethoprim removal at CASP was rather low.

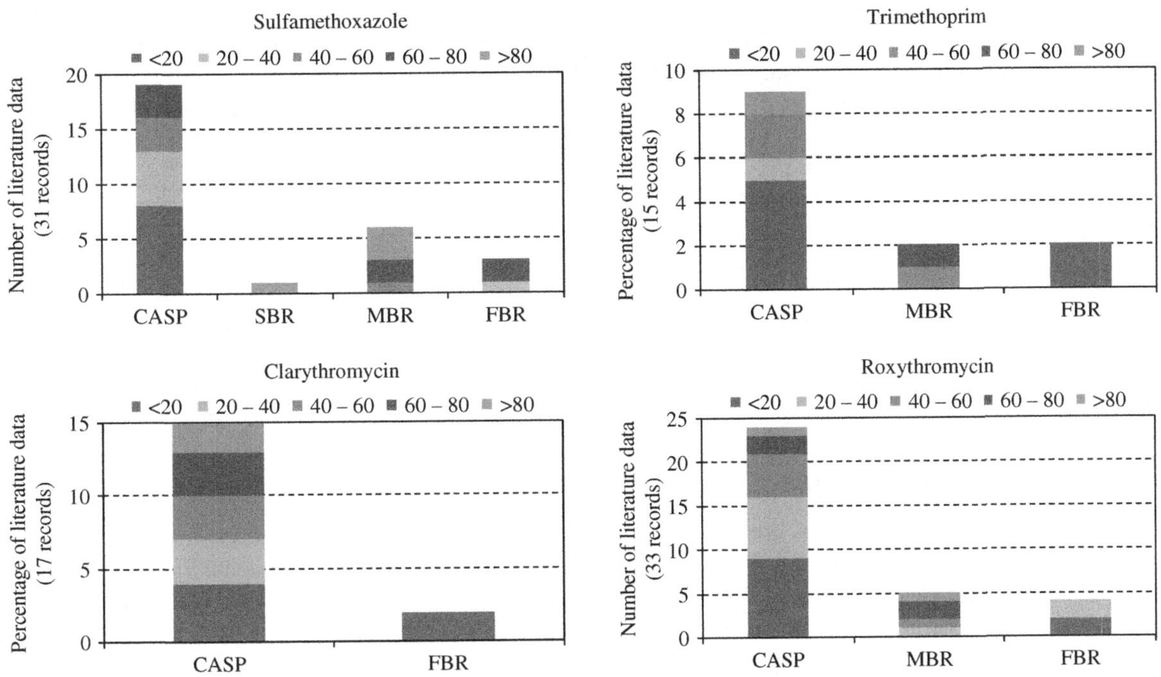

Figure 5.3 Reactor configuration influence on the removal of antibiotics. (Clara *et al.*, 2005b; Gobel *et al.*, 2007; Snyder *et al.*, 2007; Radjenovic *et al.*, 2007; 2009; Reif *et al.*, 2008; and other). *Continued over.*

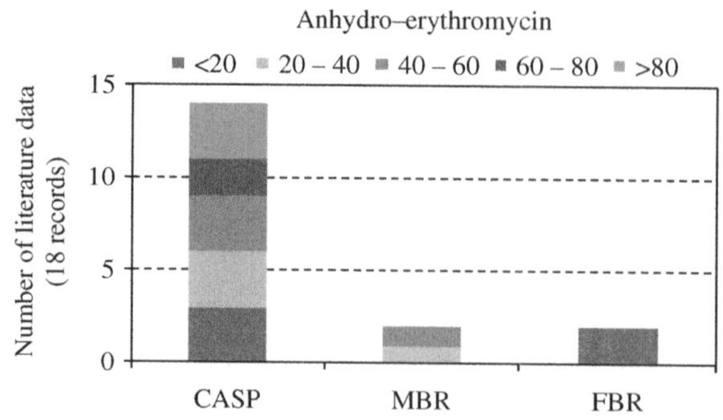

Figure 5.3 *Continued.* Reactor configuration influence on the removal of antibiotics. (Clara *et al.*, 2005b; Gobel *et al.*, 2007; Snyder *et al.*, 2007; Radjenovic *et al.*, 2007; 2009; Reif *et al.*, 2008; and other).

Clarythromycin and (Anhydro-)erythromycin have almost the same degradation pattern; their elimination at CASP covers the whole range of the removal rates and shows quite poor susceptibility of these antibiotics at CASP Figure 5.4. Roxithromycin is hardly removed at CASP and FBR. Most of the records concerning the elimination of this compound at CASP present rather low removal rates. In case of MBR, the degradation pattern seems to head for an increase in degradation of Roxithromycin (Clara *et al.*, 2005b).

Norfloxacin was reported to be quite persistent compound. Ternes and Joss (2006) show that the elimination of this PP at CASP was below 20%. No other data of this compound was available.

There is a very few data regarding ***Ciprofloxacin*** and ***Ofloxacin*** (Vieno *et al.*, 2007; Ternes and Joss, 2006). According to the available data both compounds are removed above 80% and 60% at CASP and CASP-OD, respectively.

There is no data concerning the removal of ***Tetracycline*** and ***Clindamycin*** available, although these compounds were found in the WWTP effluents (Batt *et al.*, 2006).

5.2.4 Sludge adsorption

The K_d value for primary sludge and secondary sludge were found only for ***norfloxacin*** and ***ciprofloxacin*** and they were equal to 2.5 and 2.6 L/kgSS for primary sludge, respectively, and 37 and 26 L/kgSS for secondary sludge, respectively. The changes of the K_d value for primary sludge and secondary sludge is caused mainly by different compounds of the sludges. According to the literature data, mentioned values of K_d value can not be negligible for these compounds and their elimination is connected with the sorption.

The K_d value for and secondary sludge for ***sulfamethoxazole, clarithromycin, azithromycin*** and ***erythromycin*** is below 0.4 L/kgSS, therefore the sorption processes do not play a main role in degradation of mentioned compounds (Drillia *et al.*, 2005; Carballa *et al.*, 2008; Ternes *et al.*, 2004).

The log K_d value for activated sludge were measured and or calculated for several antibiotics (Table 5.1)

Table 5.1 K_d values of some antibiotics.

Compounds	Log K_d	Ref
Sulfamethoxazole, roxythromycin	1.13–2.7	Drillia *et al.*, 2005; Gobel *et al.*, 2005; Carballa *et al.*, 2008
Clarithromycin, azithromycin, trimethroprin, clarythromycin, sulfapyridine	2.19–2.6	Gobel *et al.*, 2005
Ciprofloxacin, norfloxacin, tetracycline	4.2–4.3	Golet *et al.*, 2003; Kim *et al.*, 2005

5.2.5 Advanced technologies

Sulfonamide and macrolides antibiotics are reported to be very well removed by ozonation with reagent dose above 0.2 mg O_3/L (Huber *et al.*, 2003; Lin *et al.*, 2009). However, ozonation can cause a slight acute toxicity increment in the first stage of the process (Dantas *et al.*, 2008). The use of combined ozone systems (O_3/H_2O_2, O_3/UVA/TiO_2) can contribute to improve the efficiency of the process, by decreasing time of exposure, employing less amount of O_3, and by mitigating the increase of the toxicity (Lin *et al.*, 2009; Beltran *et al.*, 2008). Nano- and ultrafiltration was not efficient in the removal of this antibiotic.

Reyes *et al.* (2006) studied the elimination of Tetracycline using a combination of UV and TiO_2. No significant differences were observed between efficiency of 1 g/L and 0.5 g/L of TiO_2 – in both cases the removal exceeded 85%. Ozonation of oxytetracyclin is efficient by an increase of the time of exposure (1h) and pH (to promote O_3 decomposition) but the first by-products after partial ozonation (5–30 min) were found more toxic than the parent compound (Li *et al.*, 2008).

Finally, concerning quinolone antibiotics, the behaviour is the same. Ofloxacin was completely degraded by means of UV/TiO_2 with 20 mg of O_2 and pH = 3, which increases adsorption potential of this compound. Levofloxacin ozonation is influenced by the pH (degradation rate constants two times faster at pH 10 than at pH 7 and 3) and H_2O_2 amount (De Witte *et al.*, 2009).

The chlorination of trimethropin is not very efficient. Indeed, TMP structure is not substantially degraded upon reactions with HOCl. Instead, a wide variety of (multi)chlorinated and hydroxylated products are formed (Dodd and Huang, 2007).

Removal of antibiotics by reverse osmosis/nanofiltration (RO/NF) from a model wastewater of a manufacturing plant producing pharmaceuticals for veterinary use was also investigated. The rejection of the examined antibiotics by the selected RO and the tight NF membranes is acceptably high, exceeding in most cases 98.5%. The loose NF membrane retains the smaller antibiotics molecules less effective (Kosutic *et al.*, 2007).

5.3 ANTICONVULSANTS

5.3.1 Sludge Retention Time (SRT)

Carbamazepine is the most studied anticonvulsant, therefore the most of collected literature data concerns this compound in particular (Stumpf *et al.*, 1999; Clara *et al.*, 2004; Miao *et al.*, 2005; Lindqvist *et al.*, 2005; Joss *et al.*, 2005; Lishman *et al.*, 2006; Kim *et al.*, 2007; Vieno *et al.*, 2007; Paxéus 2004; and other).

Moreover, Carbamazepine is known to be very persistent – 43 out of 48 records present the removal rate <20% (Figure 5.4). Therefore it can be stated that SRT has no influence of the removal efficiency of this compound.

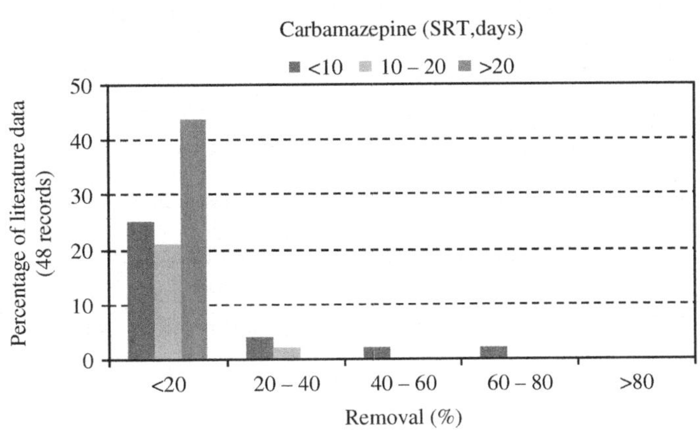

Figure 5.4 SRT influence on the removal of anticonvulsants.

Phenytoin is also poorly eliminated in wastewater treatment, with the removal efficiency ~40% (Yu *et al.*, 2006). No dependency on the SRT could be estimated, as no data with SRT range other than 8–10 days is available.

Two other anticonvulsants, reported by Yu *et al.* (2006), ***Gabapentin*** and ***Valproic acid*** shows very high removal rate (>99%). However, as mentioned above, the data concerns only nutrient removal sludge retention time.

5.3.2 Hydraulic Retention Time (HRT)

Carbamazepine, like in case of SRT, is the most commonly studied compound among anticonvulsants (Clara *et al.*, 2005b; Kreuzinger *et al.*, 2004; Nakada *et al.*, 2006; Zhang *et al.*, 2008 and more). According to the literature data it can be stated, that likewise SRT, the hydraulic retention time has no influence on the removal of Carbamazepine. There seems to be a slight increase of the removal efficiency within the HRT range from 12–25 hours (Figure 5.5), however, the elimination is still insufficient (20–40%).

There is lack of data regarding ***Valproic acid***, ***Gabapentin*** and ***Phenytoin***.

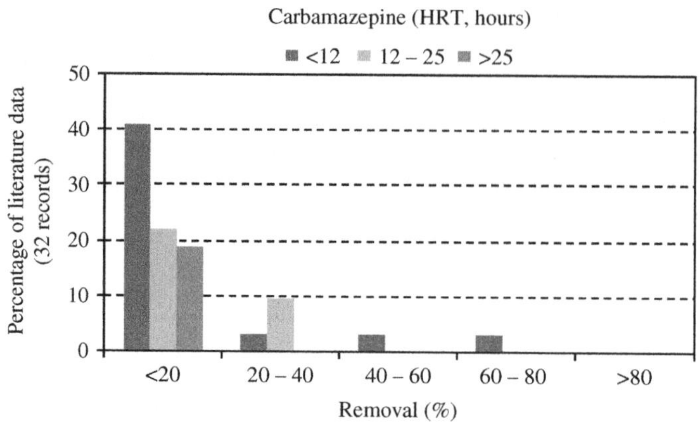

Figure 5.5 HRT influence on the removal of anticonvulsants.

5.3.3 Reactor configuration

Carbamazepine is one of the most persistent compounds detected in the wastewater effluents.

Figure 5.6 shows very low or no removal or all treatments found in the literature. Therefore it can be stated, that reactor configuration has no influence on the removal of Carbamazepine.

Gabapentin and Valproic acid are completely removed at CASP (Yu *et al.*, 2006). Additionally, Yu *et al.* (2006) reported also the removal of Phenytoin at CASP, equal to 44%. Regrettably no other treatment was mentioned.

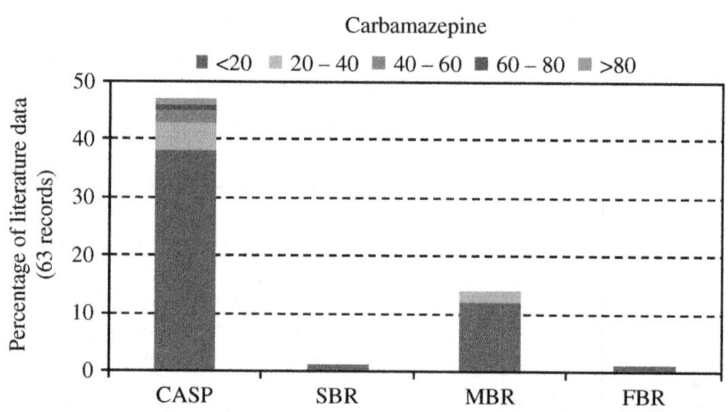

Figure 5.6 Reactor configuration influence on the removal of anticonvulsants.

5.3.4 Sludge adsorption

The K_d value for primary sludge and secondary sludge were found for carbamazepine. The K_d value for primary sludge varies from 1 to 57 L/kgSS, for secondary sludge is equal to 1.2 L/kgSS. There are some data concerning concentration of carbamazepine in primary and secondary sludge coming from 0, 3 g/L and 20 μg/L (Oppel *et al.*, 2004; Ternes *et al.*, 2004; Clara *et al.*, 2004; Drillia *et al.*, 2005b; Scheytt *et al.*, 2005; Urase and Kikuta, 2005; Carballa *et al.*, 2008; Chenxi *et al.*, 2008).

5.3.5 Advanced technologies

Carbamazepine is a fast-reaction pharmaceutical, with no dependency on pH According to (Huber *et al.*, 2003). However, with lower doses of O_3 the removal efficiency decrease significantly. Catalytic oxidation is reporter to remove efficiently carbamazepine. Molinari *et al.* (2006) reported complete removal of Carbamazepine by UV/TiO_2. The adsorption advantage is related to the solubility of the drug; at pH around 3 and 11 Carbamazepine was completely soluble, therefore adsorption percentage on TiO_2 particles was low.

Combination of UV with H_2O_2 resulted in very effective removal of Carbamazepine (Vogna *et al.*, 2004). He experiment was performed in the presence of humic acids, which decreased reactivity of studied compound. The reason is that humic acids act as OH radicals scavengers.

Vogna *et al.* (2004) also confirmed, that Carbamazepine is non-pH dependent. Finally, The catalytic ozonation of carbamazepine in aqueous solution reduced the total organic carbon to about one third of its initial value for reactions carried out at 25°C in slightly acidic conditions with pH 5 and an initial concentration of 15 mg/L (Rosal *et al.*, 2008).

5.4 ANTIINFLAMMATORIES (ANALGESICS)

5.4.1 Sludge Retention Time (SRT)

Ibuprofen, ***Ketoprofen*** and ***Indomethacin*** show the removal rates >80% for the sludge retention time typical for nutrient removal (Figure 5.7). However, the efficiency of the elimination of the compounds seems to drop down with the decrease of the SRT below 10 days. On the other hand, the elimination observed for SRT > 20 is rather complete (Clara *et al.*, 2005a; Lishman *et al.*, 2006; Radjenovic *et al.*, 2007; 2009; Yu *et al.*, 2009). On the contrary, ***Naproxen*** is better removed with SRT < 10 days (e.g., Lishman *et al.*, 2006; Radjenovic *et al.*, 2007; 2009) and tends to be more persistent with sludge retention time above 20 days (Joss *et al.*, 2005).

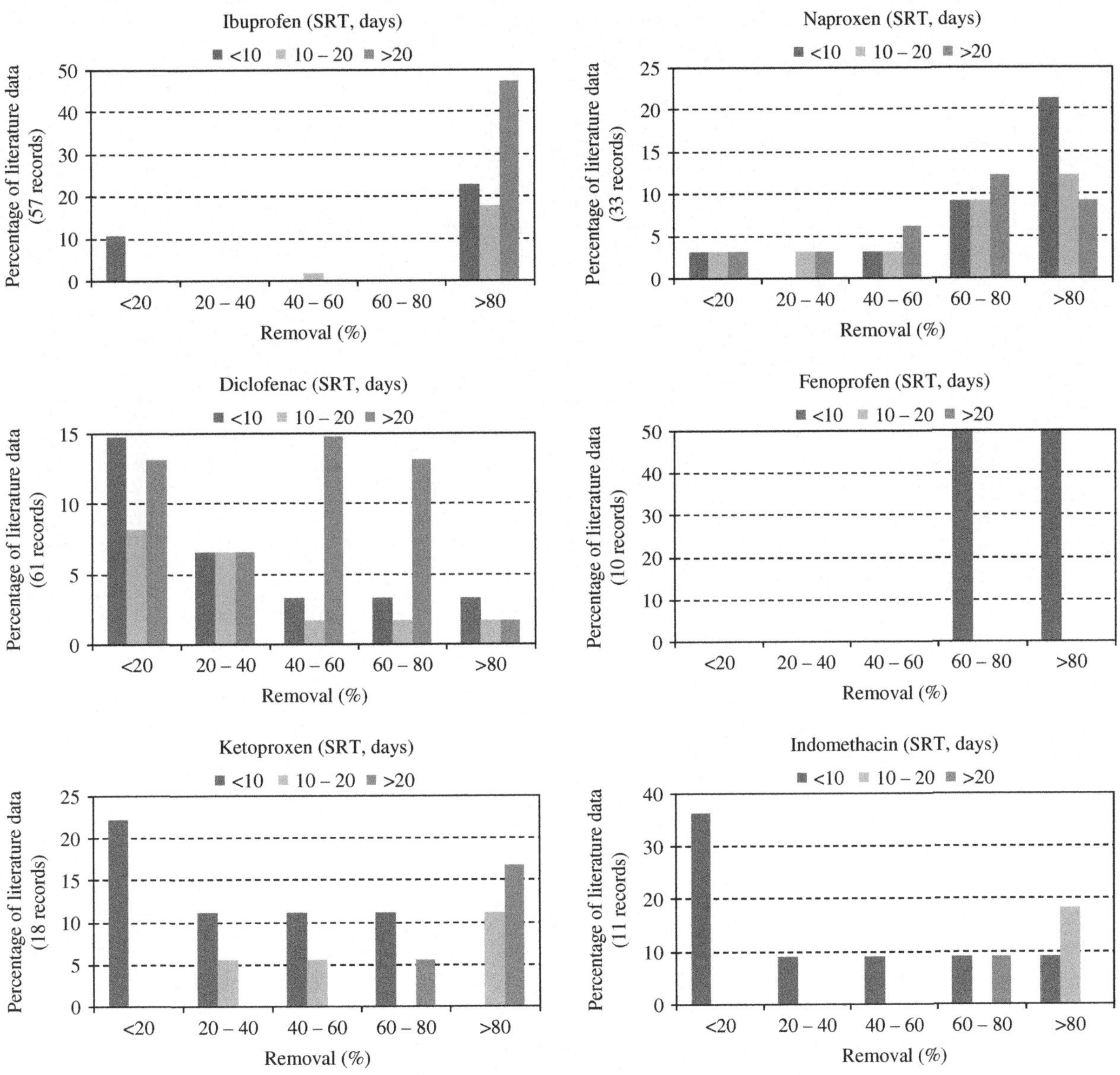

Figure 5.7 SRT influence on the removal of antiinflammatories (analgesics).

Diclofenac is very poorly removed in STPs with SRT < 20 days (Figure 5.9). Increase of the sludge retention time above 20 days seems to improve the removal efficiency of Diclofenac, however the elimination rate is still in the range from 0 to 80%. Only Lishman *et al.* (2006) reported complete removal of this compound.

Nakada *et al.* (2006) presented data concerning removal rates of ***Fenoprofen*** at five STPs with sludge retention times from 3.8 to 8.4 days. In all cases the compound was eliminated in the range of 65–95%.

Insufficient data prevents the determination of SRT impact on the removal of some analgesic; Yu *et al.* (2006) reported ***Acetaminophen*** to be completely removed with sludge retention time between 8 ad 10 days, Jones *et al.* (2007) presented data concerning ***Mefenamic acid***, ***Paracetamol*** and ***Salbutamol*** – all compounds show very high removal rates (>85%) at SRT equal to13 days.

5.4.2 Hydraulic Retention Time (HRT)

Paracetamol and ***Salbutamol*** were reported to be very degradable (Jones *et al.*, 2007), although the data concerned only HRT equal to 13.5 hours. Therefore no precise conclusions could be made.

Ibuprofen is very effectively removed with no dependency on hydraulic retention time (Figure 5.8). On the contrary, ***Naproxen*** shows very good removal rates exceeding 60%, with slight, but noticeable increase with longer HRT. Likewise, ***Diclofenac*** proves to be more susceptible with hydraulic retention time above 25 hours. However, the most commonly reported removal rate of this compound fits in the range of 0–60%.

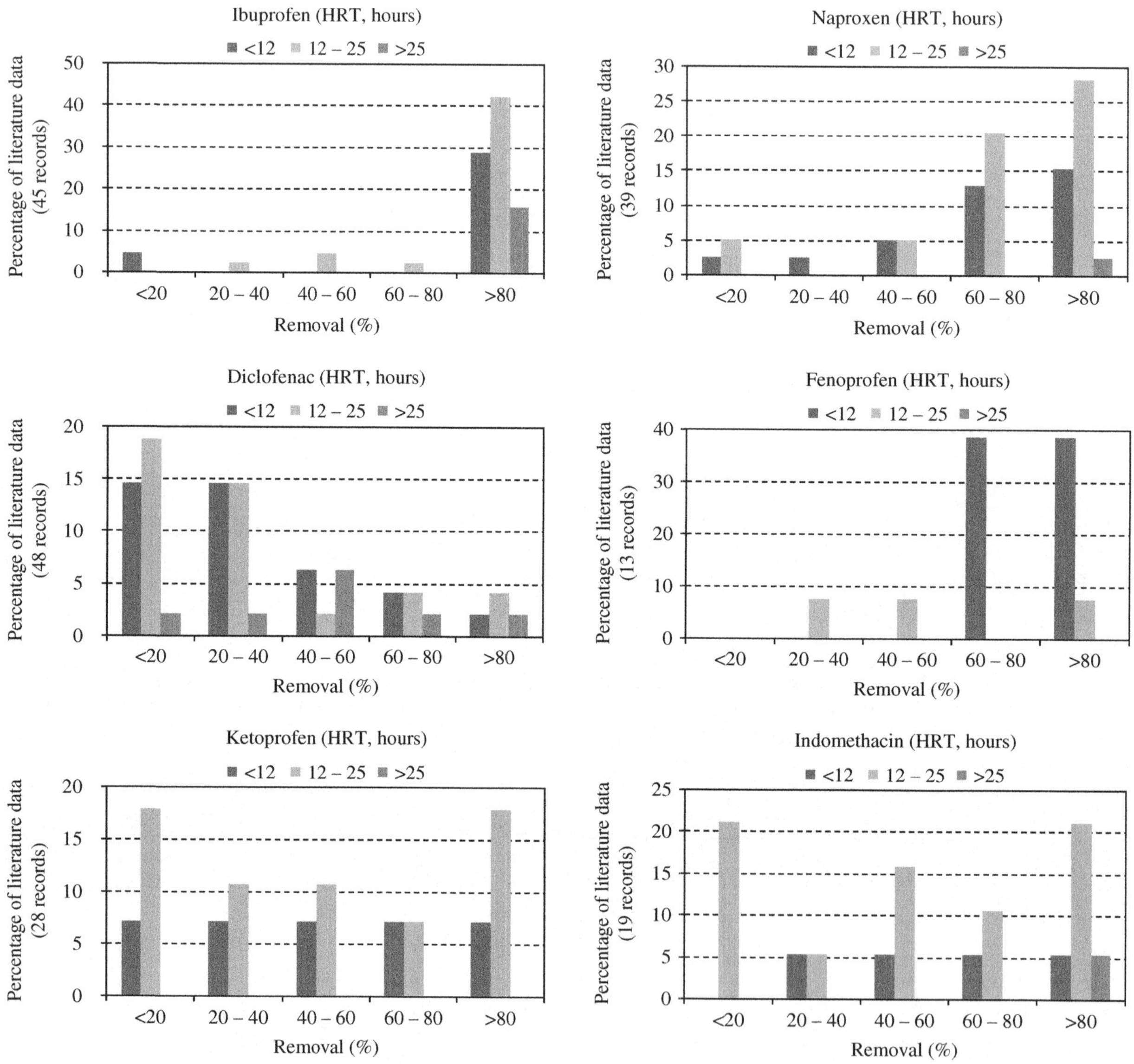

Figure 5.8 HRT influence on the removal of antiinflammatories (analgesics).

On the contrary, ***Fenoprofen*** seems to show higher removal rates with HRT < 12 hours, while at the range of 20–95% the elimination pattern is indefinite.

Ketoprofen, ***Indomethacin*** and ***Mefenamic acid*** cover the whole range of the removal rates for the HRT below 25 hours (Figure 5.8). Therefore it can be stated that HRT has no impact on the elimination pattern of these compounds. Though, except from Indomethacin, no data regarding longer hydraulic retention time was obtained.

Urase *et al.* (2005) reported no removal of ***Propyphenazone*** at HRT equal to 1 day. No other information regarding this compound was found.

No data concerning Flurbiprofen and Acetaminophen affiliation with HRT could be obtained.

5.4.3 Reactor configuration

Ibuprofen shows very good removal at CASP, MBR, FBR and Lagoon, within the range of 90% to complete removal (Figure 5.9).

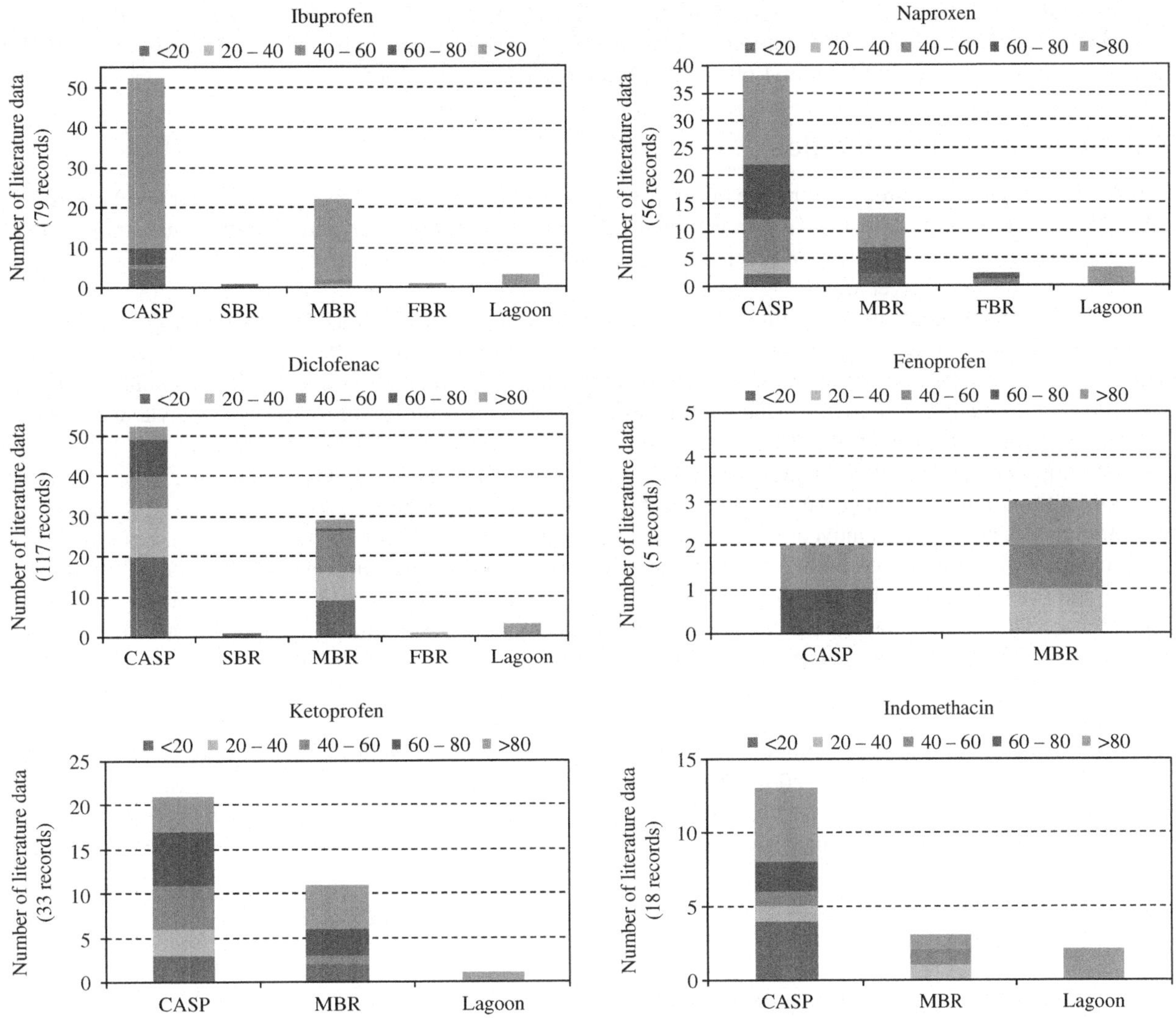

Figure 5.9 Reactor configuration influence on the removal of antiinflammatories (Stumpf *et al.*, 1999; Clara *et al.*, 2005; Urase *et al.*, 2005; Bernhard *et al.*, 2006; Kimura *et al.*, 2005; 2007; and other).

Likewise, Naproxen was reported to be well degraded at CASP and Lagoon, and eliminated at the medium range at CASP (Figure 5.9). MBR and FBR show rather high removal rates, with CASP-TF being the most efficient (Urase *et al.*, 2005; Kimura *et al.*, 2007; Snyder *et al.*, 2007).

Diclofenac usually was reported to be quite persistent, taking into consideration CASP and MBR, while very low or no removal was reported in case of SBR and FBR (Figure 5.9). Stumpf *et al.* (1999) show the results of elimination of Diclofenac for CASP-TF to be very high, exceeding 75%.

There is a limited amount of records regarding Fenoprofen removal. Nakada *et al.* (2006) reported the elimination of this compound at the range of 65–95% for CASP. On the other hand, Urase *et al.* (2005) shows the removal of Diclofenac for MBR to fit in the range between 22 and 92%.

Ketoprofen was reported to be effectively removed at CASP-TF and CASP-P (Figure 5.9). MBR and CASP show the variety of the removal rates, ranging from no removal to 100% (Lishman *et al.*, 2006; Lindqvist *et al.*, 2005).

As presented at Figure 5.9 ***Indomethacin*** shows either high (> 80% for CASP-TF ad Lagoon) or average elimination rates (CASP and MBR).

No data could be found concerning ***Flurbiprofen***.

Elimination of the compounds in pilot wastewater treatment plant was ⩾87% for ibuprofen, naproxen and ketoprofen but only 49–59% for diclofenac. Results after one month of operation revealed 30% elimination with no sign of adaptation by the biomass (Kosjek *et al.*, 2007).

5.4.4 Sludge adsorption

The K_d measured for *ibuprofen* and naproxen is below 20 L/gSS (Ternes *et al.*, 2004; Joss *et al.*, 2006; Carballa *et al.*, 2008) and it can reach higher value according to the experimental condition of measurement or estimation (Stuer lauridsen *et al.*, 2000; Carballa *et al.*, 2008). The values of this coefficient for *diclofenac* is more variable according to the works from 37–72 L/kgSS (Carballa *et al.*, 2008) up to 459 L/kgSS (Ternes *et al.*, 2004).

In order to compare, sediment Distribution coefficients (K_d values) determined by the batch experiments varied from, 0.55 to 4.66 for diclofenac, and 0.18 to 1.69 for ibuprofen (Scheytt *et al.*, 2005).

5.4.5 Advanced technologies

Ibuprofen is very effectively removed by means of O_3/H_2O_2 with O_3 dose above 3.7 mg/L (Zwiener and Frimmel, 2000) while UV/H_2O_2 allow only 20 to 40% of removal by using 0.29 and 1 mM of H_2O_2 respectively (Yuan *et al.*, 2009). It is significant, that satisfactory degradation efficiency was obtained when the applied ozone concentration was equal to the DOC value of the studied river water.

Huber *et al.* (2003) compared the efficiency of ozonation with and without H_2O_2. As could be predicted, using O_3/H_2O_2 combination increased the removal of Ibuprofen about approximately 40%.

Boyd *et al.* (2005) showed the removal of ***Naproxen*** by means of coagulation/flocculation, combined with O_3 and Cl_2. Complete elimination of this compound was achieved. Naproxen was also rapidly transformed via direct photolysis in model river water (Packer *et al.*, 2003), totally removed by means of UV/TiO_2 (Molinari *et al.*, 2006), catalytic ozonation (Rosal *et al.*, 2008; 2009) and ozonation (Benitez *et al.*, 2009). Experimental conditions, in particular pH and ozone concentration are important parameter to reach high level of degradation.

Diclofenac was effectively removed by direct photolysis (Packer *et al.*, 2003), ozonation with O_3 doses above 0.5 mg O_3/L (Huber *et al.*, 2003). Combining the process with H_2O_2 resulted in complete degradation of the compound (Zwiener and Frimmel, 2000). In turn, Vogna *et al.* (2004) examined the efficiency of UV and UV/H_2O_2 on the removal of Diclofenac. This compound can be degraded either by dechlorination or mineralization. In case of the combined process most of the UV radiation was absorbed by H_2O_2 and photolysis of Diclofenac was not a main contributing reaction pathway. However, in the absence of H_2O_2 photolytic substrate decay and chloride release were markedly slowed down.

Finally, Fenton photo-assisted treatment was able to achieve total mineralization of Diclofenac solutions (COD from 12 to 4000 mg O_2/l) (Ravina *et al.*, 2002; Badawy *et al.*, 2009). This process was also successfully applied to the degradation of paracetamol (Badawy *et al.*, 2009) even when present in complex samples, such as STP effluent (Trovo *et al.*, 2009). Moreover, the use of solar irradiation allowed a complete degradation in a shorter time than that obtained with black-light irradiation.

5.5 β-BLOCKERS

5.5.1 Sludge Retention Time (SRT)

All compounds studied within this group of PPs show rather average removal rates within the SRT up to 20 days (Figure 5.10). No data concerning SRT $>$ 20 days is available. In detail, Atenolol is been reported to be removed between 60–80%, with two records out of ten showing the elimination of 37% (Vieno *et al.*, 2007). Acebutolol is usually removed in the range of 20–60% with fewer records relevant to the lower elimination. On the contrary, removal of Sotalol fits in the range of 40–80%, with more records regarding the lower removal. Additionaly, Sotalol seems to be slightly better removed with SRT between 10 and 20 days.

Metoprolol was reported o be the most persistent β-blocker with removal efficiencies of 2–34% (Vieno *et al.*, 2007).

5.5.2 Hydraulic Retention Time (HRT)

From 8 to 10 records were found regarding hydraulic retention time impact on the removal of ***Atenolol***, ***Metoprolol***, ***Acebutolol*** and ***Sotalol***. For all compounds, however, there seems to be an increasing tendency of the elimination pattern with lower HRT values (Figure 5.11). Overall removal rates are ordered as follows: 0–40%, 20–60%, 40–80% for Metoprolol, Acebutolol and Sotalol, respectively, and concern the HRT $<$ 25 hours. With the same value of hydraulic retention time Atenolol removal rate ranges from 0 to 80%. However, there is a lack of data regarding HRT $>$ 25 hours.

Propranolol and ***Bisoprolol***, although found in the STPs' effluents (e.g., Bendz *et al.*, 2005; Miege *et al.*, 2006), were not studied according to the HRT impact on their removal rates.

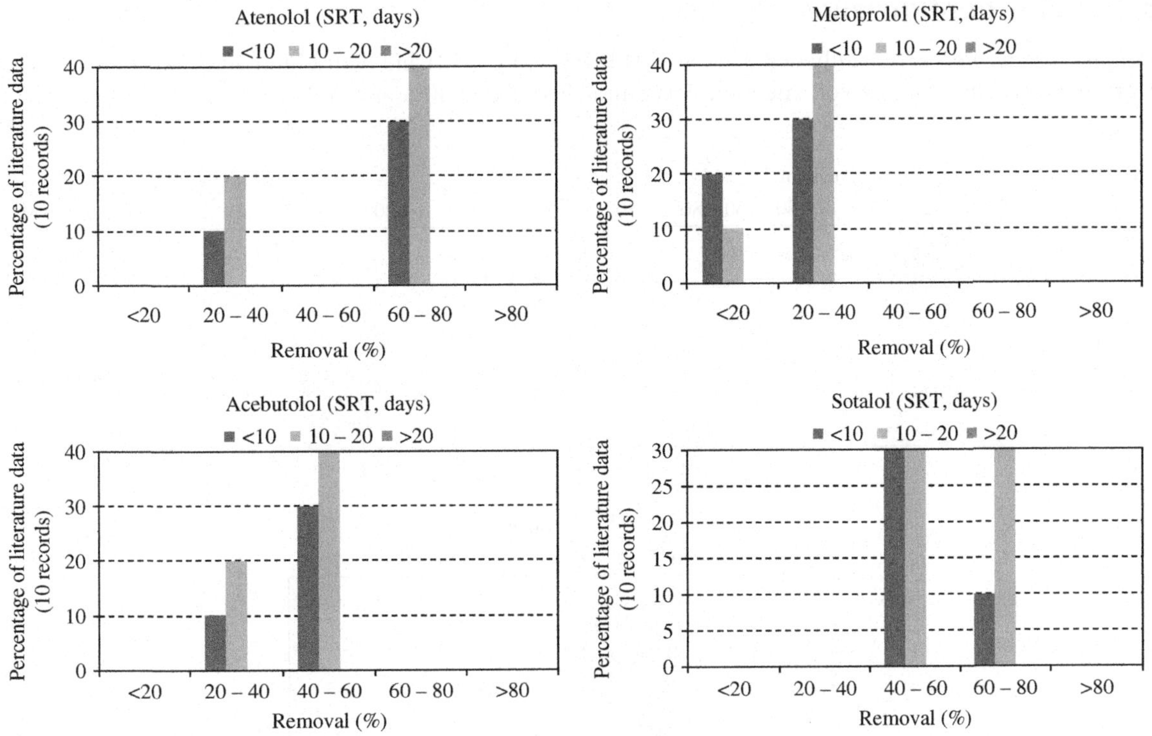

Figure 5.10 SRT influence on the removal of β-blockers.

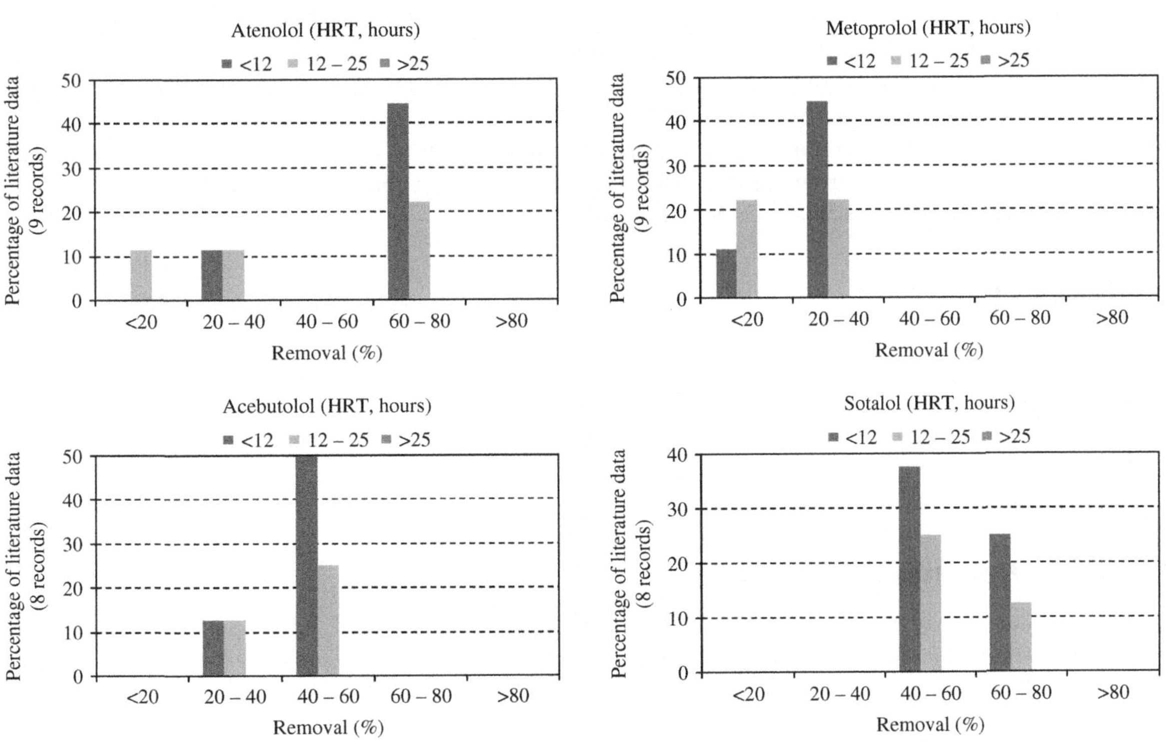

Figure 5.11 HRT influence on the removal of β-blockers.

5.5.3 Reactor configuration

Atenolol and ***Metoprolol*** were hardly degraded at CASP-P (Figure 5.12) and their elimination was rather low at CASP. However, inversely than Metoprolol, Atenolol was eliminated above 80% at CASP.

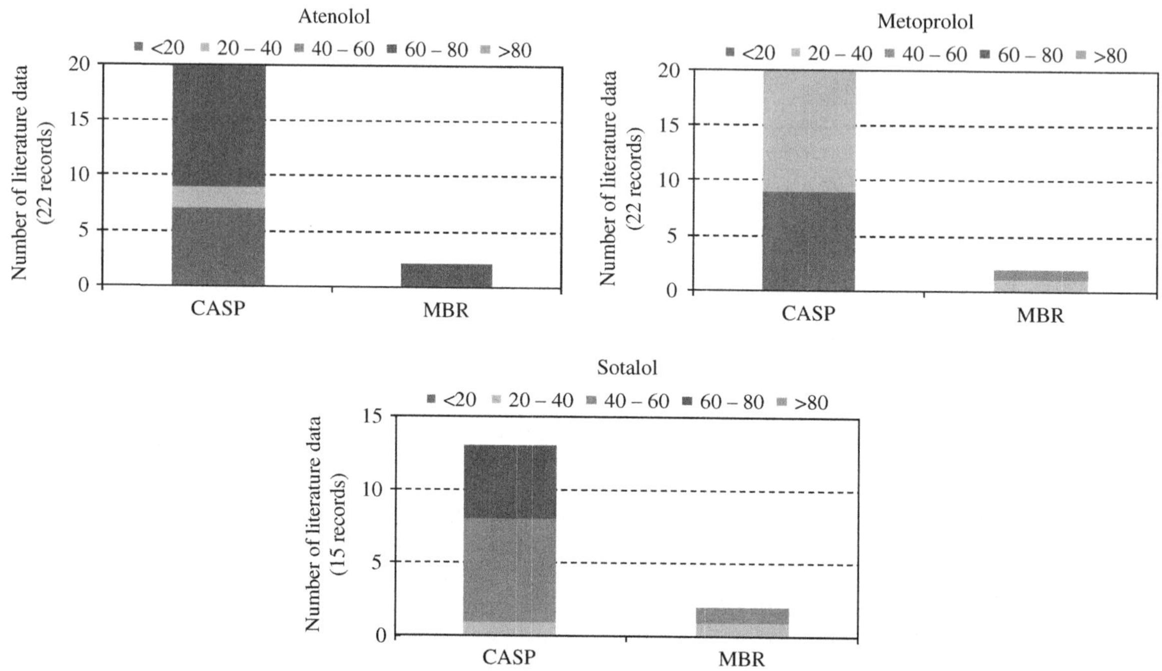

Figure 5.12 Reactor configuration influence on the removal of β-blockers.

In case of ***Acebutolol*** and ***Sotalol*** Vieno *et al.* (2007) reported the removal rate for CASP and CASP-OD. Both compounds were quite susceptible during conventional treatment. However, their behaviour was significantly different in case of CASP (Figure 5.12).

Bendz *et al.* (2005) reported the elimination of ***Propranolol*** at CASP to be approximately 30%. Few data of this compound was available and they are moreover very inconsistent. Elimination rates ranging from less than 10% up to 96% are reported (Ternes, 1998; Blac, 2003; Maurer *et al.*, 2007).

5.5.4 Sludge adsorption

There is rare information regarding sorption of these compounds in the available literature. Ternes *et al.* (2004) estimate a concentration close to 0.3 g/L in primary sludge for propranolol, sotalol and atenolol. More recently, Maurer *et al.* (2007) determined the sorption coefficients (KD, standard deviations in brackets) as 0.04(70.035), 0.04(70.033), 0.00(70.023) and 0.32(70.058) L/g $_{COD}$, for sotalol, atenolol, metoprolol and propranolol, respectively.

5.6 HORMONES

5.6.1 Sludge Retention Time (SRT)

Hormones are known to be well removed during wastewater treatment (Clara *et al.*, 2005a; Servos *et al.*, 2005; Lishman *et al.*, 2006; Johnson *et al.*, 2005; and more). However, there is data concerning average or low removal, especially for SRT < 10 days, while Kreuzinger *et al.* (2004) and Schlüsener (2005) reported even no removal. Altogether, it can be stated that the longer sludge retention time, the better removal rate of hormones. Among all chosen compounds the most persistent seems to be ***17α-ethinylestradiol*** with 12 records out of 45 showing the elimination below 40% (Figure 5.13).

Esperanza *et al.* (2007) studied the removal of ***Testosterone***, ***Androstenedione*** and ***Progesterone*** at pilot scale STP with SRT equal to 6 days. All compounds were completely removed during the treatment.

There is no data regarding to sludge retention time influence on ***16α-hydroxyestrone*** removal.

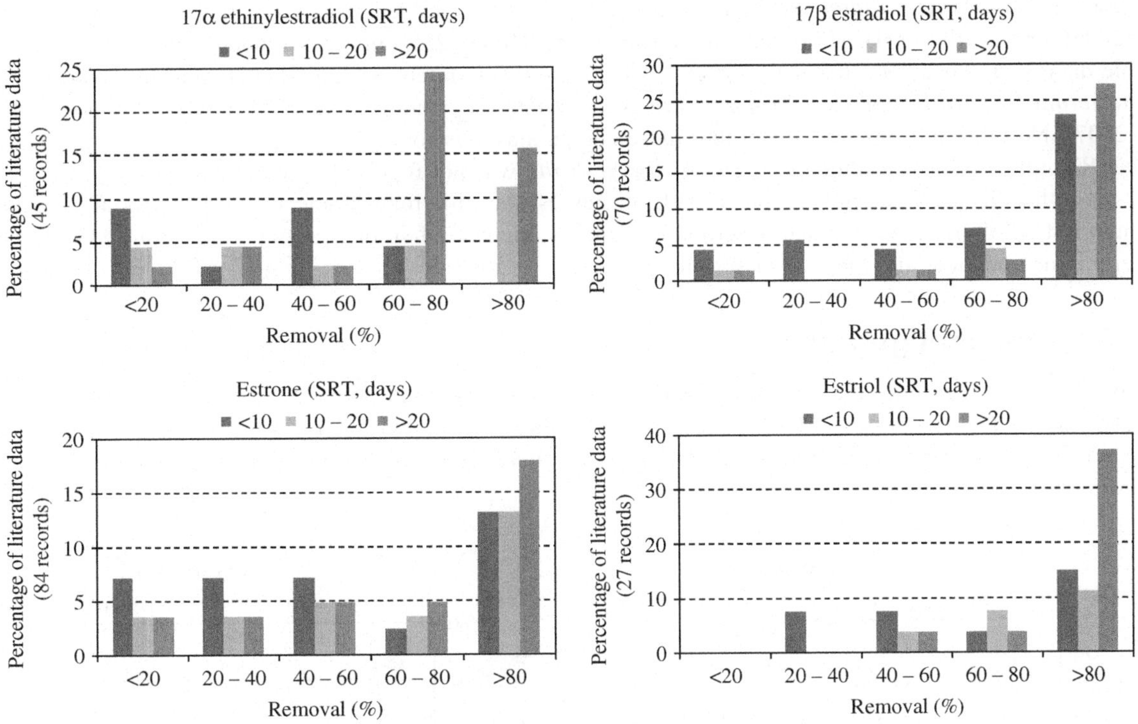

Figure 5.13 SRT influence on the removal of hormones.

5.6.2 Hydraulic Retention Time (HRT)

Most of the data concerning elimination of hormones during wastewater treatment presents high removal rates. However, majority of records regards ***17β-estradiol***, ***Estriol*** and ***Estrone*** (Figure 5.14). These compounds tend to be very well removed at the whole range of hydraulic retention time, so no influence of this parameter can be observed.

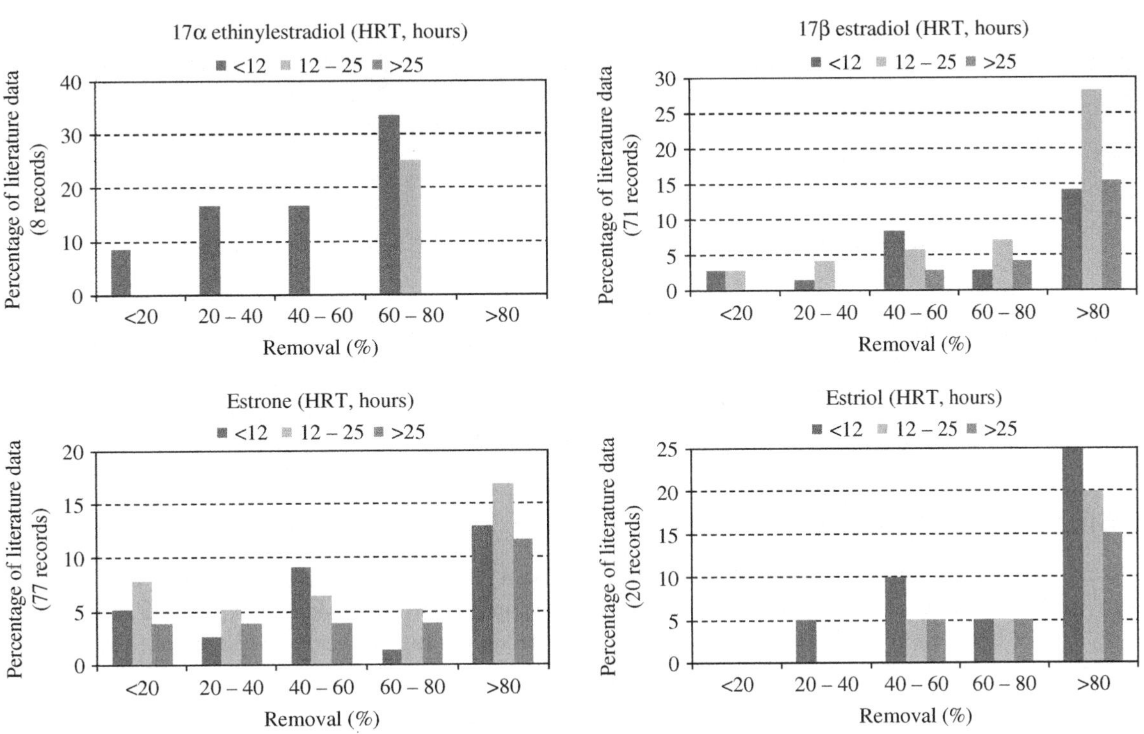

Figure 5.14 HRT influence on the removal of hormones.

In case of ***17α-ethinylestradiol*** the highest removal rate was obtained with HRT range of 12–25 hours (Urase *et al.*, 2005; Joss *et al.*, 2004). However, no data regarding HRT > 25 hours was found. Therefore, according to the available data it can be stated, that with higher values of hydraulic retention time, better removal could be obtained.

Testosterone, ***Androsterone*** and ***Progesterone*** were completely removed at HRT equal to 6 hours (Esperanza *et al.*, 2007). No other information concerning these compounds was found.

The HRT influence on the removal of ***16α-hydroxyestrone*** was not reported.

On the other hand, Cajthaml *et al.* (2009) discusses the different efficiencies of bacterial degradation of 17α-ethinylestradiol under aerobic and anaerobic conditions, the role of sulfate-, nitrate-, and iron-reducing conditions in anaerobic degradation, and the role of sorption.

5.6.3 Reactor configuration

Majority of records regarding the removal of hormones shows very good degradability of these compounds. That is especially true for ***17β-estradiol*** and ***Estrone*** (Figure 5.15) – both compounds were eliminated above 80% at MBR, FBR and Lagoon. CASP seems to be also very efficient, however there are records showing average or even poor removal of these hormones.

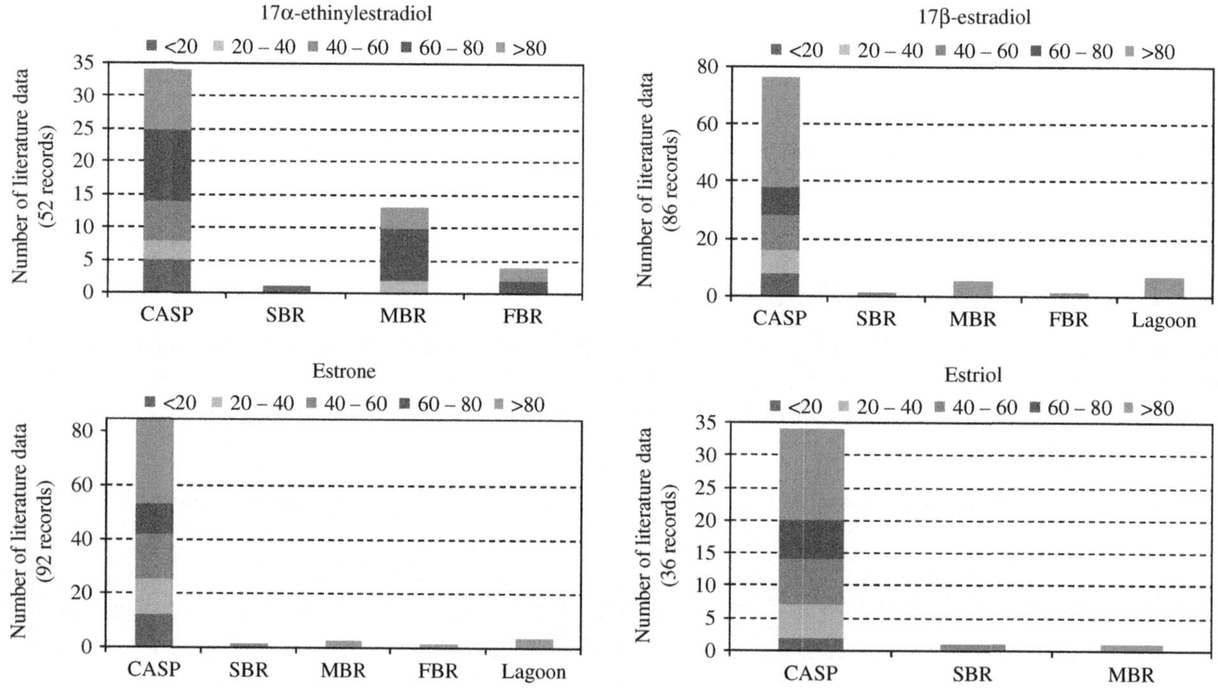

Figure 5.15 Reactor configuration influence on the removal of hormones.

17α-ethinylestradiol is effectively removed by MBR and FBR (Figure 5.15). Very low removal was observed for SBR (Clara *et al.*, 2005a). In case of CASP, this compound fits in the whole range of the removal rates, meaning that the elimination depends on the configuration of particular facility. The similar situation occurs in case of ***Estriol*** and ***16α-hydroxyestrone***. Yet, approximately 45% of records regarding CASP show the removal of Estriol at the range above 80%.

5.6.4 Sludge adsorption

A significant higher K_d value was reported by Ternes *et al.* (2004) for ***ethinylestradiol*** in primary sludge as well as for secondary sludge in comparison to the other examined substances (primary sludge: $K_d = 278$ L/kgSS; secondary sludge: $K_d = 249$ L/kgSS). Carballa *et al.* (2008) observed the same ranges for ***ethinylestradiol*** (120–707 L/Kg), but also for Estrone (150–590 L/Kg) and E2 (200–680 L/Kg).

5.6.5 Advanced technologies

Zhang and Zhou (2005) reported the removal of ***Estrone*** (E1) and ***17β-estradiol*** (E2) by means of granular activated carbon (GAC) in comparison with other adsorbents. Adsorption capacity of GAC was decreased with an increase in adsorbent concentration and by the presence of surfactants or humic acids. Moreover, the adsorption of E1 and E2 on the carbonaceus adsorbent is more rapid than on GAC. In both cases, pH seemed to be an important parameter, as the elimination rate increased at pH above 8.

The degradation of E1 and E2 in aqueous solution can be achived by ozonation. The efficiency is increased at pH values ranged of 3.0–9.0 with a maximum output at pH 9.0. E1 could reach to 94% after 8 min and even E2 could be completely degraded after only 4 min. Moreover, with increasing O_3 dosage, the degradation efficiency increased (Lin *et al.*, 2009). O_3/H_2O_2 process is also effective for E2 and EE2 removal in aqueous solution. Estrogens removals were higher than 96% for all ozone doses and pH values investigated and addition of H_2O_2 to the ozonation process did not improve estrogens removal. Increase of ozone dosage increases estrogenic activity removal, even if it was not completely removed in pHs 7 and 11 but totally at pH 3 (Maniero *et al.*, 2008). Photocatalysis (Belgiorno *et al.*, 2007) as well as chlorination (Auirol *et al.*, 2006) is also well efficient on the degradation of E2.

Nano- and ultrafiltration was presented in case of 17α-ethinylestradiol and Estriol removal (Yoon *et al.*, 2006; 2007). For both methods the elimination rate was rather poor and did not exceed 40–60%.

Progesterone, Testosterone and Androstenedione were effectively eliminated by means o nanofiltration and poorly or not removed by ultrafiltration (Yoon *et al.*, 2007).

5.7 TRANQUILIZERS

5.7.1 Sludge Retention Time (SRT)

Very few data regarding on the removal of tranquilizers is available. Kreuzinger *et al.* (2004) reported the elimination of ***Diazepam*** at STPs with sludge retention times ranging from 0.7 to 275 days. In all cases the removal rate did not surpass 25% with no removal for the SRT equal to 0.7 days (Figure 5.16).

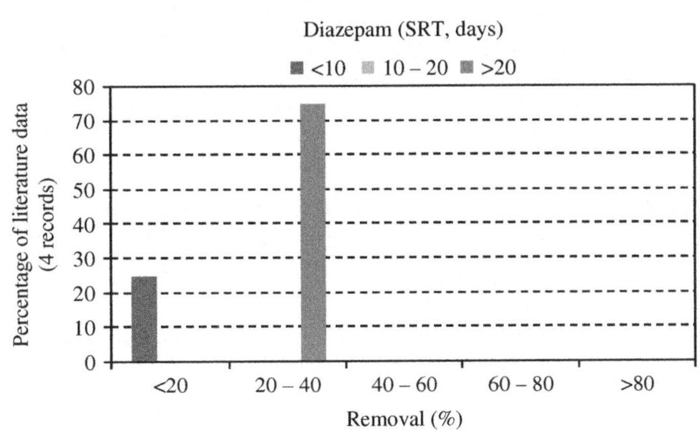

Figure 5.16 SRT influence on the removal of tranquilizers.

5.7.2 Hydraulic Retention Time (HRT)

No data containing information about HRT of wastewater treatment plants in case of ***Diazepam*** removal (or any other tranquilizer) was found.

5.7.3 Reactor configuration

The only data concerning compounds from this group was given by Kreuzinger *et al.* (2004) and regarded the elimination of ***Diazepam*** at CASP. No removal higher than 25% was reported. The lack of data prevents the comparison of different wastewater treatment technologies.

5.7.4 Sludge adsorption

There are information concerning *diazepam* concentration in sludge: onto primary sludge the amount of compound was equal to 580 μg/L (K_d = 125 L/kgSS), onto secondary – around 241 μg/L (K_d = 21 L/kgSS) (Carballa *et al.*, 2007b; Ternes *et al.*, 2004).

5.7.5 Advanced technologies

Diazepam removal varied from 24 to 65% using 2 mg O_3 L^{-1} (Huber *et al.*, 2003). In case of ultrafiltration no removal was reported, however nanofiltration resulted in the elimination from 20–90% (Yoon *et al.*, 2006).

5.8 X-RAY CONTRAST MEDIA (CM)

5.8.1 Sludge Retention Time (SRT)

Contrast media are known to be the most persistent compounds. However, the data concerning these PPs is strongly limited. Among all CM reported to be found in STPs' effluents ***Iopromide*** is the only compound described in the literature in regard to the impact of the sludge retention time on the removal. Figure 5.17 shows that SRT < 10 days results in no removal, while the increase of the sludge retention time allows to eliminate Iopromide up to 95% (Joss *et al.*, 2005).

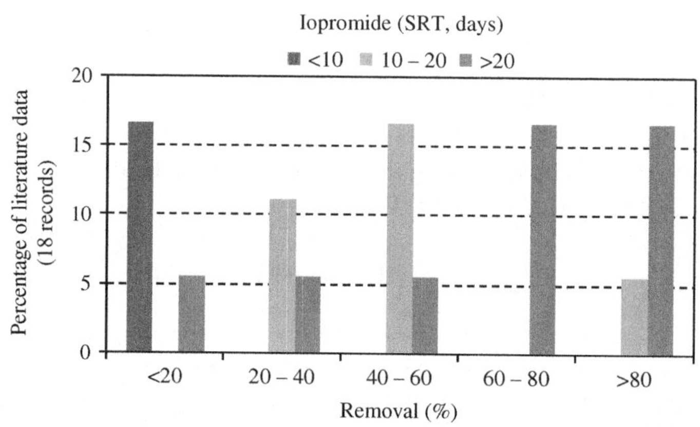

Figure 5.17 SRT influence on the removal of X-ray contrast media.

5.8.2 Hydraulic Retention Time (HRT)

There is a lack of data regarding influence of the HRT on contrast media elimination during wastewater treatment. However, Carballa *et al.* (2004) and Clara *et al.* (2005b) reported no removal of ***Iopromide*** with HRT ranging from 2 to 29 hours (Figure 5.18). Based on this information in can be stated, that contrast media tend to be very persistent regardless of the hydraulic retention time.

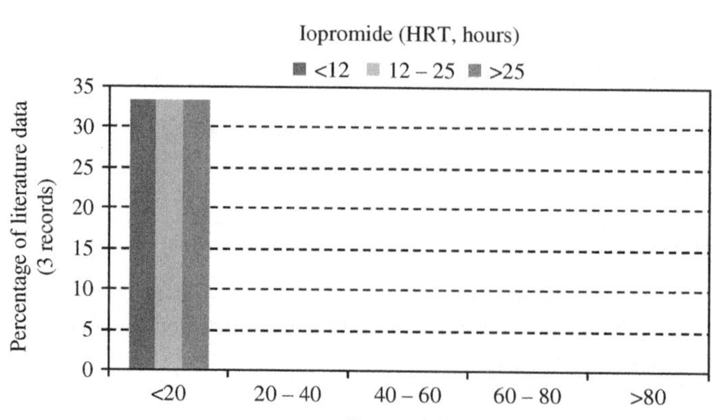

Figure 5.18 HRT influence on the removal of X-ray contrast media.

5.8.3 Reactor configuration

Ternes and Hirsch (2000) reported no removal of ***Diatrizoate*, *Iomeprol*, *Iopamidol*, *Iothalamic acid*** and ***Ioxithalamic acid*** at CASP. No further information according to this compounds' elimination was found.

Iopromide was hardly removed at CASP-C (Figure 5.19), on average eliminated by MBR, and quite well removed in FBR. In case o CASP it can be stated, that Iopromide is rather persistent, however there was an evidence of the removal rates even above 95% (Joss *et al.*, 2005; Kreuzinger *et al.*, 2004).

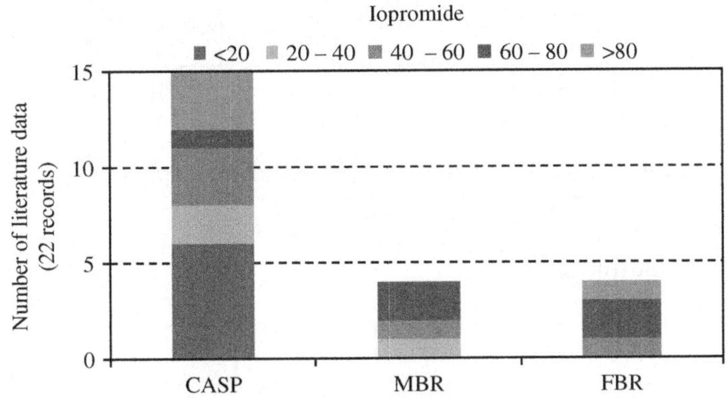

Figure 5.19 Reactor configuration influence on the removal of X-ray contrast media.

5.8.4 Sludge adsorption

Available literature delivers data concerning K_d value for ***Iopromide*** <5 L/kgSS for primary sludge and 11 L/kgSS for secondary sludge (Ternes *et al.*, 2004). Concentration between 0.3 and 2.5 g/L are described in primary sludge (Haiβ and Kuemmerer, 2006; Ternes *et al.*, 2007; Carballa *et al.*, 2008).

5.8.5 Advanced technologies

Doll *et al.* (2003, 2005) studied the removal of ***Iomeprol*** by means of UV/TiO_2. The results varied from 12–80%, which can be explained with the usage of different UV lamps: Hombikat UV100 was more effective for photocatalytic oxidation of Iomeprol than P25. Similar results were obtained for ***Iopromide***. Moreover, ozonation was not able to fully mineralized Iomeprol at pH 7 and 9 but allow the production of unknown oxidation by-products (Steitz *et al.*, 2008).

Kim *et al.* (2007) reported very high removal of Iopromide using granular activated carbon (GAC). The advantage of this treatment technology is that low concentrations of organic matter in lake or river water results in less competition for the binding of particles to activated carbon.

5.9 LIPID REGULATORS

5.9.1 Sludge Retention Time (SRT)

Lipid regulators show a variety of removal rates depending on the sludge retention time. Both ***Bezafibrate*** and ***Gemfibrozil*** are removed above 80% with SRT > 20 days, well removed with sludge age ranging from 10 to 20 days, and poorly removed with SRT < 10 days (Figure 5.20).

No data is available regarding SRT influence on the removal of ***Clofibric acid*** and ***Fenofibric acid***.

5.9.2 Hydraulic Retention Time (HRT)

Bezafibrate shows the highest removal rate at the HRT < 25 hours (Figure 5.21) ranging from 77–100%. The exception is a record presenting the removal of 36%, which was reported for the carbon removal WWTP with HRT of 2 hours (Clara *et al.*, 2005b). Lower degradation, equal to 54%, was reported also by Clara *et al.* (2005b) at the HRT of 29 hours. On the contrary, hydraulic retention time seems to have no influence on the elimination pattern of ***Gemfibrozil***, which in most cases is removed above 60%. ***Clofibric acid*** shows a tendency of decreasing removal efficiency as HRT increase (Figure 5.21).

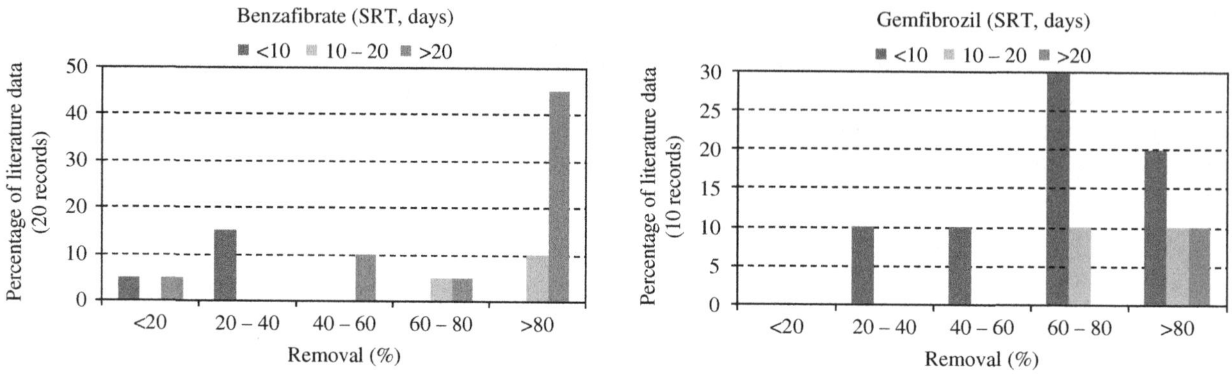

Figure 5.20 SRT influence on the removal of lipid regulators.

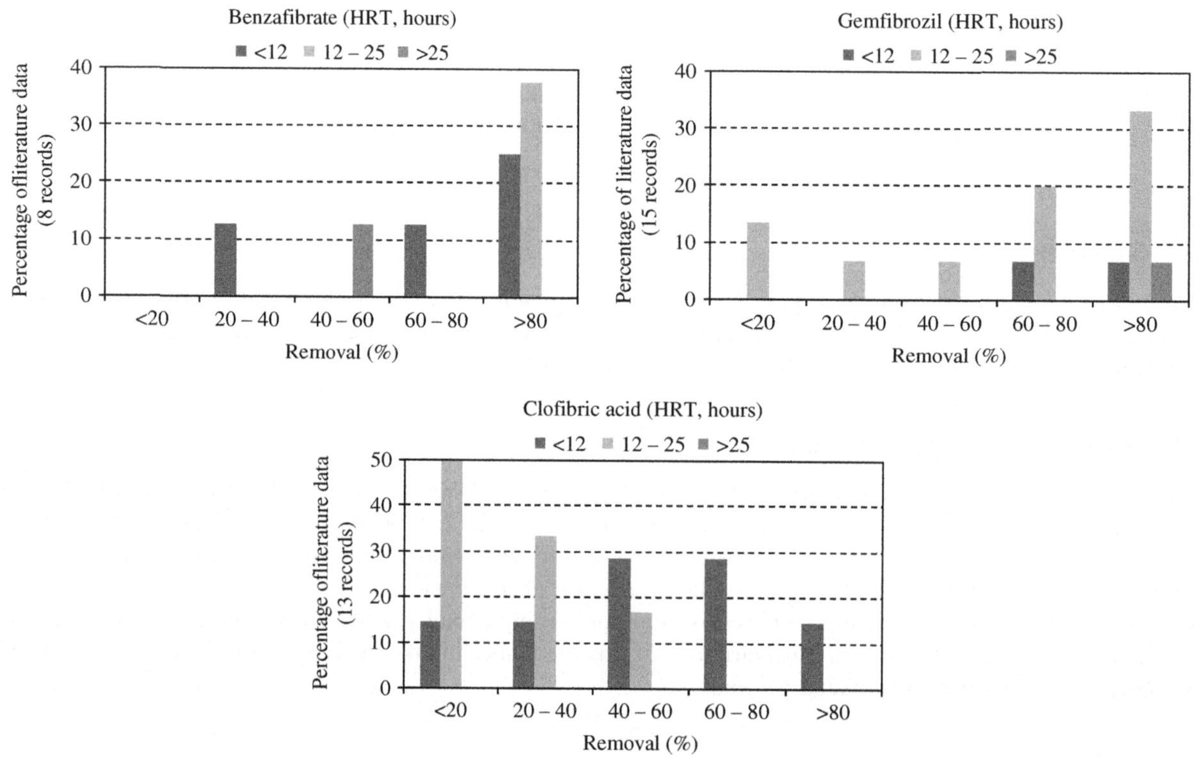

Figure 5.21 HRT influence on the removal of lipid regulators.

5.9.3 Reactor configuration

The most efficient technology in eliminating ***Bezafibrate*** seems to be MBR, with the removal rates above 60% (Figure 5.22). Furthermore, CASP shows rather high degradation of this compound (most of the records concern the elimination of more than 80%). Poor degradation of Bezafibrate can be observed for CASP-TF and CASP-C.

On the contrary, MBR is not effective in the removal of ***Gemfibrozil*** and ***Clofibric acid***. Although there is data concerning the elimination above 80% of Gemfibrozil, the majority of records show rather low removal rates (<20%). CASP and CASP-TF are characterized by average elimination of both compounds.

Additionally, Stumpf *et al.* (1999) presented the removal efficiency of CASP-TF of ***Fenofibric acid*** to be around 40%. There is a lack of information regarding other treatment technologies.

5.9.4 Sludge adsorption

K_d values determined for ***clofibric acid*** indicate, that this compound is not sorbed in significant degree onto primary sludge (K_d less than 30 L/kgSS), as well as onto secondary sludge (K_d = 4.8 L/kgSS) (Joss *et al.*, 2006; Ternes *et al.*, 2007; 2004).

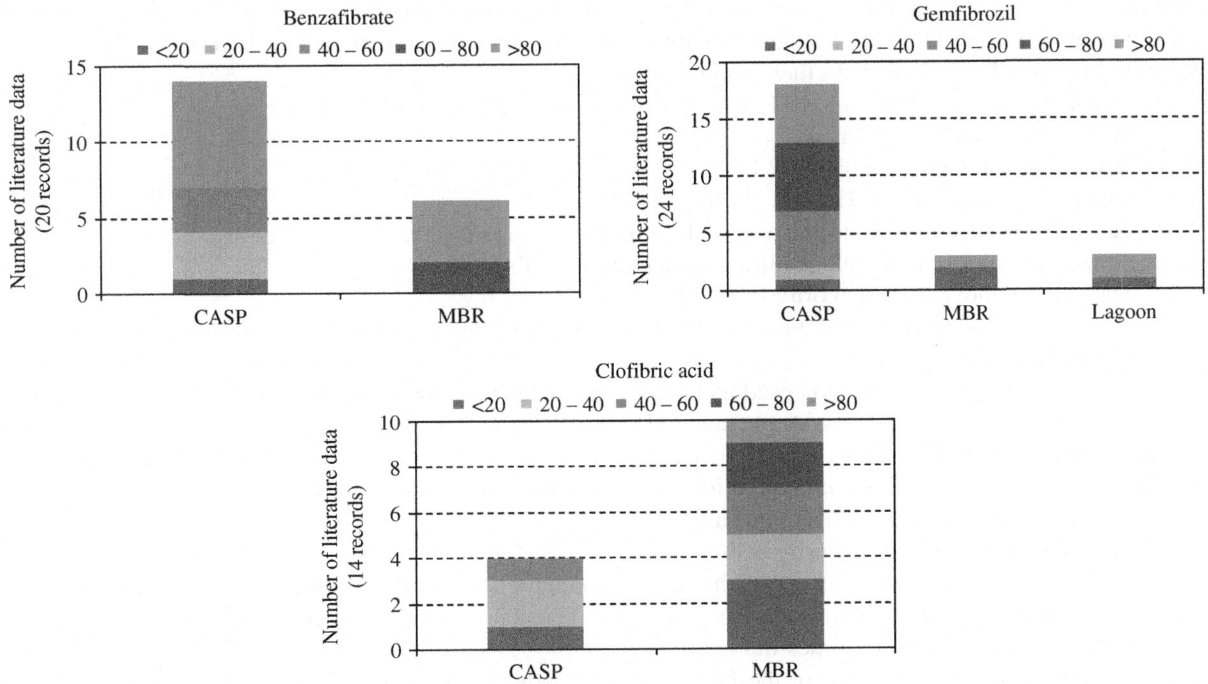

Figure 5.22 Reactor configuration influence on the removal of lipid regulators.

5.9.5 Advanced technologies

Huber *et al.* (2003) presented the influent of ozone dose on the removal of ***Bezafibrate***. Doses from 0.1–2 mg O_3/L were examined. Satisfactory elimination was achieved with 0.5 mg O_3/L. This results were confirmed by Dantas *et al.* (2007). Moreover, combining O_3 with H_2O_2 results in very high removal, exceeding 95%. Photo-Fenton was also successfully applied to the degradation of benzafibrate (Trovo *et al.*, 2009) and shows that the use of solar irradiation allowed a complete degradation in a shorter time than that obtained with black-light irradiation.

In case of ***Clofibric acid*** combination of UV and TiO_2 was reported to be highly efficient (Molinari *et al.*, 2006; Doll *et al.*, 2003). According to Molinari, adsorption advantage can be related to the drug solubility in water and increase with the decrease of pH. On the other hand, Zwiener and Frimmel (2000) compared the efficiency of Clofibric acid removal from Milli-Q water and river water by means of O_3/H_2O_2. However the removal of this compound was quite low for Milli-Q water (~46%), it decreased significantly in the river water (12% of removal). This could be explained with the presence of radical scavengers, which compete with pharmaceuticals for OH radicals. On the other hand, higher concentrations of O_3/H_2O_2 increase the degradation kinetics of Clofibric acid.

Andreozzi *et al.* (2003a) reported very high removal of this compound using UV and UV/H_2O_2. However, the elimination was performed in Milli-Q water as a medium, therefore the results from the environmental waters could vary.

Yoon *et al.* (2006) treated river water with ultra- and nanofiltration to measure the removal of ***Gemfibrozil***. None of these technologies was able to eliminate the compound effectively.

5.10 CONCLUSIONS

(1) The sludge retention time (SRT) is one of the crucial parameters, which influence on the design, operation and control of wastewater treatment plants. According to the available data it can be stated, that PPs can be divided into three major groups:
- Compounds with the optimum SRT range for which their removal is the most effective, e.g.: the majority of antibiotics and antiinflammatories is well removed with SRT typical for the nutrient removal (10–20 days);
- Compounds on which SRT has no impact and are characterized by:
    - very low removal rates (anticonvulsants),
    - average removal rates (β-blockers),
    - satisfactory removal rates (majority of hormones);
- Compounds with visible influence of SRT on their removal rate, e.g. lipid regulators.

(2) Hydraulic retention time (HRT) is also one of the most fundamental parameters for wastewater treatment plants performance. Likewise in case of sludge retention time, all studied compounds can be divided into two groups according to the HRT influence on their removal rates:
- Compounds with the optimum HRT, which results in the highest elimination rate, e.g.:
  - Antibiotics – HRT $<$ 12 hours,
  - Antiinflammatories – HRT $>$ 12 hours,
  - Lipid regulators – 12 $<$ HRT $<$ 25 hours;
- Groups of compounds with no impact of HRT on their removal rate, characterized by:
  - very low or no removal (e.g. contrast media, anticonvulsants),
  - very high removal (e.g. majority of hormones).

Therefore the optimum range of hydraulic retention time for PPs elimination during wastewater treatment would be from 12 to 25 hours.

(3) Because of the variety of the common used technology and reactor configuration, several statements could be made:
- For CASP and MBR there is the whole range of removal rates. Therefore, the elimination of PPs depends rather on parameters like SRT, HRT, flow rate etc.
- There is no impact of reactor configuration on some compounds removal, which are characterized by:
  - low or no removal (e.g. anticonvulsants),
  - very high removal (e.g. majority of hormones),
  - in some cases only slight impact of the reactor configuration could be observed (e.g. contrast media).

Lack of data limits proper comparison of the removal efficiencies of most compounds according to a variety of treatment technologies. Table 5.2 resumes the main results.

(4) It can be concluded, that behaviour of the PPs in sewage sludge is the issue, which demands more investigations. In most cases there are some single records, which give very general overview of the sorption and desorption processes of PPs.

According to the literature data, for compounds showing sorption coefficient below 500 L/kgSS onto primary sludge and 300 L/kgSS onto secondary sludge can be negligible. Notwithstanding, during sludge treatment processes compounds with lower K_d values should be also taken into consideration, because the sorbed amount do not depend only on the K_d, but also on the sludge concentration. Therefore, based on the outcome of the POSEIDON project it can be stated, that the limit of the relevance of the sorption coefficient is around 1 L/kgSS.

(5) No data regarding the influence of red-ox conditions on the removal of PPs could be found. In most cases the treatment takes place in overall conditions (including aerobic, anoxic and anaerobic).

(6) No influence of the climatic zones on the removal of PPs was observed. Comparison of the data on the removal of PPs concerning countries from Europe, North and South America, Asia and Australia revealed no influence of the climatic zone on the PPs elimination efficiency (data not shown). It is more likely that the removal rates of pharmaceutical compounds depends on their consumption pattern, therefore as a next step this data will be analyzed. It is more likely that the removal rates of pharmaceutical compounds depends on their consumption pattern.

(7) Basing on the available data, it can be concluded, that the current municipal STPs are not able to guarantee a complete elimination of PPs. However, there are some possibilities of the enhancing PPs removal at the existing facilities:
- the optimum SRT range within 10–20 days,
- the optimum range of HRT for PPs elimination varies from 12 to 25 hours.

Therefore, the pre- or post-treatment in some cases is effective, e.g. reverse osmosis, however, removed pollutants are concentrated in the waste stream, which also must be treated itself; the treatment by means of AOPs can lead to creation toxic by-products.

(8) Source control and source separation should be implemented in order to reduce or minimize the introduction of pharmaceutical compounds to the environment in the following cases:
- if the treatment before the dilution in the sewage system is more effective,
- if the losses during the transport via sewer system are to be avoided.

Table 5.2 Comparison of the reactor configuration impact on the removal of PPs.

	CASP	SBR	MBR	FBR	Lagoon
		ANTIBIOTICS			
Sulfamethoxazole	**x**	**xxx**	**xxx**	**xxx**	*n.a.*
Trimethoprim	**x**	*n.a.*	*n.a.*	*n.a.*	*n.a.*
Clarythromycin	overall	*n.a.*	*n.a.*	*n.a.*	*n.a.*
Roxythromycin	**x**	*n.a.*	**xx**	**x**	*n.a.*
(Anhydro-) erythromycin	overall	*n.a.*	*n.a.*	*n.a.*	*n.a.*
Ciprofloxacin	**xxx**	*n.a.*	*n.a.*	*n.a.*	*n.a.*
Ofloxacin	**xxx**	*n.a.*	*n.a.*	*n.a.*	*n.a.*
Norfloxacin	**x**	*n.a.*	*n.a.*	*n.a.*	*n.a.*
Sulfapyridine	**xx**	*n.a.*	*n.a.*	*n.a.*	*n.a.*
		ANTICONVULSANTS			
Carbamazepine	**x**	**x**	**x**	**x**	*n.a.*
Gabapentin	**xxx**	*n.a.*	*n.a.*	*n.a.*	*n.a.*
Valproic acid	**xxx**	*n.a.*	*n.a.*	*n.a.*	*n.a.*
Phenytoin	**xx**	*n.a.*	*n.a.*	*n.a.*	*n.a.*
		ANTIINFLAMMATORIES			
Ibuprofen	**xxx**	**x**	**xxx**	**xxx**	**xxx**
Naproxen	**xx/xxx**	*n.a.*	**xxx**	**xx/xxx**	**xxx**
Diclofenac	**x/xx**	**x**	**x/xx**	**x**	**xxx**
Fenoprofen	**xxx**	*n.a.*	**xx/xxx**	*n.a.*	*n.a.*
Ketprofen	overall	*n.a.*	overall	*n.a.*	**xxx**
Indomethacin	**xx**	*n.a.*	**xx**	*n.a.*	**xxx**
		β-BLOCKERS			
Atenolol	**xx**	*n.a.*	*n.a.*	*n.a.*	*n.a.*
Metoprolol	**x**	*n.a.*	*n.a.*	*n.a.*	*n.a.*
Acebutolol	**xx**	*n.a.*	*n.a.*	*n.a.*	*n.a.*
Sotalol	**xx**	*n.a.*	*n.a.*	*n.a.*	*n.a.*
Propranolol	*n.a.*	*n.a.*	*n.a.*	*n.a.*	*n.a.*
		HORMONES			
17α-ethinylestradiol	overall	**x**	**xx/xxx**	**xxx**	*n.a.*
17β-estradiol	**xxx**	**xx**	**xxx**	**xxx**	**xxx**
Estrone	**xx/xxx**	**xx**	**xxx**	**xxx**	**xxx**
Estriol	**xx/xxx**	**xx**	**xxx**	*n.a.*	*n.a.*
16α-hydroxyestrone	overall	*n.a.*	*n.a.*	*n.a.*	*n.a.*
Testosterone	**xxx**	*n.a.*	*n.a.*	*n.a.*	*n.a.*
Androstenedione	**xxx**	*n.a.*	*n.a.*	*n.a.*	*n.a.*
Progesterone	**xxx**	*n.a.*	*n.a.*	*n.a.*	*n.a.*
		Tranquilizers			
Diazepam	**x**	*n.a.*	*n.a.*	*n.a.*	*n.a.*
		X-RAY CONTRAST MEDIA			
Diatrizoate	**x**	*n.a.*	*n.a.*	*n.a.*	*n.a.*
Iohexol	**x**	*n.a.*	*n.a.*	*n.a.*	*n.a.*
Iomeprol	**x**	*n.a.*	*n.a.*	*n.a.*	*n.a.*
Iopamidol	**x**	*n.a.*	*n.a.*	*n.a.*	*n.a.*
Iopromide	overall	*n.a.*	**xx**	**xx/xxx**	*n.a.*
Iothalamic acid	**x**	*n.a.*	*n.a.*	*n.a.*	*n.a.*
Ioxithalamic acid	**x**	*n.a.*	*n.a.*	*n.a.*	*n.a.*
		LIPID REGULATORS			
Bezafibrate	**xx/xxx**	*n.a.*	**xxx**	*n.a.*	*n.a.*
Gemfibrozil	**xx**	*n.a.*	**x/xxx**	*n.a.*	**xxx/x**
Clofibric acid	**x/xx**	*n.a.*	overall	*n.a.*	*n.a.*
Fenofibric acid	*n.a.*	*n.a.*	*n.a.*	*n.a.*	*n.a.*

x - very low or no removal; **xx** - average removal; **xxx** - good or very good removal; overall - removal range from 0 - 100%; n.a. - data not available.

REFERENCES

Andreozzi, R., Caprio, V., Marotta, R. and Radovnikovic, A. (2003a) Ozonation and H_2O_2/UV treatment of clofibric acid in water: a kinetic investigation. *Journal of Hazardous Materials* **103**(3), 233–246.

Auirol, M., Filali-Meknassi, Y., Tyagi, R.D., Adams, C.D. and Surampalli, R.Y. (2006) Endocrine disrupting compounds removal from wastewater, a new challenge. *Process Biochemistry* **41**(3), 525–539.

Badawy M.I., Wahaab R.A. and El-Kalliny A.S. (2009) Fenton-biological treatment processes for the removal of some pharmaceuticals from industrial wastewater. Journal of Hazardous Materials, 167, Issues 1–3, 567–574.

Batt, A.L., Kim, S. and Aga, D.S. (2007) Comparison of the occurrence of antibiotics in four full-scale wastewater treatment plants with varying designs and operations. *Chemosphere* **68**(3), 428–435.

Batt, A.L., Bruce, I.B. and Aga, D.S. (2006) Evaluating the vulnerability of surface waters to antibiotic contamination from varying wastewater treatment plant discharges. *Environmental Pollution* **142**(2), 295–302.

Bedner, M. and MacCrehan, W.A. (2006) Reactions of the amine-containing drugs fluoxetine and metoprolol during chlorination and dechlorination processes used in wastewater treatment. *Chemosphere* **65**(11), 2130–2137.

Belgiorno, V., Rizzo, L., Fatta, D., Della Rocca, C., Lofrano, G., Nikolaou, A., Naddeo, V. and Meric, S. (2007) Review on endocrine disrupting-emerging compounds in urban wastewater: occurrence and removal by photocatalysis and ultrasonic irradiation for wastewater reuse. *Desalination* **215**(1–3), 166–176.

Beltran, F.J., Aguinaco, A., García-Araya, J.F. and Oropesa, A. (2008) Ozone and photocatalytic processes to remove the antibiotic sulfamethoxazole from water. *Water Research* **42**(14), 3799–3808.

Bendz, D., Paxéus, N.A., Ginn, T.R. and Loge, F.J. (2005) Occurrence and fate of pharmaceutically active compounds in the environment, a case study: Höje River in Sweden. *Journal of Hazardous Materials* **122**(3), 195–204.

Benitez F.J., Acero J.L., Real F.J. and Roldán G. (2009) Ozonation of pharmaceutical compounds: Rate constants and elimination in various water matrices. Chemosphere **77**(1), 53–59.

Bernhard, M., Muller, J. and Knepper, T.P. (2006) Biodegradation of persistent polar pollutants in wastewater: Comparison of an optimised lab-scale membrane bioreactor and activated sludge treatment. *Water Research* **40**, 3419–3428.

BLAC (2003) Arzneimittel in der Umwelt – Auswertung der Untersuchungsergebnisse. (Pharmaceuticals in the environment – Analysis of the investigation results). Freie und Hansestadt Hamburg, Behorde fur Umwelt und Gesundheit im Auftrag des Bund/Lánderausschusses fur Chemikaliensicherheit (BLAC) (in German). Document download: http://blak-uis.server.de/servlet/is/2146/P-2c.pdf (24 April 2006).

Boyd, G.R., Zhang, S. and Grimm, D.A. (2005) Naproxen removal from water by chlorination and biofilm processes. *Water Research* **39**(4), 668–676.

Cajthaml, T., Křesinová, Z., Svobodová, K., Sigler, K. and Řezanka, T. (2009) Microbial transformation of synthetic estrogen 17α-ethinylestradiol. *Environmental Pollution* **157**(12), 3325–3335.

Carballa, M., Omil, F., Lema, J.M., Llompart, M., Garcia-Jares, C., Rodriguez, I., Gomez, M. and Ternes, T. (2004) Behavior of pharmaceuticals, cosmetics and hormones in a sewage treatment plant. *Water Research* **38**, 2918–2926.

Carballa, M., Omil, F. and Lema, J.M. (2005) Removal of cosmetic ingredients and pharmaceuticals in sewage primary treatment. *Water Research* **39**(19), 4790–4796.

Carballa, M., Fink, G., Omil, F., Lema, J.M. and Ternes, T. (2008) Determination of the solid–water distribution coefficient (K_d) for pharmaceuticals, estrogens and musk fragrances digested sludge. *Water Research* **41**, 287–295.

Carballa, M., Omil, F. and Lema, J.M. (2007a) Calculation methods to perform mass balances of micropollutants in sewage treatment plants, application to pharmaceutical and personal care products (PPCPs). *Environ. Sci. Technol* **41**, 884–890.

Carballa, M., Omil, F., Ternes, T. and Lema, J.M. (2007b) Fate of pharmaceutical and personal care products (PPCPs) during anaerobic digestion of sewage sludge. *Water Research* **41**, 2139–2150.

Chenxi, W., Spongberg, A.L. and Witter, J.D. (2008) Determination of the persistence of pharmaceuticals in biosolids using liquid-chromatography tandem mass spectrometry. *Chemosphere* **73**(4), 511–518.

Clara, M., Strenn, B. and Kreuzinger, N. (2004) Carbamazepine as a possible anthropogenic marker in the aquatic environment: investigations on the behaviour of Carbamazepine in wastewater treatment and during groundwater infiltration. *Water Research* **38**(4), 947–954.

Clara, M., Kreuzinger, N., Strenn, B., Gans, O. and Kroiss, H. (2005a) The solids retention time – a suitable design parameter to evaluate the capacity of wastewater treatment plants to remove micropollutants. *Water Research* **39**(1), 97–106.

Clara, M., Strenn, B., Gans, O., Martinez, E., Kreuzinger, N. and Kroiss, H. (2005b) Removal of selected pharmaceuticals, fragrances and endocrine disrupting compounds in a membrane assisted bioreactor and conventional wastewater treatment plants. *Water Research* **39**, 4797–4807.

Dantas, R.F., Canterino, M., Marotta, R., Sans, C., Esplugas, S. and Andreozzi, R. (2007) Bezafibrate removal by means of ozonation: Primary intermediates, kinetics, and toxicity assessment. *Water Research* **41**(12), 2525–2532.

Dantas, R.F., Contreras, S., Sans, C. and Esplugas, S. (2008) Sulfamethoxazole abatement by means of ozonation. *Journal of Hazardous Materials* **150**, 790–794.

De Wever, H., Weiss, S., Reemtsma, T., Vereecken, J., Müller, J., Knepper, T., Rörden, O., Gonzalez, S., Barcelo, D. and Hernando, M.D. (2007) Comparison of sulfonated and other micropollutants removal in membrane bioreactor and conventional wastewater treatment. *Water Research* **41**(4), 935–945.

De Witte, B., Van Langenhove, H., Hemelsoet, K., Demeestere, K., De Wispelaere, P., Van Speybroeck, V. and Dewulf, J. (2009) Levofloxacin ozonation in water: Rate determining process parameters and reaction pathway elucidation. *Chemosphere* **76**(5), 683–689.

Dodd, M.C. and Huang, C.H. (2007) Aqueous chlorination of the antibacterial agent trimethoprim: Reaction kinetics and pathways. *Water Research* **41**(3), 647–655.

Doll, T.E., Frimmel, F.H. (2003) Fate of pharmaceuticals – photodegradation by simulated solar UV – light. *Chemosphere* **52**, 1757–1769.

Doll, T.E. and Frimmel, F.H. (2005) Removal of selected persistent organic pollutants by heterogenous photocatalysis in water. *Catalysis Today* **101**, 195–202.

Drillia, P., Dokianakis, S.N., Fountoulakis, M.S., Kornaros, M., Stamatelatou, K. and Lyberatos, G. (2005) On the occasional biodegradation of pharmaceuticals in the activated sludge process: The example of the antibiotic sulfamethoxazole. *Journal of Hazardous Materials* **122** (3), 259–265.

Drillia, P., Stamatelatou, K. and Lyberatos, G. (2005b) Fate and mobility of pharmaceuticals in solid matrices. *Chemosphere* **60**, 1034–1044.

Esperanza, M., Suidan, M.T., Vega, R.M., Gonzalez, C., Sorial, G.A., McCauley, P. and Brenner, R. (2007) Fate of sex hormones in two pilot-scale municipal wastewater treatment plants: Conventional treatment. *Chemosphere* **66**, 1535–1544.

Göbel, A., McArdell, C.S., Joss, A., Siegrist, H. and Giger, W. (2007) Fate of sulfonamides, macrolides, and trimethoprim in different wastewater treatment technologies. *Science of the Total Environment* **372** (2–3), 361–371.

Göbel, A., Thomsen, A., McArdell, C.S., Joss, A. and Giger, W. (2005) Occurrence and sorption behaviour of sulfonamides, macrolides, and trimethoprim in activated sludge treatment. *Environmental Science Technology* **39**, 3981–3989.

Golet, E.M., Xifra, I., Siegrist, H., Alder, A.C. and Giger, W. (2003) Environmental exposure assessment of fluoroquinolone antibacterial agents from sewage to soil. *Environ. Sci. Technol.* **37**, 3243–3249.

González, S., Müller, J., Petrovic, M., Barceló, D. and Knepper, T.P. (2006) Biodegradation studies of selected priority acidic pesticides and diclofenac in different bioreactors. *Environmental Pollution* **144**(3), 926–932.

Ghosh, G.C., Okuda, T., Yamashita, N. and Tanaka, H. (2009) Occurrence and elimination of antibiotics at four sewage treatment plants in Japan and their effects on bacterial ammonia oxidation. *Water Sci. Technol* **59** (4), 779–786.

Haiß, A. and Kummerer, K. (2006) Biodegradability of the X-ray contrast compound diatrizoic acid, identification of aerobic degradation products and effects against sewage sludge micro-organisms. *Chemosphere* **62**, 294–302.

Huber, M.M., Korhonen, S., Ternes, T.A. and von Gunten, U. (2005) Oxidation of pharmaceuticals during water treatment with chlorine dioxide. *Water Research* **39**(15), 3607–3617.

Huber, M.M., Canonica, S., Park, G.-Y. and von Gunten, U. (2003) Oxidation of pharmaceuticals during ozonation and advanced oxidation processes. *Environ. Sci. Technol* **37**, 1016–1024.

Johnson, A.C., Aernib, H.R., Gerritsenc, A., Gibertd, M., Gigerb, W., Hyllande, K., Jurgensa, M., Nakarif, T., Pickeringg, A., Suterb, M.J.F., Svensonh, A. and Wettstein, F.E. (2005) Comparing steroid estrogen, and nonylphenol content across a range of European sewage plants with different treatment and management practices. *Water Research* **39**, 47–58.

Jones, O.A.H., Voulvoulis, N. and Lester, J.N. (2007) The occurrence and removal of selected pharmaceutical compounds in a sewage treatment works utilising activated sludge treatment. *Environmental Pollution* **145**(3), 738–744.

Joss, A., Andersen, H., Ternes, T., Richle, P.R. and Siegrist, H. (2004) Removal of Estrogens in Municipal Wastewater Treatment under Aerobic and Anaerobic Conditions: Consequences for Plant Optimization. *Environmental Science and Technology* **38**(11), 3047–3055.

Joss, A., Keller, E., Alder, A.C., Göbel, A., McArdell, C.S., Ternes, T. and Siegrist, H. (2005) Removal of pharmaceuticals and fragrances in biological wastewater treatment. *Water Research* **39**, 3139–3152.

Joss, A., Zabczynski, S., Göbel, A., Hoffmann, B., Löffler, D., McArdell, C.S., Ternes, T.A., Thomsen, A. and Siegrist, H. (2006) Biological degradation of pharmaceuticals in municipal wastewater treatment: Proposing a classification scheme. *Water Research* **40**(8), 1686–1696.

Kim, S., Eichhorn, P., Jensen, J.N., Weber, A.S. and Aga, D.S., (2005) Removal of antibiotics in wastewater: effect of hydraulic and solid retention times on the fate of tetracycline in the activated sludge process. *Environ. Sci. Technol.* 39, 5816–5823.

Kim, S.D., Cho, J., Kim, I.S., Vanderford, B.J. and Snyder, S.A. (2007) Occurrence and removal of pharmaceuticals and endocrine disruptors in South Korean surface, drinking, and waste waters. *Water Research* **41**(5), 1013–1021.

Kim, Y., Pipes, W.O. and Chung, P.-G. (1998) Control of activated sludge bulking by operating clarifiers in a series. *Water Science and Technology* **38**(8–9), 1–8.

Kimura, K., Hara, H. and Watanabe, Y. (2005) Removal of pharmaceutical compounds by submerged membrane bioreactors (MBRs). *Desalination* **178**(1–3), 135–140.

Kimura, K., Hara, H. and Watanabe, Y. (2007) Elimination of Selected Acidic Pharmaceuticals from Municipal Wastewater by an Activated Sludge System and Membrane Bioreactors. *Environ. Sci. Technol.* **41**, 3708–3714.

Klavarioti M., Mantzavinos D., Kassinos D. (2009) Removal of residual pharmaceuticals from aqueous systems by advanced oxidation processes. *Environment International* **35**(2), 402–417.

Kosjek, T., Heath, E. and Kompare, B. (2007) Removal of pharmaceutical residues in a pilot wastewater treatment plant. *Analytical and Bioanalytical Chemistry* **387**(4), 1379–1387.

Kosutic, K., Dolar, D., Asperger, D. and Kunst, B. (2007) Removal of antibiotics from a model wastewater by RO/NF membranes. *Separation and Purification Technology* **53**(3), 244–249.

Kreuzinger, N., Clara, M., Strenn, B. and Kroiss, H. (2004) Relevance of the sludge retention time (SRT) as design criteria for wastewater treatment plants for the removal of endocrine disruptors and pharmaceuticals from wastewater. *Water Science and Technology* **50**(5), 149–156.

Li, K., Yediler, A., Yang, M., Schulte-Hostede, S. and Wong, M.H. (2008) Ozonation of oxytetracycline and toxicological assessment of its oxidation by-products. *Chemosphere* **72**(3), 473–478.

Lin, A.Y.-C., Lin, C.-F., Chiou, J.-M. and Hong, P.K.A. (2009) O_3 and O_3/H_2O_2 treatment of sulfonamide and macrolide antibiotics in wastewater. *Journal of Hazardous Materials* **171**, 452–458.

Lin, A.Y.-C., Yu, T.-H. and Lateef, S.K. (2009a) Removal of pharmaceuticals in secondary wastewater treatment processes in Taiwan. *J. Hazard. Mater* **167**(1–3), 1163–1169.

Lin, Y., Peng, Z. and Zhang, X. (2009) Ozonation of estrone, estradiol, diethylstilbestrol in waters. *Desalination*, In Press, Corrected Proof, Available online.

Lindqvist, N., Tuhkanen, T. and Kronberg, L. (2005) Occurrence of acidic pharmaceuticals in raw and treated sewages and in receiving waters. *Water Research* **39**(11), 2219–2228.

Lishman, L., Smyth, S.A., Sarafin, K., Kleywegt, S., Toito, J., Peart, T., Lee, B., Servos, M., Beland, M. and Seto, P. (2006) Occurrence and reductions of pharmaceuticals and personal care products and estrogens by municipal wastewater treatment plants in Ontario, Canada. *Science of the Total Environment* **367**(2–3), 544–558.

Maniero, M.G., Bila, D.M. and Dezotti, M. (2008) Degradation and estrogenic activity removal of 17β-estradiol and 17α-ethinylestradiol by ozonation and O_3/H_2O_2. *Science of The Total Environment* **407**(1), 105–115.

Maurer, M., Escher, B.I., Richle, P., Schaffner, C. and Alder, A.C. (2007) Elimination of b-blockers in sewage treatment plants. *Water Research* **41**(7), 1614–1622.

Miao, X.S., Yang, J.J. and Metcalfe, C.D. (2005) Carbamazepine and its metabolites in wastewater and in biosolids in a municipal wastewater treatment plant. *Environmental Science and Technology* **39**(19), 7469–7475.

Miège, C., Favier, M., Brosse, C., Canler, J.P. and Coquery, M. (2006) Occurrence of betablockers in effluents of wastewater treatment plants from the Lyon area (France) and risk assessment for the downstream rivers. *Talanta* **70**(4), 739–744.

Molinari, R., Pirillo, F., Loddo, V. and Palmisano, L. (2006) Heterogeneous photocatalytic degradation of pharmaceuticals in water by using polycrystalline TiO_2 and a nanofiltration membrane reactor. *Catalysis Today* **118**(1–2), 205–213.

Nakada, N., Tanishima, T., Shinohara, H., Kiri, K. and Takada, H. (2006) Pharmaceutical chemicals and endocrine disrupters in municipal wastewater in Tokyo and their removal during activated sludge treatment. *Water Research* **40**(17), 3297–3303.

Oppel, J., Broll, G. Löffler, D., Meller, M., Rombke, J. and Thernes, T.A. (2004) Leaching behaviour of pharmaceuticals in soil-testing-systems: a part of an environmental risk assessment for groundwater protection. *Science of the Total Environment* **328**, 265–273.

Packer, J., Werner, J., Latch, D., McNeill, K. and Arnold, W. (2003) Photochemical fate of pharmaceuticals in the environment: Naproxen, diclofenac, clofibric acid and ibuprofen. *Aquatic Sciences* **65**, 342–351.

Paxeus, N. (2004) Removal of selected non-steroidal anti-inflammatory drugs (NSAIDs), gemfibrozil, carbamazepine, beta-blockers, trimethoprim and triclosan in conventional wastewater treatment plants in five EU countries and their discharge to the aquatic environment. *Water Science and Technology* **50**(5), 253–260.

Radjenovic, J., Petrovic, M. and Barceló, D. (2007) Analysis of pharmaceuticals in wastewater and removal using a membrane bioreactor. *Anal Bioanal Chem* **387**, 1365–1377.

Radjenovic, J., Petrovic, M. and Barceló, D. (2009) Fate and distribution of pharmaceuticals in wastewater and sewage sludge of the conventional activated sludge (CAS) and advanced membrane bioreactor (MBR) treatment. *Water Research* **43**(3), 831–841.

Ravina, M., Campanella, L. and Kiwi, J. (2002) Accelerated mineralization of the drug diclofenac via Fenton reactions in a concentric photo – reactor. *Water Research* **36**, 3553–3560.

Reyes, C., Fernández, J., Freer, J., Mondaca, M.A., Zaror, C., Malato, S. and Mansilla, H.D. (2006) Degradation and inactivation of tetracycline by TiO_2 photocatalysis. *Journal of Photochemistry and Photobiology A: Chemistry* **184**, 141–146.

Reif R., Suárez., Omil F. and Lema J.M. (2008) Fate of pharmaceuticals and cosmetic ingredients during the operation of a MBR treating sewage. *Desalination* **221**(1–3), 511–517.

Rosal, R., Rodríguez, A., Gonzalo, M.S. and García-Calvo, E. (2008) Catalytic ozonation of naproxen and carbamazepine on titanium dioxide. *Applied Catalysis B: Environmental* **84**(1–2), 48–57.

Rosal, R., Rodríguez, A., Perdigón-Melón, J.A., Petre A., García-Calvo, E., Gómez, M.J., Agüera, A. and Fernández-Alba, A.R. (2010) Occurrence of emerging pollutants in urban wastewater and their removal through biological treatment followed by ozonation. *Water Research* **44**(2), 578–588.

Scheytt, T., Mersmann, P., Lindstadt, R. and Heberer, T. (2005) Determination of sorption coefficients of pharmaceutically active substances carbamazepine, diclofenac, and ibuprofen, in sandy sediments. *Chemosphere* **60**, 245–253.

Schlüsener, M. (2005) Fate, pathways and methods for the determination of selected antibiotics and steroid hormones in the environment. PhD Thesis, Universität Duisburg-Essen, Campus Essen, Institut für Umweltanalytik und Angewandte Geochemie.

Servos, M.R., Bennie, D.T., Burnison, B.K., Jurkovic, A., McInnis, R., Neheli, T., Schnell, A., Seto, P., Smyth, S.A., Ternes, T.A. (2005) Distribution of estrogens, 17β-estradiol and estrone, in Canadian municipal wastewater treatment plants. *Science of Total Environment* **336**, 155–170.

Snyder, S.A., Adham, S., Redding, A.M., Cannon, F.S., DeCarolis, J., Oppenheimer, J., *et. al.* (2007) Role of membranes and activated carbon in the removal of endocrine disruptors and pharmaceuticals. *Desalination* **202**(1–3), 156–181.

Seitz, W., Jiang, J.-Q., Schulz, W., Weber, W.H., Maier, D. and Maier, M. (2008) Formation of oxidation by-products of the iodinated X-ray contrast medium iomeprol during ozonation. *Chemosphere* **70**(7), 1238–1246.

Stuer-Lauridsen, F., Birkved, M., Hansen, L.P., Holten-Lutzhoft, H.C. and Halling-Sorensen, B. (2000) Environmental risk assessment of human pharmaceuticals in Denmark after normal therapeutic use. *Chemosphere* **40**, 783–793.

Stumpf, M., Ternes, T.A., Wilken, R.-D., Rodrigues, S.V. and Baumann, W. (1999) Polar drug residues in sewage and natural waters in the state of Rio de Janeiro, Brazil. *The Science of the Total Environment* **225**, 135–141.

Ternes, T.A. (1998) Occurrence of drugs in German sewage treatment plants and rivers. *Water Res.* **32**(11), 3245–3260.

Ternes, T. and Hirsch, R. (2000) Occurrence and behavior of X-ray contrast media in sewage facilities and the aquatic environment. *Environment Science and Technology* **34**, 2741–2748.

Ternes, T.A., Stüber, J., Herrmann, N., McDowell, D., Ried, A., Kampmannb, M. and Teiser, B. (2003) Ozonation: a tool for removal of pharmaceuticals, contrast media and musk fragrances from wastewater? *Water Research* **37**(8), 1976–1982.

Ternes, T.A., Herrmann, N., Bonerz, M., Knacker, T., Siegrist, H. and Joss, A. (2004) A rapid method to measure the solid-water distribution coefficient (K_d) for pharmaceuticals and musk fragrances in sewage sludge. *Water Research* **38**, 4075–4084.

Ternes T. and Joss A. (2006) Human Pharmaceuticals, Hormones and Fragrances - The Challenge of Micropollutants in Urban Water Management Water Intelligence Online © IWA Publishing.

Ternes, T.A., Bonerz, M., Herrmann, N., Teiser, B. and Andersen, H.R. (2007) Irrigation of treated wastewater in Braunschweig, Germany: An option to remove pharmaceuticals and musk fragrances. *Chemosphere* **66**, 894–904.

Trovó, A.G., Melo, S.A.S. and Nogueira, R.F.P. (2008) Photodegradation of the pharmaceuticals amoxicillin, bezafibrate and paracetamol by the photo-Fenton process – Application to sewage treatment plant effluent. *Journal of Photochemistry and Photobiology A: Chemistry* **198**(2–3), 215–220.

Urase, T., Kagawa, C. and Kikuta, T. (2005) Factors affecting removal of pharmaceutical substances and estrogens in membrane separation bioreactors. *Desalination* **178**(1–3), 107–113.

Urase, T. and Kikuta, T. (2005) Separate estimation of adsorption and degradation of pharmaceutical substances and estrogens in the activated sludge process. *Water Research* **39**(7), 1289–1300.

Vieno, N., Tuhkanen, T. and Kronberg, L. (2007) Elimination of pharmaceuticals in sewage treatment plants in Finland. *Water Research* **41**(5), 1001–1012.

Vogna, D., Marotta, R., Napolitano, A., Andreozzi, R. and d'Ischia, M. (2004) Advanced oxidation of the pharmaceutical drug diclofenac with UV/H_2O_2 and ozone. *Water Research* **38**(2), 414–422.

Yoon, Y., Westerhoff, P., Snyder, S.A. and Wert, E.C. (2006) Nanofiltration and ultrafiltration of endocrine disrupting compounds, pharmaceuticals and personal care products. *Journal of Membrane Science* **270**, 88–100.

Yoon, Y., Westerhoff, P., Snyder, S.A., Wert, E.C. and Yoon, J. (2007) Removal of endocrine disrupting compounds and pharmaceuticals by nanofiltration and ultrafiltration membranes. *Desalination* **202**(1–3), 16–23.

Yu, J.T., Bouwer, E.J. and Coelhan, M. (2006) Occurrence and biodegradability studies of selected pharmaceuticals and personal care products in sewage effluent. *Agricultural Water Management* **86**(1–2), 72–80.

Yu, T.-H., Lin, A.Y.-C., Lateef, S.K., Lin, C.-F. and Yang, P.-Y. (2009) Removal of antibiotics and non-steroidal anti-inflammatory drugs by extended sludge age biological process. *Chemosphere* **77**(2), 175–181.

Yuan, F., Hu, C., Hu, X., Qu, J. and Yang, M. (2009) Degradation of selected pharmaceuticals in aqueous solution with UV and UV/H_2O_2. *Water Research* **43**(6), 1766–1774.

Zhang, Y. and Zhou, J.L. (2005) Removal of estrone and 17β-estradiol from water by adsorption. *Water Research* **39**(16), 3991–4003.

Zhang, Y., Geißen, S.-U., Gal C. (2008) Carbamazepine and diclofenac: Removal in wastewater treatment plants and occurrence in water bodies. *Chemosphere* **73**(8), 1151–1161.

Zwiener, C. and Frimmel, F.H. (2000) Oxidative treatment of pharmaceuticals in water. *Water Research* **34**(6), 1881–1885.

Zwiener, C., Seeger, S., Glauner, T. and Frimmel, F.H. (2002) Metabolites from the biodegradation of pharmaceutical residues of ibuprofen in biofilm reactors and batch experiments. *Anal. Bioanal. Chem* **372**, 569–575.

Chapter 6

Biological monitoring and endpoints

Alistair Boxall and Richard Greenwood

6.1 INTRODUCTION

Pharmaceuticals compounds are either designed to be highly active and interact with receptors in humans and animals or they are toxic towards health threatening organisms such as bacteria, fungi and parasites. Many lower animals have receptor systems similar to humans and animals moreover many of the groups of organisms that affect human and animal health and which are targeted by pharmaceuticals, such as bacteria and fungi, play a critical role in the functioning of ecosystems. It is therefore possible that pharmaceuticals may cause effects on aquatic and terrestrial organisms. For human medicines in particular, releases to the environment are likely to be almost continuous, and so organisms will be exposed for long durations. Because of this, over the past few years there has been a spectacular increase in the amount of research being performed into the potential effects and risks of pharmaceuticals in the natural environment. Over the past ten years or so, a wealth of information has been published in the scientific and grey literature on the effects of both human and veterinary medicines on a range of organisms and endpoints. This Chapter provides an overview of the published data and draws on a database that was developed during the KNAPPE project on the ecotoxicity and properties of pharmaceuticals. The parameters that went into the database included drug names and classes, species (and phylogeny), exposure duration and concentrations, endpoints measured [main ones being median effective concentration (EC50), median lethal concentration (LC50), no observable effect concentration (NOEC), lowest observable effect concentration (LOEC), lowest observable adverse effect concentration (LOAEL), no observable adverse effect concentration (NOAEL)), laboratory or field, eukaryotic versus prokaryotic, physical/chemical properties (CAS number, chemical formula, molecular weight (MW), octanol-water partition coefficient (Kow) (estimated and experimental), organic carbon-water partition coefficient (Koc), octanol-air partition coefficient (Koa), sodium adsorption ratio (SAR)], and exposure conditions (marine, freshwater, brackish).

6.2 DATA AVAILABILITY AND DISTRIBUTION

Over the past ten years a large body of data has been generated on the effects of pharmaceuticals on organisms within the environment. Around 20,000 study endpoints for both human and veterinary pharmaceuticals are available (see Table 6.1). The high number of datapoints is due to the fact that many pharmaceuticals also are commonly used in other sectors, e.g. as pesticides. The data include both acute and chronic endpoints and the top three groups of study organisms are fish, water flea and other small crustaceans. The number of studies performed per class of pharmaceutical compound is shown in the Table 6.2. The top three drug classes studied were cholinesterase inhibitors (used in veterinary medicine), lysozyme inhibitors and antiscabies agents (e.g., lindane). When all data are considered, the vast majority of studies indicate that in general pharmaceuticals are not highly toxic to aquatic organisms (Figure 6.1). A small proportion of compounds are, however, highly toxic to some organisms and endpoints.

©2010 IWA Publishing. *Pharmaceuticals in the Environment: Current knowledge and need assessment to reduce presence and impact*. Edited by Benoit Roig. ISBN: 9781843393146. Published by IWA Publishing, London, UK.

Table 6.1 Top ten most studied organism types covered in the database.

Test class	Number of studies
Fish	12137
Water Flea	2374
Small Crustacea	2140
Insect	1647
Algae	1403
Mollusc	1173
Protozoa	455
Amphibian	368
Worm	361
Large Crustacea	275
Rotifer	258

Table 6.2 Top ten most studied drug classes.

Drug class	Number of studies
Cholinesterase inhibitors	2943
Lysozyme inhibitor	2435
Antiscabies agent	2102
Surface-active agents	1201
Estrogens	1186
Anti-infective agents	1109
Pentaerythritol tetranitrate reductase inhibitor	752
Anti-menopausal agents	678
D-alanyl-D-alanine carboxypeptidase inhibitor	628
Alpha-amylase	575
Anti-bacterial agents	432

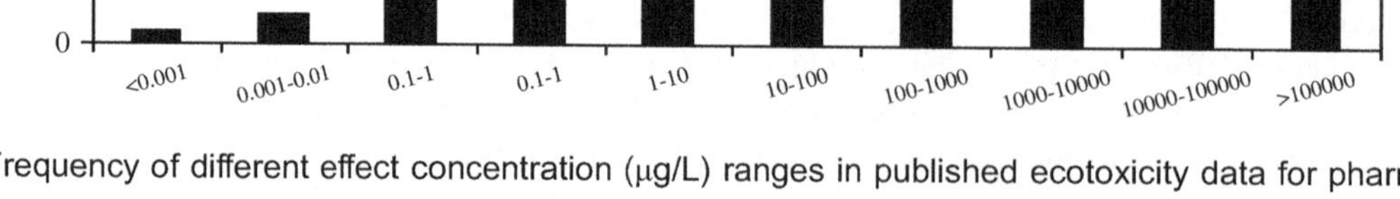

Figure 6.1 Frequency of different effect concentration (μg/L) ranges in published ecotoxicity data for pharmaceuticals.

6.3 TRADITIONAL ECOTOXICITY STUDIES

Traditional ecotoxicology studies were categorized into two main classes – acute or chronic. Acute studies were defined as those <72 h for algae, <96 h for Daphnia and <96 h for fish; exposures of longer duration were categorized as chronic for these taxa. Lastly, we must consider the endpoints that each study considers.

An overview of the OECD guidelines and associated endpoints is shown in Table 6.3. The end point for acute toxicity studies is always mortality; however, the end points for chronic studies are subject to variability and include various measurable signs of damage to the organism. The choice of end point on the outcome of EC/LC50 studies can yield different values due to variability in the sensitivity of each end point to the effect of the compound. This is highlighted in work undertaken by Gulden *et al.* (2005), who studied the impact of the choice of end points on the toxic response of fish cell lines to various pharmaceutical compounds. The authors noted that generally, the EC50 values for the inhibition of cell growth were lower than those for reduction of final cell protein, indicating that cell proliferation is a more sensitive end point for cytotoxic action of chemicals than cell survival.

Table 6.3 Overview of the OECD guidelines and associated end points.

Organisms	Type	Duration (d)	Endpoint	OECD Guideline
Fish	Acute	1, 2, 3 and 4	Mortality	203
Daphnia	Acute	1	Immobilisation	202
Daphnia	Chronic	14	Reproduction	202
Algae	Chronic	1, 2 and 3	Growth inhibition	201
Juvenile Fish	Chronic	14–28	Mortality	210
Daphnia	Chronic	21	Reproduction	211

(OECD-201, (1984); OECD-202, (1984); OECD-203, (1992); OECD-210, (1992); OECD-211, (1998)).

Analysis of the data bases reveals that acute toxicity bioassays account for a larger proportion of the studies than chronic. This slew of the data in favour of acute studies can probably be attributed to the shorter time and associated lower cost required to perform acute studies. Table 6.4 and Table 6.5 give an overview of the different organisms and species used in chronic and acute toxicity tests.

6.4 LIMITATIONS OF TRADITIONAL ECOTOXICITY STUDIES

6.4.1 Validity of chronic and acute data

Figure 6.2 summarizes the numbers of acute toxicity studies for algae (EC50), cladocerans (LC50) and fishes (LC50) within several concentration ranges. The data demonstrate that, in general, pharmaceuticals are not highly acutely toxic to aquatic organisms with the majority of compounds having an LC50 or EC50 value of 1000 μg/L or higher. There are however selected substances that have LC50 or EC50 values in the low μg/L range.

Figure 6.3 shows the number of longer term/chronic toxicity tests algae (EC50), *Daphnia* (LOEC) and fishes (LOEC). This figure shows that concentrations of API causing chronic effects in these three trophic groups are generally high although there are some instances where traditional chronic effects are observed in the sub $\mu g.L^{-1}$ range.

The validity of both acute and chronic studies as a means of assessing the toxicity of a compound is questionable due to the unrealistic nature of exposure to the toxicant. Firstly the exposure concentrations which are studied for both acute and chronic assays often differ by several orders of magnitude from those that would be encountered in the environment. Pharmaceutical concentrations detected in the field are generally below 1000 ng/L (cf Chapter 4). However, analysis of the literature reveals that most studies report EC50 and LC50 values at much higher concentrations. Fent (2006) reviewed the ecotoxicity of human pharmaceuticals and concluded that for most human medicines analyzed, acute effects to aquatic organisms are unlikely, except for spills. For investigated pharmaceuticals chronic LOECs in standard laboratory organisms are about two orders of magnitude higher than maximal concentrations in STP effluents. For diclofenac, the LOEC for fish toxicity was in the range of waste water concentrations, whereas the LOEC of propranolol and fluoxetine for zooplankton and benthic organisms were near to maximal measured STP effluent concentrations. In surface water, concentrations are lower and so, therefore, are the environmental risks (Fent, 2006).

In order to establish whether there is any evidence of risk of pharmaceuticals to aquatic systems, the traditional effects data were compared with occurrence data in surface water from the comprehensive review of Monteiro and Boxall (2009). Maximum Environmental Concentrations (MECs) were used in the analysis – these were the highest concentrations measured in any of the reviewed published monitoring studies. If the concentration was below the limit of quantification, the MEC was set equal to the limit of quantification.

MECs are compared with effects data in Figure 6.4 and Figure 6.5.

None of the acute standard endpoints was higher than the MEC for this active ingredient. However, using an assessment factor of 1000, which is typically used for acute endpoints if data for all three trophic levels are available, would lead to risk quotients greater than 1 for clarithromycin, 19-norethisterone, sulfamethoxazole, and acetaminophen.

Only 10 active ingredients had a long-term endpoint that was publicly available. MECs were higher than long-term standard tests for 17α-ethinylestradiol and 17α/β estradiol. Assuming that the TGD assessment factor (i.e. 10) for long-term studies could be applied, there would also be an indication of a risk to surface water organisms for ibuprofen and sulfamethoxazole.

Table 6.4 Biomonitoring of chronic effect of pharmaceutical products.

Taxonomic group	Species	Substance (therapeutic class)	References
Algae	*Chlorella sp.* (green)	Metronidazole (anti-protozoal)	Lanzky and Halling-Sørensen, 1997
	Cyclotella meneghiniana (diatom)	Oxofloxacin, Sulfamethoxazole (antibiotic), Carbamazepine (antiepileptic), Clofibric acid (antihyperlipoproteinemic), Diclofenac (antiinflammatory), Propranolol (ß-blocker)	Ferrari *et al.*, 2003, 2004
	Desmodesmus subspicatus (green)	Carbamazepine (antiepileptic), Clofibric acid (antihyperlipoproteinemic), Captopril (antihypertensive), Diclofenac, Ibuprofen, Naproxen (antiinflammatory), Propranolol, Metoprolol (ß-blocker)	Cleuvers, 2003
	Pseudokirchneriella subcapitata (green)	Oxofloxacin, Sulfamethoxazole (antibiotic); Carbamazepine (antiepileptic), Clofibric acid (antihyperlipoproteinemic), Diclofenac (antiinflammatory), Propranolol (ß-blocker), Bezafibrate, Fenofibrate, Gemfibrozil (lipid regulator), Ivermectin (parasiticide)	Ferrari *et al.*, 2003, 2004; Isidori *et al.*, 2007; Garric *et al.*, 2007
	Synechococcus leopolensis (cyanobacteria)	Oxofloxacin, Sulfamethoxazole (antibiotic), Carbamazepine (antiepileptic), Clofibric acid (antihyperlipoproteinemic), Diclofenac (antiinflammatory), Propranolol (ß-blocker)	Ferrari *et al.*, 2003; 2004
Fish	*Carassius carassius* (Crucian Carp)	Methyltestosterone (androgen)	Fujioka, 2002
	Danio rerio (zebra fish)	Carbamazepine (antiepileptic), Clofibric acid (antihyperlipoproteinemic), Diclofenac (antiinflammatory)	Ferrari *et al.*, 2003
	Oncorhynchus mykiss (rainbow trout)	Ethinylestradiol (oestrogen)	Purdom *et al.*, 1994
	Oryzias latipes (Japanese rice fish)	Methyltestosterone (androgen), Flutamide (antiandrogen), 17-β-estradiol, Diethylstilbestrol, Ethinylestradiol (oestrogen), Propranolol (ß-blocker), Ibuprofene (anti-inflammatory)	Huggett *et al.*, 2002; Flippin *et al.*, 2007
	Pimephales promela (fathead minnow)	Methyltestosterone (androgen), Ethinylestradiol (oestrogen)	Länge *et al.*, 2001; Zerulla *et al.*, 2002; Ankley *et al.*, 2001

(continued)

Table 6.4 *(continued)*

Taxonomic group	Species	Substance (therapeutic class)	References
Invertebrate	*Bithynia tentaculata (snail)*	Ethinylestradiol (oestrogen)	Belfroid and Leonards, 1996
	Brachionus calyciflorus (rotifer)	Oxofloxacin, Sulfamethoxazole (antibiotic), Carbamazepine (antiepileptic), Clofibric acid (antihyperlipoproteinemic), Diclofenac (antiinflammatory), Propranolol (ß-blocker), Bezafibrate, Fenofibrate, Gemfibrozil (lipid regulator)	Ferrari *et al.*, 2003, 2004; Isidori *et al.*, 2007
	Ceriodaphnia dubia (water flea)	Oxofloxacin, Sulfamethoxazole (antibiotic), Citalopram, Fluoxetine, Fluvoxamine, Paroxetine, Sertraline (antidepressant), Carbamazepine (antiepileptic), Clofibric acid (antihyperlipoproteinemic), Diclofenac (antiinflammatory), Propranolol (ßblocker), Bezafibrate, Fenofibrate, Gemfibrozil (lipid regulator)	Ferrari *et al.*, 2003, 2004; Henri *et al.*, 2004; Brooks *et al.*, 2003; Isidori *et al.*, 2007
	Chironomus riparius (larva)	Carbamazepine (antiepileptic), Clofibric acid (antihyperlipoproteinemic)	Nentwig *et al.*, 2004
	Daphnia magna (waterflea)	Clofibric acid (antihyperlipoproteinemic), Fluoxetine (antidepressant), Diethylstilbestrol, Ethinylestradiol, Faslodex (oestrogen), Iopromide (X-ray contrast medium), Ibuprofen (antiinflammatory), Bezafibrate, Fenofibrate, Gemfibrozil (lipid regulator), Ivermectin (parasiticide)	Köpf, 1995; Baldwin *et al.*, 1995; Schweinfurth *et al.*, 1996; Flaherty and Dodson, 2005; Heckmann *et al.*, 2007; Clubbs and Brooks, 2007; Garric *et al.*, 2007
	Hyalella azteca (amphipod)	Fluoxetine (antidepressant), Ethinylestradiol (oestrogen), Propranolol (ß-blocker), Acetaminophen, Diclofenac, Ibuprofen, Naproxen, Salicylic acid (antiinflammatory), Gemfibrozil (lipid regulator), Triclosan (antibacterial)	Brooks *et al.*, 2003; Vandenburg *et al.*, 2003; Huggett *et al.*, 2002; Borgmann *et al.*, 2007
	Hydra vulgaris (cnidarian)	Diazepam (anxiolytic); Amlodipine (Ca channel blocker), Digoxin (cardiotonic), Bendroflumethiazide, Furosemide (diuretic), Acetaminophen, Aspirin, Ibuprofen (antiinflammatory), Atenolol (ß-blocker)	Pascoe *et al.*, 2003
	Hydra attenuata (cnidarian)	Caffeine, ibuprofen, naproxen (antiinflammatory) oxytetracycline, novobiocin, trimethoprim, sulfamethoxazole, sulfapyridine (antibiotic), carbamazepine (anticonvulsant), gemfibrozil, bezafibrate (lipid regulator)	Quinn *et al.*, 2008
	Lumbriculus variegates (oligochaete worm)	Carbamazepine (antiepileptic), Clofibric acid (antihyperlipoproteinemic)	Nentwig *et al.*, 2004

(continued)

Table 6.4 (*continued*)

Taxonomic group	Species	Substance (therapeutic class)	References
Invertebrate	*Lymnaea stagnalis* (snail)	Methyltestosterone (androgen), Ethinylestradiol (oestrogen)	Czech *et al.*, 2001; Belfroid and Leonards, 1996
	Marisa cornuarietis (snail)	Methyltestosterone (androgen), Ethinylestradiol (oestrogen)	Schulte-Oehlmann *et al.*, 2004
	Nitocra spinepes (copepod)	Diethylstilbestrol, Ethinylestradiol, Estradiol (oestrogen)	Breitholtz and Bengtsson, 2001
	Planorbis carinatus (snail)	Ibuprofen (antiinflammatory)	Pounds *et al.*, 2004
	Thamnocephalus platyurus (shrimp)	Bezafibrate, Fenofibrate, Gemfibrozil (lipid regulator)	Isidori *et al.*, 2007
	Tisbe battagliai (copepod)	Diethylstilbestrol (oestrogen)	Hutchinson, 2002
	Vibrio fisheri (marine bacteria)	Bezafibrate, Fenofibrate, Gemfibrozil (lipid regulator)	Isidori *et al.*, 2007
	Lemna gibba (duckweed)	Trimethoprim, Neomycin, Streptomycin, Cephalexin, Ciprofloxacin, Levofloxacin, Lomefloxacin, Norfloxacin, Oxofloxacin, Erythromycin, Lincomycin, Roxithromycin, Tylosin, Amoxicillin, Sulfadimetoxine, Sulfamethazine, Sulfamethoxazole, Sulfochlorpyridazine, Chlortetracycline, Doxycycline, Oxytetracycline, Tetracycline (antibiotic), Fluoxetine (antidepressant), Metformin (Antidiabetic), Carbamazepine (antiepileptic), Atorvastatin (anti-hyperlipidemic), Caffeine (CNS stimulant), Acetaminophen, Ibuprofen (antiinflammatory)	Brain *et al.*, 2004
	Lemna minor (duckweed)	Carbamazepine (antiepileptic) ; Clofibric acid (antihyperlipoproteinemi), Captopril (antihypertensive), Diclofenac Ibuprofen, Naproxen (antiinflammatory), Metoprolol, Propranolol (ß-blocker)	Cleuvers, 2003
Algae	*Chlorella vulgaris* (green)	Gemfibrozil (lipid regulator), Fluoxetine, Fluvoxamine, Sertraline (SSRIs)	Zurita *et al.*, 2007; Johnson *et al.*, 2007
	Desmodesmus subspicatus (green algae)	Aspirin (antiinflammatory)	Cleuvers, 2003
	Microcystis aeruginosa (cyanobacteria)	Penicillin, Amoxicillin, Chlorotetracycline, Oxytetracycline, Streptomycin (antibiotic)	Holten Lutzhoft *et al.*, 1999; Halling-Sorensen *et al.*, 2000

Table 6.5 Biomonitoring of acute effects of pharmaceutical products.

Taxonomic group	Species	Substance (therapeutic class)	References
Algae	*Pseudokirchneriella subcapitata* (green)	Naproxen (antiinflammatory), Bezafibrate, Fenofibrate, Gemfibrozil (lipid regulator), Fluoxetine, Fluvoxamine, Sertraline (SSRIs)	Isidori *et al.*, 2005a; 2007; Johnson *et al.*, 2007
	Selenastrum capricornutum (cyanobacteria)	Chlorotetracycline, Metronidazole, Oxytetracycline (antibiotic)	Lanzky and Halling-Sørensen, 1997; Holten Lutzhoft *et al.*, 1999; Halling-Sorensen *et al.*, 2000
	Synechococcus leopolensis (cyanobacteria)	Carbamazepine (antiepileptic)	Ferrari *et al.*, 2003
	Tetraselmis chuii (green)	Oxytetracycline, Florfenicol (antibiotic)	Ferreira *et al.*, 2007
Fish	*Brachydanio rerio* embryos (zebrafish)	Clofibric acid (lipid regulator)	Henschel *et al.*, 1997
	Carassius auratus (goldfish)	Clofibric acid (lipid regulator)	Mimeault *et al.*, 2005
	Cyprinodon variegatua (sheephead minnow)	17β-ethinylestradiol (estrogen)	Zillioux *et al.*, 2001
	Danio rerio embryos (zebrafish)	Erythromycin, Oxytetracyclin, Sulfamethoxazole, Ofloxacin, Lincomycin, Clarithromycin (antibiotic), Clofibric acid (lipid regulator); Carbamazepine (anticonvulsant), Diclofenac (antiinflammatory), Propranolol (ß-blocker), Mianserin (antidepressant), Ethinylestradiol (oestrogen)	Isidori *et al.*, 2005b; Ferrari *et al.*, 2004; Fraysse *et al.*, 2006; van der Ven *et al.*, 2006; Segner *et al.*, 2003; Orn *et al.*, 2003
	Lepomis macrochirus (bluegill sunfish)	Ibuprofen (antiinflammatory)	Webb, 2004
	Oncorhynchus mykiss (rainbow trout)	Penicillin, Amoxicillin (antibiotic), Fluoxetine (antidepressant), Diclofenac (antiinflammatory), Carbamazepine (anticonvulsant), 17β-ethinylestradiol	Laville *et al.*, 2004; Jos *et al.*, 2003; Jobling *et al.*, 1996
	Oryzias latipes (Japanese rice fish)	Estadiol, Estriol (oestrogen), sulfamethoxazole, sulfachlorpyridazine, sulfathiazole, sulfamethazine, sulfadimethoxine, trimethoprim (antibiotic), acetaminophen (antiinflammatory), carbamazepine (antiepileptic), cimetidine (antihistaminic), diltiazem (Ca channel blocker)	Nimrod and Benson, 1998, Kang *et al.*, 2002; Metcalfe *et al.*, 2001; Balch *et al.*, 2004
	Pimephales promelas – eggs (fathead minnow)	Fluoxetine (SSRI); 17β-ethinylestradiol (estrogen)	Brooks *et al.*, 2003; Lange *et al.*, 2001; Jobling *et al.*, 2004
	Salvelinus namaycush (lake trout)	Erythromycin (antibiotic)	Marking *et al.*, 1988
	Pomatoschistus minitus (sand goby)	Ethinylestradiol (estrogen)	Robinson *et al.*, 2003

(*continued*)

Table 6.5 (*continued*)

Taxonomic group	Species	Substance (therapeutic class)	References
Invertebrate	*Artemia parthenogenica* (shrimp)	Oxytetracycline, Florfenicol (antibiotic)	Ferreira *et al.*, 2007
	Brachionus calyciflorus (rotifer)	Erythromycin, Oxytetracyclin, Sulfamethoxazole, Ofloxacin, Lincomycin, Clarithromycin (antibiotic), Carbamazepine (antiepileptic), Naproxen (antiinflammatory), Bezafibrate, Fenofibrate, Gemfibrozil (lipid regulator)	Ferrari *et al.*, 2004; Isidori *et al.*, 2005a, 2005b, 2007
	Ceriodaphnia dubia (water flea)	Erythromycin, Oxytetracyclin, Sulfamethoxazole, Ofloxacin, Lincomycin, Clarithromycin (antibiotic); Carbamazepine (antiepileptic), Naproxen (antiinflammatory); Fluoxetine (SSRI), Bezafibrate, Fenofibrate, Gemfibrozil (lipid regulator)	Ferrari *et al.*, 2004; Brooks *et al.*, 2003, Isidori *et al.*, 2005a; 2005b; 2007
	Daphnia magna (water flea)	Bacitracin, Ciprofloxacin, Clarithromycin, Erythromycin, Lincomycin, Metronidazole, Ofloxacin, Oxytetracycline, Sulfamethoxazole, Sulfachlorpyridazine, Sulfathiazole, Sulfamethazine, Sulfadimethoxine, Trimethoprim (antibiotic), Clofibrate, Bezafibrate, Fenofibrate, Gemfibrozil (lipid regulator); Carbamazepine (antiepileptic), Acetaminophen, Diclofenac (antiinflammatory), Fluoxetine (antidepressant), Cimetidine (antihistaminic), Diltiazem (Ca channelblocker)	Kuhn *et al.*, 1989; Brambilla *et al.*, 1994; Halling-Sorensen *et al.*, 2000; Wollenberger *et al.*, 2000; Ferrari *et al.*, 2003; Cleuvers, 2004; Flaherty and Dodson, 2005; Kim *et al.*, 2007; Isidori *et al.*, 2005b; 2007; Zurita *et al.*, 2007
	Hydra attenuata (cnidarian)	caffeine, ibuprofen, naproxen (antiinfl) oxytetracycline, novobiocin, trimethoprim, sulfamethoxazole, sulfapyridine (antibiotic), carbamazepine (anticonvulsant), gemfibrozil, bezafibrate (lipid regulator)	Quinn *et al.*, 2008
	Elliptio complanata (freshwater mussel)	Oxytetracycline (antibiotic), Clofibric acid (lipid regulator), Carbamazepine (antiepileptic)	Gagne *et al.*, 2006
	Gammarus pulex (crustacean)	Fluoxetine (SSRI), Ibuprofen (antiinflammatory)	De Lange *et al.*, 2006
	Marisa cornuarietis (prosobranch snail)	17β-ethinylestradiol (oestrogen)	Schulte-Oehlmann *et al.*, 2004
	Sphaerium striatinum (fingernail clam)	Fluoxetine (antidepressant)	Fong, 1998
	Spirostomum ambiguum (protozoan)	Fluoxetine (antidepressant)	Nalecz-Jawecki, 2007
	Streptocephalus probiscideus (fairy shrimp)	Acetaminophen (antiinflammatory)	Kuhn *et al.*, 1989

(*continued*)

Table 6.5 (*continued*)

Taxonomic group	Species	Substance (therapeutic class)	References
Invertebrate	*Thamnocephalus platyurus* (crustacean)	Erythromycin, Oxytetracyclin, Sulfamethoxazole, Ofloxacin, Lincomycin, Clarithromycin (antibiotic), Fluoxetine (antidepressant), Naproxen (antiinflammatory), Bezafibrate, Fenofibrate, Gemfibrozil (lipid regulator)	Isidori *et al.*, 2005a; 2007; Nalecz-Jawecki, 2007
	Vibrio fischeri (marine bacterium)	Clarithromycin, Erythromycin, Oxytetracyclin, Ofloxacin, Lincomycin, sulfamethoxazole, sulfachlorpyridazine, sulfathiazole, sulfamethazine, sulfadimethoxine, trimethoprim (antibiotic) ; acetaminophen (antiinflammatory), carbamazepine (antiepileptic), cimetidine (antihistaminic), diltiazem (Ca channel blocker), Bezafibrate, Fenofibrate, Gemfibrozil (lipid regulator)	Kim *et al.*, 2007; Isidori *et al.*, 2005b; 2007
Plant	*Lemna gibba* (duckweed)	Penicillin, Amoxicillin, Ciprofloxacin, Erythromycin (antibiotic)	Brain *et al.*, 2008

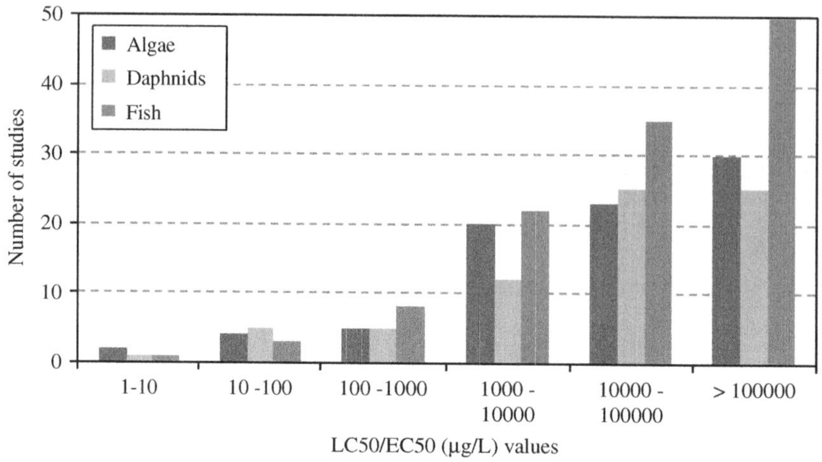

Figure 6.2 Distribution of traditional acute LC50 and EC50 values for pharmaceuticals to fish, daphnids and algae.

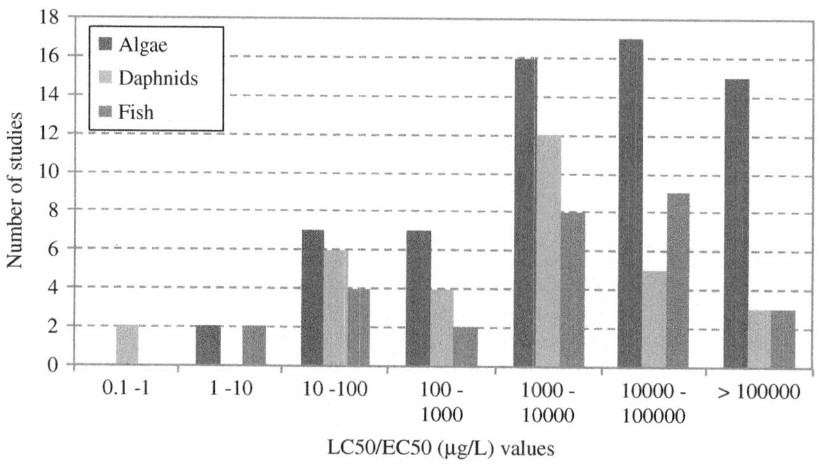

Figure 6.3 Distribution of traditional chronic LC50/EC50 values for pharmaceuticals to fish, daphnids and algae.

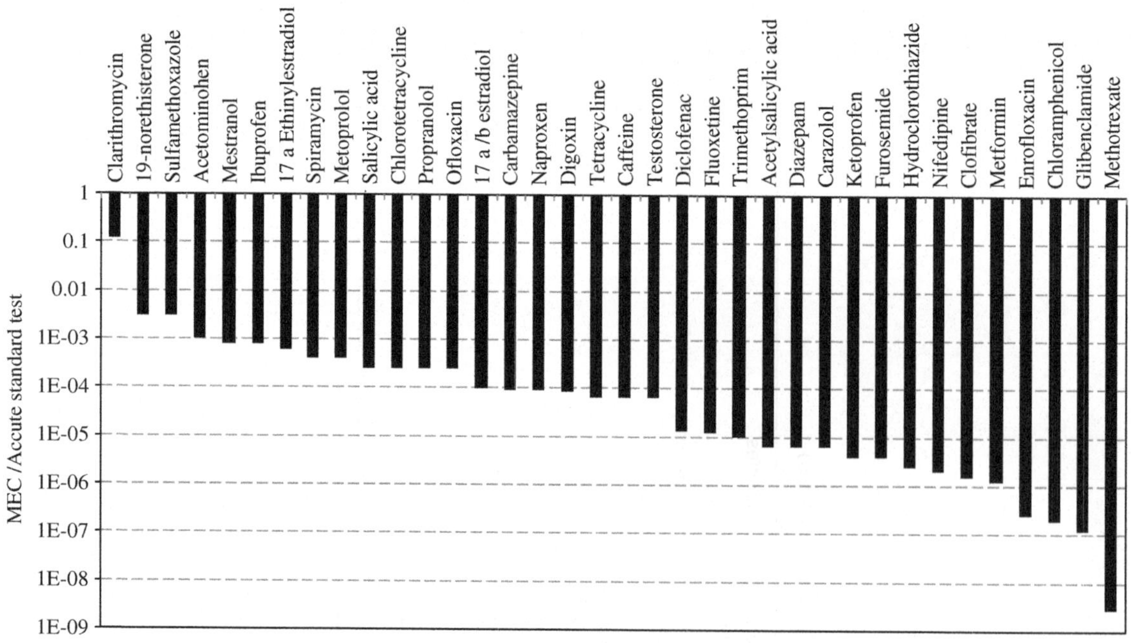

Figure 6.4 MEC:Traditional acute EC50/LC50 values for a range of pharmaceuticals.

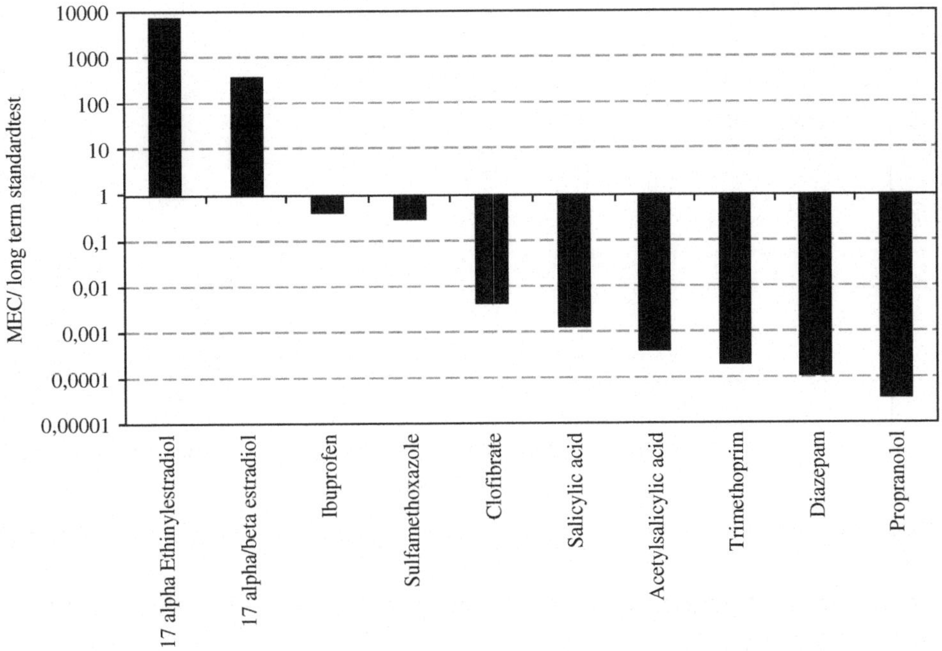

Figure 6.5 MEC:Traditional chronic EC50/LC50 values for a range of pharmaceuticals.

6.4.2 Effects of mixtures

Other differences between laboratory and field exist that add further doubt to the validity of acute and chronic studies. Laboratory-based studies generally study single substance toxicity as opposed to the mixture of pharmaceuticals an organism would be exposed to in the natural environment. Instead, the typical exposure is towards mixtures pharmaceuticals, industrial pollutants, personal care products and pesticides. Even if only the pharmaceuticals are considered, they usually do not occur as individual, isolated compounds in the environment. For example, analysis of estuarine waters around the UK has revealed measurable concentrations of pharmaceutical mixtures, including mefenamic acid, diclofenac and propranolol at concentrations of between 16 and 195 ng/L (Thomas and Hilton, 2004). Heberer (2002) reviewed the occurrence and fate of pharmaceutical products and noted that in some studies more than 80 pharmaceutical compounds were detected. Fent (2006) noted that there are only a few studies dealing with the effects of mixtures of pharmaceuticals. However, Cleuvers (2004) reported the toxicity of a mixture of the antiinflammatory drugs diclofenac, ibuprofen, naproxen, and acetylsalicylic acid on *Daphnia sp*. The authors observed that toxicity of the mixture was considerable, even at concentrations at which the single substances showed no, or only very slight effects. Cleuvers (2005) reported that the toxicity of a mixture of three β-blockers to *Daphnia sp.* could be predicted accurately using the concept of concentration addition (CA). This concept implies that the toxic unit of a mixture – the ratio between the total mixture concentration and a pre-defined effect concentration of the mixture, which is most often its EC50 – equals the total sum of the toxic units of its components:

$$\sum_{i=1}^{n} \frac{c_i}{EC50_i} = \frac{c_{Mix}}{EC50_{Mix}} \tag{6.1}$$

where c_i gives the concentration (or dose) of the *i*th component in a n-compound mixture and EC50 *i* denotes the concentration of that substance which provokes 50% effect if applied singly; cMix is the total concentration of the mixture and EC50Mix gives the mixture concentration that provokes 50% total effect.

It should be noted, that CA implicitly assumes that all EC50 values have been recorded in the same assay. The validity of CA has been demonstrated in a range of different assays and pharmaceuticals, e.g. for antibiotics, estrogens, anti-inflammatory drugs, ß-blockers or selective serotonin re-uptake inhibitors (SSRIs), a group of commonly used antidepressants.

Figure 6.6 illustrate the application of the method to SSRIs. Data were obtained on acute ecotoxicity to algae, fish and daphnids and predicted concentrations were obtained for Europe and the US and measured concentrations were obtained for Canada. The analysis indicated that the co-occurrence of SSRIs is unlikely to cause acute effects in aquatic systems.

Due to a lack of data, it was not possible to apply the approach to chronic endpoints. This showed that in the future consideration should also be given to the effects of mixtures of different chemical classes (i.e. mixtures of different pharmaceutical classes as well as mixtures of pharmaceuticals with other types of chemical (e.g. pesticides).

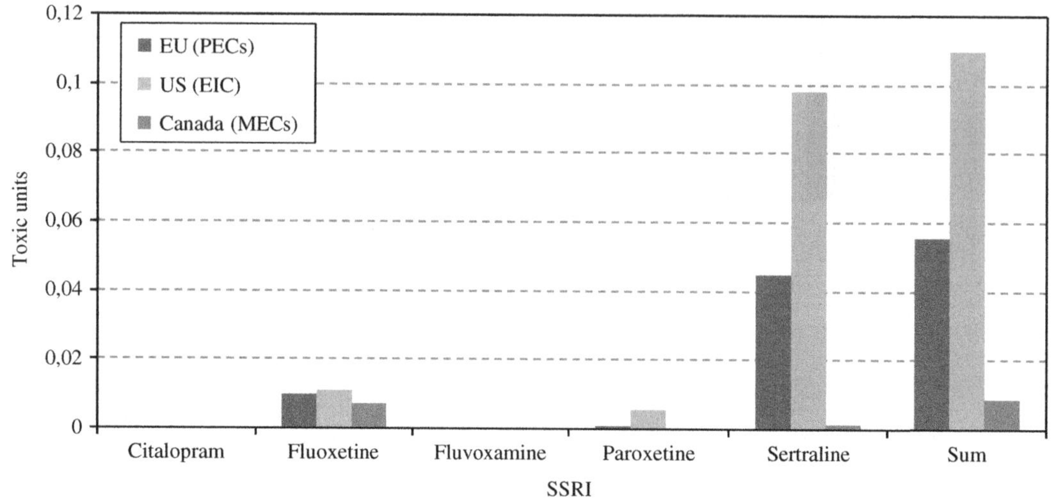

Figure 6.6 Toxic units for individual SSRIs and SSRI mixtures in the EU, US and Canada.

6.4.3 Can the effects of pharmaceuticals be predicted using qsars or acute/chronic ratios?

For other classes of chemicals, extrapolation approaches (e.g. use of acute/chronic ratios) and QSARs have been proposed for estimating ecotoxicity values. In order to test whether these approaches might be suitable for use on pharmaceuticals, the workshop evaluated a range of different approaches for ecotoxicity estimation. Example outputs from these analyses are shown in Figure 6.7 and Figure 6.8. It can be concluded that currently available QSAR approaches and the application of typical acute: chronic ratios for estimating chronic toxicity from acute toxicity are probably inappropriate for use on pharmaceuticals. Further development work is therefore required in this area.

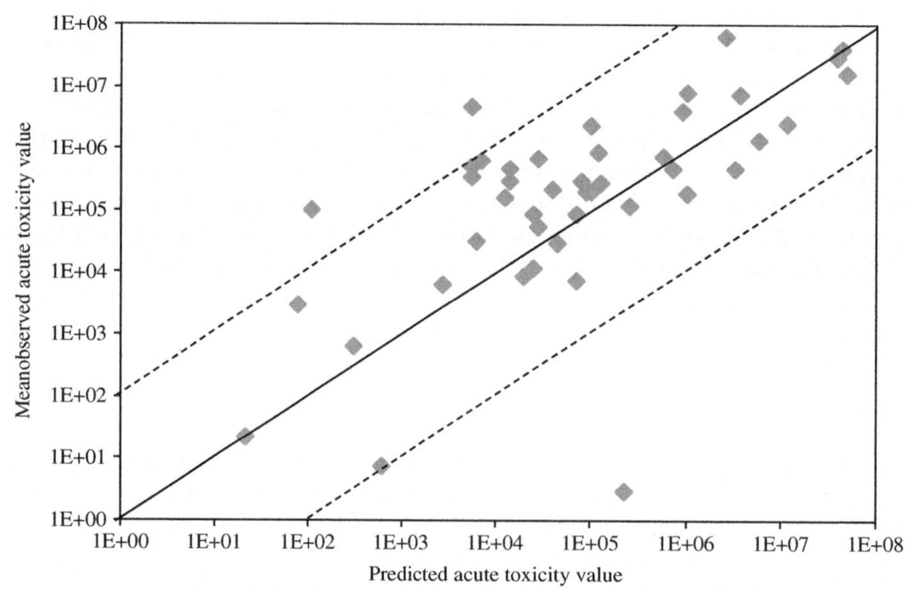

Figure 6.7 QSAR Predicted versus mean observed acute toxicity values for *Pimephales promelas*.

6.4.4 Do metabolites pose a risk?

Following administration, a pharmaceutical may be metabolised and consequently, a mixture of the metabolite(s) and the parent compound may be released to the environment. Once in the environment, both the metabolites and the parent compound may be further degraded by abiotic and biotic processes. The fate and effects of all the transformation products (i.e. metabolites and degradation products) may differ from those of the parent compound which means that their environmental risks could be very different from those of the parent compound. With the exception of a few drugs, limited information is available in the public domain on the ecotoxicity of metabolites and transformation products.

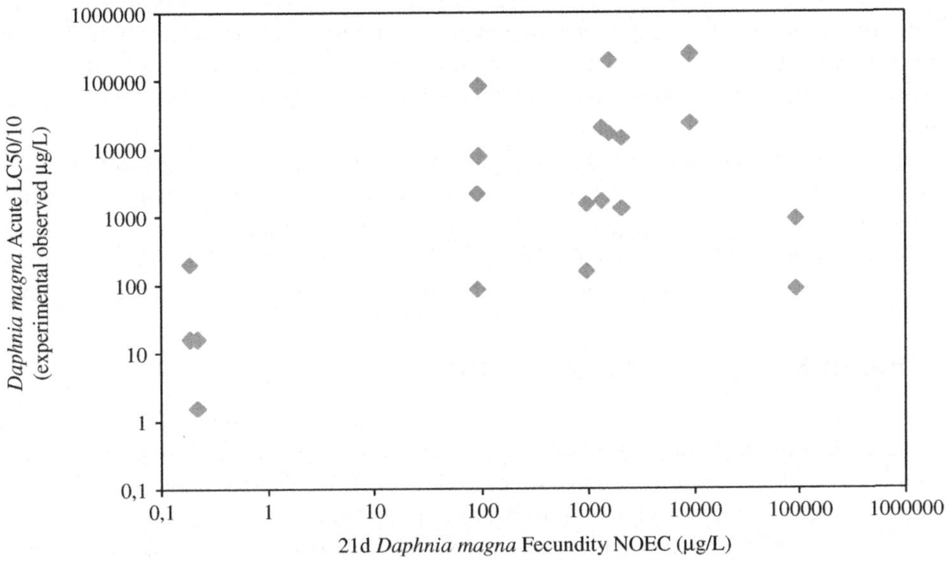

Figure 6.8 Comparison of measured chronic ecotoxicity values with estimated values obtained using acute data and an acute: chronic ratio of ten.

Ways in which transformation products of most concern could be identified without the need for extensive testing have been explored. Data is available on the ecotoxicity of pesticide transformation products that provides useful information on which factors make a transformation product more ecotoxic than the parent compound. Most pesticide transformation products are less toxic than, or have similar toxicity to their associated parent compound. However some are more than 3 times more toxic than the parent compound. These observed increases can be due to a number of reasons:

(1) uptake of the transformation product into organisms is greater than for the parent compound, due to either an increase in lipophilicity or a change in dissociation or both;
(2) the transformation product contains the active moiety of the parent compound; and
(3) the transformation reaction results in the introduction of a toxic moiety which is not present in the parent compound.

Using the above knowledge, if the structure of a metabolite or transformation product of a pharmaceutical is known, it is possible to identify which substances are likely to be more toxic than the parent compound. The lipophilicity and dissociation behaviour of a molecule can be estimated using quantitative structure-property relationships. If the active moiety (or pharmacophore) of the parent compound is known, then a quick visual assessment of the transformation product can tell you whether this moiety is present within it. Several approaches are available to identify whether a new toxic moiety has been introduced during the transformation reaction. These approaches generally use structural alerts that are associated with a range of environmental models of action. If the transformation product contains one of these structural alerts but this alert is not present in the parent compound, then the transformation product may be more toxic than the parent compound.

6.5 NON TRADITIONAL ENDPOINTS

6.5.1 What are the limitations of the standardized test endpoints and methods?

Several studies have commented on the sufficiency of standardized ecotoxicity data for the characterisation of the hazard of pharmaceuticals (Brooks *et al.*, 2003; Ankley *et al.*, 2005). Such traditional methods include common model organisms such as *Daphnia magna* and *Pimephales promelas* that are amenable to laboratory culture. Acute toxicity tests with these organisms include mortality as the primary response, compared with growth inhibition in the green algae models (e.g., *Selanastrum capricornutum*). Traditional chronic or subchronic responses generally include survival and reproduction (and sometimes growth) for *D. magna* and survival and growth for *P. promelas* following a 21 d or 7 d exposure period, respectively. Although such bioassays have high utility for environmental safety assessment of many industrial chemicals, whole effluent toxicity and ambient toxicity studies, it is likely that short term growth and reproduction responses are not related to target-mediated responses, which may be observed at concentrations that are orders of magnitude lower then standardized chronic toxicological benchmark concentrations (TBCs).

For example, Kidd *et al.* (2007) observed that the population of *P. promelas* disappeared following treatment with 5–6 ng/L of 17α-ethinyl estradiol (EE2) in a whole lake experiment. However, an acute (48 h) *P. promelas* LC50 tests with EE2 has been reported at ~1700 ug/L, which would result in an acute-to-chronic (ACR) ratio of approximately 340000, a value markedly greater than typical uncertainty factors of 10–1000 applied to LC50 values during prospective ERAs under EMEA.

In order to explore whether the novel effects data indicate a risk to the aquatic environment, measured environmental concentrations were compared to the lowest toxic endpoints (either novel or traditional) obtained for a range of pharmaceuticals. The data used in the comparison are summarised in Table 6.6.

The lowest effect datum was orders of magnitude lower than the MEC for 17α-ethinylestradiol, acetaminophen, 17α/β estradiol, and salicylic acid indicating that these substances may pose a risk to the environment (Figure 6.9).

6.5.2 Use of biomarkers as novel end points

Biomarkers are used for the detection of a limited number of pharmaceutical products, but the numbers are very low. The range of pharmaceutical products monitored includes antiinflammatory and antibiotic products, with some studies being carried out on X-ray contrast media, antidepressants and β-blockers. The biomarkers used are expressed in living organisms such as fishes, including zebrafish and rainbow trout. Other smaller organisms, such as crustaceans, have also been employed (Nunes *et al.*, 2006). Biomarkers representative of oxidative stress have recently been described. Nunes *et al.* (2006) showed that exposure to some pharmaceutical compounds induced alterations in the cellular redox status in a crustacean, *A. parthenogenetica*. More precisely, diazepam caused a significant inhibition of soluble cholinesterases (ChE) (LOEC = 7.04 mg/L) and total glutathione-peroxidase (GPx) activities. Sodium dodecylsulphate (SDS) was responsible for a decrease in the activity of both ChE (LOEC = 8.46 mg/L) and glutathione reductase (GRed) (LOEC = 4.08 mg/L). Both fibrates (clofibrate and clofibric acid) were responsible for significant decreases in Se-dependent GPx, with LOEC values of 176.34 and 3.09 mg/L, respectively. Clofibrate also caused a slight increase in the thiobarbituric acid reactants (TBARS) content of *A. parthenogenetica* homogenates. In addition, diazepam was shown to have the capability of interfering with *A. parthenogenetica* neurotransmission, through the inhibition of ChE.

Hong *et al.* (2007) investigated the expression levels of cytochrome P450 1A, p53 and vitellogenin in the liver, kidney and gill of male Japanese Medaka, (*Oryzias latipes*). Cytochrome P450 (CYP1A) has an essential function in the biotransformation and detoxification of endogenous and exogenous compounds and can be used to indicate the cellular toxicity caused by chemicals. p53 is a biomarker for analyzing the carcinogenicity and DNA damage due to environmental toxicants, whilst vitellogenin is a precursor of yolk protein found in eggs. Its induction in males is an indicator of exposure to oestrogen or estrogen-like chemicals. Hong (2007) demonstrated that diclofenac causes a modification in the expression of these markers in an aquatic model organism, *O latipes*. The results suggested that at high concentrations (mg L-1) and even at environmentally relevant concentrations (1 μg/L), diclofenac lead to the notable induction of three biomarker genes. These levels can also possibly cause acute cellular toxicity as well as apoptotic or carcinogenic effects in medaka tissue. In particular, after exposure to 8 mg/L and 1 μg/L of diclofenac, they were highly expressed in tissue-specific gene expression patterns. The elevated expression levels of three biomarkers suggested that diclofenac has potential to cause cellular toxicity, p53-related genotoxicity and oestrogenic effects. Markers of redox activity of several pharmaceuticals (Table 6.7) in fish tissue (microsomes) have been used as biomarkers. Indeed, primary cultures of trout hepatocytes are well recognized model systems in toxicological research because these cells are the major drug-metabolizing cells in fish, and can be used conveniently to characterize cellular toxicity. Special attention was given to the redox activity of pharmaceutical products in microsomal membrane preparations by investigating the rate of oxidation of reduced nicotinamide adenine dinucleotide phosphate (NADPH) and the formation of lipid peroxidation (LPO) after a 60 minute incubation period.

In addition, Gagné *et al.* (2006) exposed primary cultures of rainbow trout hepatocytes to various drugs found in municipal effluent for 48 h at 15°C. Most pharmaceuticals accelerated the rate of NADPH oxidation in the presence of microsomes and increased LPO in microsomal membranes. Exposure of trout hepatocytes to these products leads in many cases to decreased cell viability, increased CYP3A-related monooxygenase activity (benzyl ether resorufin dealkylase), and LPO. No induction of CYP1A1-related activity (7-ethoxyresorufin O-deethylase) was observed. These results show that these products have the potential to produce a toxic response in aquatic organisms, and that the above biomarkers respond readily to several pharmaceutical products in aquatic organisms.

Table 6.6 Lowest effects data and maximum environmental concentrations used in the risk characterisation process.

Class	Compound	MEC [µg/L]	Lowest overall endpoint	Effect concentration µg/L	MEC/Lowest effect
Analgesics and antiinflammatories	Acetominophen	10.000	*Hyallela azteca* 28 d 3 generations LOEC sex ratio	0.1	100
	Acetylsalicylic Acid	0.340	*Daphnia magna/longespina* 21 d NOEC	1000	
	Diclofenac	1.200	*Lemna minor* 7 d EC50	7500	1.60×10^{-4}
	Diazepam	0.033	*Hydra* polyp regeneration LOEC	10	3.30×10^{-3}
	Ibuprofen	5.044	*Hydra* polyp regeneration LOEC	10	0.504
	Ketoprofen	0.300	*Vibrio fischeri* 10 min EC50	1.56×10^{4}	1.92×10^{-5}
	Naproxen	2.000	*Anabena flos-aquae* 72 h EbC50	1.23×10^{4}	1.63×10^{-4}
	Salicylic Acid	8.800	*Hyallela azteca* 28 d 3 generations LOEC	0.1	88
Fluoroquinolone antibiotics	Ciprofloxacin	0.030	*Microcyctus aeruginosa* X d EC50	5	0.006
	Enrofloxacin	0.020	*Penaeus vannamei* 48 h EC50	14.3	1.40×10^{-3}
	Ofloxacin	0.306	*Pseudomonas putida* EC50	10	3.06×10^{-2}
Macrolide antibiotics	Clarithromycin	0.260	*Pseudokircheriella subcapitata* 72 h EC50	2	0.13
	Roxithromycin	0.560	*Lemna gibba* 7 d EC50	1000	5.60×10^{-4}
	Spiramycin	0.740	*Microcystis aeruginosa* 7 d EC50	5	0.148
Penicillin antibiotics	Amoxycillin	0.006	*Synechococcus leopolensis* 4 d EC50	2.2	2.73×10^{-3}
Sulfonamide antibiotics	Sulfamethazine	0.220	*Lemna gibba* 7 d EC50	1000	2.20×10^{-4}
	Sulfamethoxazole	1.900	*Synechococcus leopolensis* 96 h EC50	26.8	7.09×10^{-2}
Tetracycline antibiotics	Chlorotetracycline	0.690	*Microcystis aeruginosa* 7 d EC50	50	1.38×10^{-2}
	Doxycycline	0.080	*Lemna gibba* 7 d EC50	316	2.53×10^{-4}
	Oxytetracycline	0.340	*Penaeus vannamei* 48 h EC50	61.1	5.56×10^{-3}
	Tetracycline	0.140	*Vibrio fischeri* 24 h EC50	25.1	5.58×10^{-3}
Other antibiotics	Chloramphenicol	0.060	*Penaeus stylirostrus* 48 h EC50	41.3	1.45×10^{-3}
	Trimethoprim	0.710	*Anabaena variabilis* 144 h EC50	1.10×10^{4}	6.45×10^{-5}
Anti depressants	Fluoxetine	0.012	*Pseudokircheriella subcapitata* 48 h EC50	24	5.00×10^{-4}
Antiepileptics	Carbamazepine	7.100	*Ceriodaphnia dubia* 48 h EC50	1.15×10^{4}	6.20×10^{-4}
Antineoplastic agents	Methotrexate	0.000	*Tetrahymena pyriformis* 48 h EC50	4.50×10^{4}	6.67×10^{-9}
	Tamoxifen	0.071	*Acartia tonsa* 5 d EC50	49	1.45×10^{-3}
ß-blockers	Atenolol	0.241	*Hydra vulgaris* 7 d polypstructure LOEC	1.00×10^{4}	2.41×10^{-5}
	Carazolol	0.110	*Daphnia magna* 48 h EC50	1.48×10^{4}	7.43×10^{-6}
	Metoprolol	2.200	*Desmodesmus subspicatus* 72 h EC50	7300	3.01×10^{-4}
	Propanolol	0.590	*Streptocephalus proboscideus* 24 h EC50	1870	3.16×10^{-4}

(*continued*)

Table 6.6 (*continued*)

Class	Compound	MEC [µg/L]	Lowest overall endpoint	Effect concentration µg/L	MEC/Lowest effect
Hormones and steroids	17 Alpha Ethinylestradiol	0.831	*Pimephales promelas* 308 d EC50	1.66×10^{-3}	501
	17-Alpha/Beta Estradiol	0.200	*Oncorhynchus mykiss*, 21 d EC 50	0.015	13.3
	19-Norethisterone	0.872	*Pseudokirchneriella subcapitata* 72 h EC50	216	4.04×10^{-3}
	Mestranol	0.407	*Pseudokirchneriella subcapitata* 72 h EC50	500	8.14×10^{-4}
	Estrone	1.600	*Tisbe battagliani* 10 d LC50	100	0.016
	Testosterone	0.214	*Acartia tonsa* 5 d EC50	1500	1.43×10^{-4}
Lipid regulators	Clofibrate	0.040	*Daphnia magna* 21 d EC50	106	3.77×10^{-4}
	Gemfibrozil	1.550	*Vibrio fischeri* 10 min EC50	1.88×10^{4}	8.24×10^{-5}
Other	Caffeine	6.000	*Artemisia salina* 1 d LC50	5.27×10^{4}	1.14×10^{-4}
	Cotinine	0.900	*Lemna gibba* 7 d EC50	1000	9.00×10^{-4}
	Digoxin	0.260	*Hydra vulgaris* polyp regeneration	10	0.026
	Diltiazem	0.049	*Vibrio fischeri* 24 h EC50	1.52×10^{5}	3.22×10^{-7}
	Furosemide	0.255	*Ceriodaphnia dubia* 48 h EC50	156	1.63×10^{-3}
	Glibenclamide	0.012	Fish, Daphnia and Algae	1.00×10^{5}	1.20×10^{-7}
	Hydrochlorothiazide	0.236	Fish, Daphnia and Algae	1.00×10^{5}	2.36×10^{-6}
	Metformin	0.150	*Anabaena* sp. EC50	1.00×10^{5}	1.50×10^{-6}
	Nifedipine	0.010	*Danio rerio* 96 h LC50	5770	1.73×10^{-6}

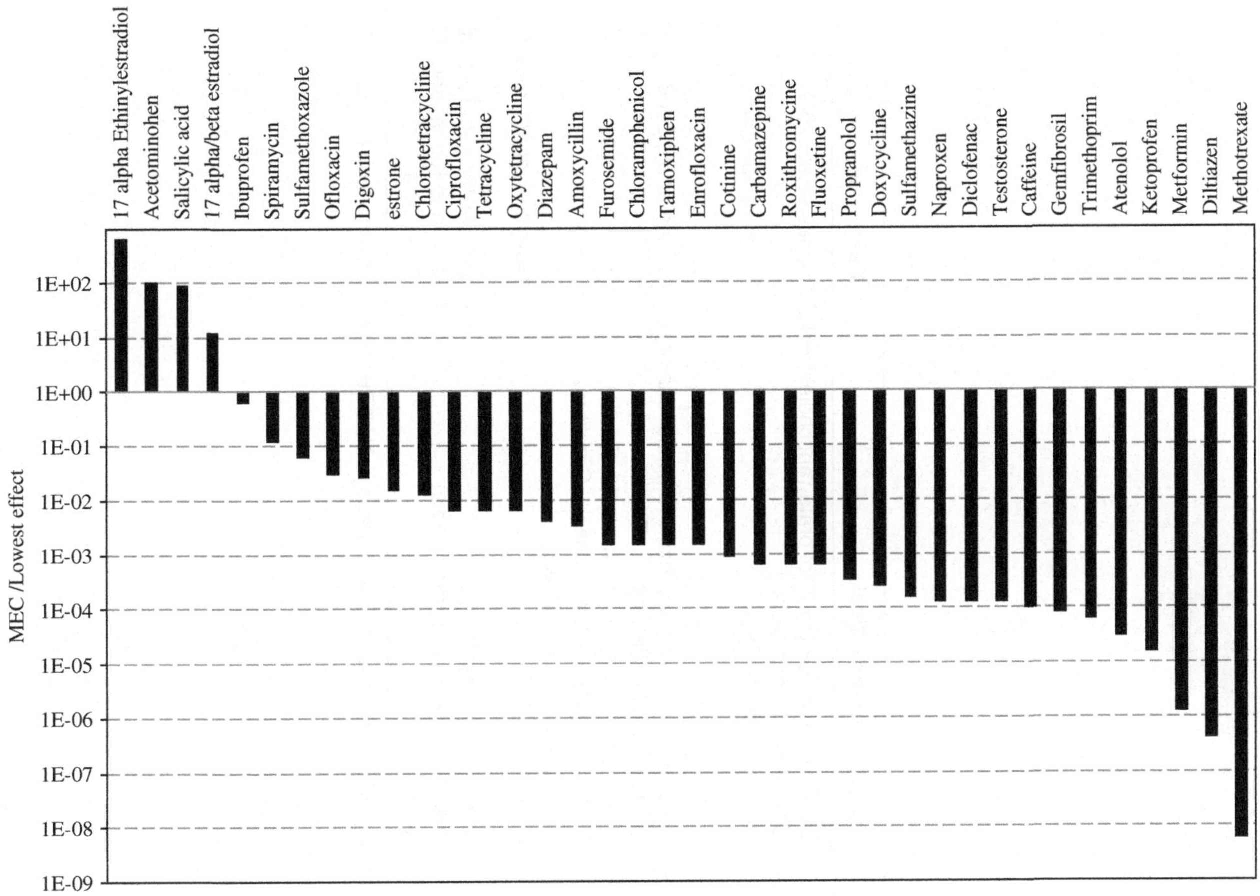

Figure 6.9 MEC:lowest ecotoxicity values for a range of pharmaceuticals.

Markers of oestrogenic activity of pharmaceutical products (such as ethynylestradiol – EE2) have been investigated in fish. Research carried out by Martyniuk (2007) showed that vitellogenin 1 and 3 (*vtg1*; *vtg3*), oestrogen receptor alpha (*esr1*), and apolipoprotein A1 (*apoA1*) mRNA are particularly expressed when adult zebrafish are exposed to a concentration of >10 ng/L EE2. Real-time RT-PCR confirmed that *vtg1* was induced 700- fold, *vtg3* was induced 100-fold and *esr1* was induced 20-fold. Van der Ven *et al.* (2006) also used induction of vitellogenin to demonstrate the oestrogenic activity of mianserin (antidepressant) in the aquatic environment. In this study, the possible mode of action through which the antidepressant mianserin can interfere with normal neuro-endocrine signalling and reproduction in a teleost was further investigated by analyzing differential gene expression in brain and gonad tissue.

Medical knowledge can also be used to help in the environmental risk assessment of pharmaceutical products. Owen and co-workers (2007) developed a comparative review of pharmacological and toxicological data for mammals and aquatic life in order to assess the possible effects of ß-blocker residues in surface waters. They found that some ß-adrenergic receptors have been characterised in fish, but identified a need for more work to extend knowledge in this area. They advocated a comparative approach that made maximum use of the extensive and well-established background knowledge available from mammalian studies alongside research in fish and other aquatic organisms. This should reduce the work needed to identify relevant biomarkers of β-blocker toxicity in aquatic organisms.

Table 6.7 Markers of redox activity of several pharmaceuticals in fish tissue (microsomes).

Biomarker	Species	Target pharmaceutical	Reference
Oxidative stress: GPx, Gred, SOD, GSTs? Lipid peroxidation, Cholinesterase	*Artemia parthenogenetica* (Brine shrimp)	Diazepam (Anxiolytic), Clofibrate (anticholesterol), Clofibric acid (metabolite)	Nunes *et al.*, 2006
P450 1A, p53, vitellogenin (parameters indicating biotransformation, genotoxicity and reproductive and endocrine effects, respectively	Oryzias latipes (Japanese rice fish)	Diclofenac (Antiinflamm)	Hong *et al.*, 2007
NADH, LPO	Oncorhynchus mykiss – hepatocytes (Rainbow trout)	Caffeine, Ibuprofen, Naproxen (antiinflammatory), Oxytetracycline, Novobiocin, Trimethoprim, Sulfamethoxazole, Sulfapyridine (antibiotic), Carbamazepine (anticonvulsant), Gemfibrozil, Bezafibrate (lipid regulator)	Gagné *et al.*, 2006
Vitellogenin 1 and 3, oestrogen receptor alpha, apolipoprotein A1	*Danio rerio* (Zebrafish)	Ethynylestradiol	Martyniuk *et al.*, 2007
Vitellogenin, zona pellucida protein	*Danio rerio* (Zebrafish)	Mianserin (antidepressant)	van der Ven *et al.*, 2006
Quinines, polyamines, phospholipids, ATP		Diatrizoate (X-ray contrast media)	Haiß and Kummerer, 2006
ß-adrenergic receptors	Fish	ß-blockers	Owen *et al.*, 2007

GPx: glutathione-peroxidase; Gred: glutathione reductase; SOD: total superoxide dismutase; GSTs: glutathione-S-transferases NADPH: nicotinamide adenine dinucleotide phosphate; LPO: lipid peroxidation.

6.6 THE USE OF NOVEL APPROACHES TO THE RAPID IDENTIFICATION OF POLLUTION INCIDENTS

Since its conception, the Internet has been used as a platform for the exchange of knowledge. Its near universal global penetration combined with the ability to update websites rapidly has facilitated the development of news sites that report current events in almost real time. These websites present a unique opportunity to identify reports of pollution incidents that may be associated with global pharmaceutical product use. However, manual monitoring of these websites is both time consuming and labour intensive and. web agents could be used to facilitate the monitoring of such sites. Web agents are complex software systems that operate on the Internet, and related corporate, government, or military intranets. They can be deployed to monitor multiple websites for the occurrence of specific terms, and, when these are found, alert the end user of their presence. Deployment of these web agents could be used to facilitate the rapid response of an organisation to both monitor and contain pharmaceutical product releases.

6.7 CONCLUSIONS

A large body of data has been generated over the past decade on the effects of pharmaceuticals on aquatic organisms. Generally, pharmaceuticals show low acute toxicity to fish, daphnids and algae. While the majority of pharmaceuticals also show low toxicity in standard chronic studies, a number of pharmaceuticals are highly toxic. A comparison of standard acute and chronic test endpoints with available monitoring data indicate that, with a few exceptions, pharmaceuticals do not pose an unacceptable risk to the environment.

The effects of a number of pharmaceuticals have been assessed using non-traditional eoctoxicity tests and endpoints. The effect endpoints for the majority of the novel tests are many orders of magnitude lower than the traditional endpoints and existing uncertainty factors do not reflect these differences. Most of the novel endpoints can be linked to important ecological functions, and so can be regarded as ecologically relevant. Even when novel data are included in the risk characterisation, most pharmaceuticals seem to pose a low risk to ecosystem health. There are, however, some exceptions that may need further assessment or control in the future.

Based on the review and analysis of the data, the following recommendations have been developed:

(1) For many pharmaceuticals, only limited data is available on their ecotoxicity. Some of this data will have been generated by pharmaceutical companies. If this data were more freely available, it could be used alongside the available monitoring data to determine whether concentrations of pharmaceuticals are of ecological concern or not.

(2) Further work is required to understand the significance of novel endpoints (including results from studies employing proteomics and metabolomics) in terms of their ecological relevance. These studies will help to establish whether or not the standard chronic tests are appropriate.

(3) Work is required to understand how risks of pharmaceuticals may change in the future as a result of climate change and pandemics.

(4) More synthesis of published and unpublished data would be worthwhile. Ecotoxicity studies should report all relevant background information (e.g., measured exposure concentrations, pH, how NOEC, EC50 were derived, and full information, not just summary statistics, and confidence limits) to facilitate this synthesis. Reported effects data should be compared with monitoring data in order to identify substances of most concern. Scientists and industry should be encouraged to share data (tables of parameters). Studies yielding 'surprising' results should be repeated.

(5) Further work is required to understand whether and how we can extrapolate from mammalian data to environmental effects. The use of contra-indications to indicate potential for ecological risks should be considered and the utility of 'omics' based approaches should be explored for risk assessment purposes. It would be helpful if case studies could be developed that demonstrate read across from mammalian data to environmental risk for a range of substances.

(6) The impacts of mixtures of (a) pharmaceuticals of the same class; (b) pharmaceuticals of different classes; and (c) pharmaceuticals and other substance types should be assessed. Modelling approaches are available from other sectors that could be used for this purpose.

(7) Long-term studies at realistic exposure concentrations and under realistic environmental conditions might be useful for some pharmaceuticals although strong consideration would need to be given on how best to interpret these studies.

(8) Most work has been performed on the aquatic environment. Further work into impacts on terrestrial systems is warranted.

ACKNOWLEDGEMENT

Thomas Backhaus (Göteborg University, Sweden), Bryan Brooks (Baylor University, USA), Daniel Gouy (Sanofi-Aventis, France), Silke Hickman (UBA, Germany), Karen Kidd (University of New Brunswick, Canada), Chris Metcalfe (Trent University, Canada), Melanie Netherton (University of York, UK), Joanne Parrott (Environment Canada, Canada), Edward Smith (Göteborg University, Sweden), Jürg Oliver Straub (Roche, Switzerland), Jason Weeks (Cefas, UK) who contributed to the workshop organised during KNAPPE project.

REFERENCES

Ankley, G.T., Jensen, K.M., Kahl, M.D., Korte, J.J. and Makynen, E.A. (2001) Description and evaluation of a short-term reproduction test with the fathead minnow (*Pimephales promelas*). *Environ. Toxicol. Chem.* **20,** 1276–1290.

Ankley, G.T., Black, M.C., Garric, J., Hutchinson, T.H. and Iguchi, T. (2005) A framework for assessing the hazard of pharmaceutical materials to aquatic species. In Human pharmaceuticals: assessing the impacts on aquatic ecosystems, Williams RT, Ed., SETAC Press; Pensacola, FL, Chapter 6, pp. 183–237;.

Balch, G.C., Mackenzie, C.A. and Metcalfe, C.D. (2004) Alterations to gonadal development and reproductive success in Japanese medaka (*Oryzias latipes*) exposed to 17α-ethinylestradiol. *Environ. Toxicol. Chem.* **23**, 782–791.

Baldwin, W.S., Milam, D.L. and Leblanc, G.A. (1995) Physiological and biochemical perturbations in Daphnia magna following exposure to the model environmental estrogen diethylstilbestrol. *Environ. Toxicol. Chem.* **14**, 945–52.

Belfroid, A. and Leonards, P. (1996) Effect of ethinyl estradiol on the development of snails and amphibians. SETAC 17th Annual Meeting, Washington, DC, USA.

Borgmann, U., Bennie, D.T., Ball, A.L. and Palabrica V. (2007) Effect of a mixture of seven pharmaceuticals on *Hyalella azteca* over multiple generations. *Chemosphere* **66**, 1278–1283.

Brain, R.A., Johnson, D.J., Richards, S.M., Sanderson, H., Sibley, P.K. and Solomon, K.R. (2004) Effects of 25 pharmaceutical compounds to *Lemna gibba* using a seven-day static-renewal test. *Environ. Toxicol. Chem.* **23**, 371–82.

Brain, R.A., Hanson, M.L., Solomon, K.R. and Brooks, B.W. (2008) Aquatic plants exposed to pharmaceuticals: effects and risks. *Rev. Environ. Contamin. Toxicol.* **192**, 67–115.

Brambilla, G., Civitarcale, C. and Migliore, L. (1994) Experimental toxicity and analysis of bacitracin, flumequine and sulphadimethoxine in terrestrial and aquatic organisms as a predictive model for ecosystem damage. *Quim Anal.* **13**(sup.1), 114–8.

Breitholtz, M. and Bengtsson, B.E. (2001) Oestrogens have no hormonal effect on the development and reproduction of the harpacticoid copepod *Nitocra spinipes. Marine Pollution Bulletin* **42**(10), 879–886.

Brooks, B.W., Foran, C.M., Richards, S.M., Weston, J., Turner, P.K. and Stanley, J.K. (2003) Aquatic ecotoxicology of fluoxetine. *Toxicol Lett.* **142**, 169–83.

Brooks, B.W., Huggett, D.B., Brain, R.A. and Ankley, G.T. (2009) Risk assessment considerations for veterinary medicines in aquatic systems. In *Veterinary Pharmaceuticals in the Environment*, (eds. K. Henderson, J. Coats), American Chemical Society, Washington, DC.

Cleuvers, M. (2003) Aquatic ecotoxicity of pharmaceuticals including the assessment of combination effects. *Toxicol Lett.* **142**, 185–194.

Cleuvers, M. (2004) Mixture toxicity of the anti-inflammatory drugs diclofenac, ibuprofen, naproxen, and acetylsalicylic acid. *Ecotoxicology and Environmental Safety* **59**, 309–315.

Cleuvers, M. (2005) Initial risk assessment for three beta-blockers found in the aquatic environment. *Chemosphere* **59**(2), 199–205.

Clubbs, R.L. and Brooks, B.W. (2007) *Daphnia magna* responses to a vertebrate estrogen receptor agonist and an antagonist: A multigenerational study. *Ecotoxicology and Environmental Safety* **67**, 385–398.

Czech, P., Weber, K. and Dietrich, D.R. (2001) Effects of endocrine modulating substances on reproduction in the hermaphroditic snail Lymnaea stagnalis L. *Aquat Toxicol.* **53**, 103–14.

De Lange, H.J., Noordoven, W., Murk, A.J., Lurling, M. and Peeters, E.T. (2006) Behavioural responses of *Gammarus pulex* (Crustacea, Amphipoda) to low concentrations of pharmaceuticals. *Aquat Toxicol.* **78**, 209–16.

Fent, K., Weston, A.A. and Caminada, D. (2006) Ecotoxicology of human pharmaceuticals. *Aquatic Toxicology* **76**, 122–159.

Ferrari, B., Mons, R., Vollat, B., Fraysse, B., Paxeus, N., Lo Giudice, R., Pollio, A. and Garric, J. (2004) Environmental risk assessment of six human pharmaceuticals: are the current environmental risk assessment procedures sufficient for the protection of the aquatic environment? *Environ Toxicol Chem.* **23**, 1344–54.

Ferrari, B., Paxéus, N., Lo Giudice, R., Pollio, A. and Garric, J. (2003) Ecotoxicological impact of pharmaceuticals found in treated wastewaters: study of carbamazepine, clofibric acid, and diclofenac. *Ecotoxicol. Environ. Saf.* **55**, 359–70.

Ferreira, C.S., Nunes, B.A., de Melo Henriques-Almeida, J.M. and Guilhermino, L. (2007) Acute toxicity of oxytetracycline and florfenicol to the microalgae *Tetraselmis chuii* and to the crustacean *Artemia parthenogenetica. Ecotoxicology and Environmental Safety* **67**, 452–458.

Flaherty, C.M. and Dodson, S.I. (2005) Effects of pharmaceuticals on Daphnia survival, growth, and reproduction. *Chemosphere* **61**, 200–207.

Flippin, J.L., Huggett, D. and Foran, C.M. (2007) Changes in the timing of reproduction following chronic exposure to ibuprofen in Japanese medaka, *Oryzias latipes. Aquat. Toxicol.* **81**, 73–78.

Fong, P. (1998) Zebra mussel spawning is induced in low concentrations of putative serotonin reuptake inhibitors. *Biol Bull.* **194**, 143–8.

Fraysse, B., Mons, R. and Garric, J. (2006) Development of a zebrafish 4-day embryo-larval bioassay to assess toxicity of chemicals. *Ecotoxicology and Environmental Safety* **63**, 253–267.

Fujioka, Y. (2002) Effects of hormone treatments and temperature on sex reversal of Nigorobuna Carassius carassius grandoculis. *Fish Sci.* **68**, 889–893.

Gagne, F., Blaise, C., Fournier, M. and Hansen, P.D. (2006) Effects of selected pharmaceutical products on phagocytic activity in *Elliptio complanata* mussels. *Comp Biochem Physiol C* **143**, 179–86.

Garric, J., Vollat, B., Duis, K., Péry, A., Junker, T., Ramil, M., Fink, G. and Ternes, T.A. (2007) Effects of the parasiticide ivermectin on the cladoceran *Daphnia magna* and the green alga *Pseudokirchneriella subcapitata*. *Chemosphere* **69**, 903–910.

Gülden, M., Mörchel, S. and Seibert, H. (2005) Comparison of mammalian and fish cell line cytotoxicity: impact of endpoint and exposure duration. *Aquatic Toxicology* **71**(3), 229–236.

Haiß, A. and Kummerer, K. (2006) Biodegradability of the X-ray contrast compound diatrizoic acid, identification of aerobic degradation products and effects against sewage sludge micro-organisms. *Chemosphere* **62**, 294–302.

Halling-Sorensen, B., Lutzhoft, H.H.C., Andersen, H.R. and Ingerslev, F. (2000) Environmental risk assessment of antibiotics: comparison of mecillinam, trimethoprim and ciprofloxacin. *J Antimicrob Chemother.* **46**(Suppl. 1), 53–8.

Heberer, T. (2002) Occurrence, fate, and removal of pharmaceutical residues in the aquatic environment: a review of recent research data. *Toxicol Lett.* **131**, 5–17.

Heckmann, L-H., Callaghan, A., Hooper, H.L., Connon, R., Hutchinson, T.H., Maund, S.J. and Sibly, R.M. (2007) Chronic toxicity of ibuprofen to *Daphnia magna*: Effects on life history traits and population dynamics. *Toxicology Letters* **172**, 137–145.

Henri, T.B., Kwon, J.W., Armbrust, K.L. and Black, M.C. (2004) Acute and chronic toxicity of five selective serotonin reuptake inhibitors in *Ceriodaphnia dubia*. *Environ Toxicol Chem.* **23**, 2229–33.

Henschel, K.P., Wenzel, A., Diedrich, M. and Fliedner, A. (1997) Environmental hazard assessment of pharmaceuticals. *Regul Toxicol Pharmacol.* **25**, 220–5.

Holten Lutzhoft, H.C., Halling-Sorensen, B. and Jorgensen, S.E. (1999) Algal toxicity of antibacterial agents applied in Danish fish farming. *Arch Environ Contam Toxicol.* **36**, 1–6.

Hong, H.N., Kim, H.N., Park, K.S., Lee, S.K. and Gu, M.B. (2007) Analysis of the effects diclofenac has on Japanese medaka (Oryzias latipes) using real-time PCR. *Chemosphere* **67**, 2115–2121.

Huggett, D.B., Brooks, B.W., Peterson, B., Foran, C.M. and Schlenk, D. (2002) Toxicity of selected beta adrenergic receptor blocking pharmaceuticals (β blockers) on aquatic organisms. *Arch Environ Contam Toxicol.* **43**, 229–35.

Hutchinson, T.H. (2002) Reproductive and developmental effects of endocrine disrupters in invertebrates: in vitro and invivo approaches. *Toxicol Lett.* **131**, 75–81.

Isidori, M., Lavorgna, M., Nardelli, A., Parrella, A., Previtera, L. and Rubino, M. (2005a) Ecotoxicity of naproxen and its phototransformation products. *Science of the Total Environment* **348**, 93–101.

Isidori, M., Lavorgna, M., Nardelli, A., Pascarella, L. and Parrella, A. (2005b) and genotoxic evaluation of six antibiotics on non-target organisms. *Science of the Total Environment* **346**, 87–98.

Isidori, M., Nardelli, A., Pascarella, L., Rubino, M. and Parrella A. (2007) Toxic and genotoxic impact of fibrates and their photoproducts on non-target organisms. *Environment International* **33**, 635–641.

Jobling, S., Sheahan, D., Osborne, J.A., Matthiessen, P. and Sumpter, J.P. (1996) Inhibition of testicular growth in rainbow trout (*Oncorhynchus mykiss*) exposed to estrogenic alkylphenolic chemicals. *Environ Toxicol Chem.* **15**, 194–202.

Johnson, D.J., Sanderson, H., Brain, R.A., Wilson, C.J. and Solomon K.R. (2007) Toxicity and hazard of selective serotonin reuptake inhibitor antidepressants fluoxetine, fluvoxamine, and sertraline to algae. *Ecotoxicol. Environ. Saf* **67**(1), 128–139.

Jos, A., Repetto, G., Rios, J.C., Hazen, M.J., Molero, M.L., del Peso, A., Salguero, M., Fernandez-Freire, P., Perez-Martin, J.M. and Camean, A. (2003) Ecotoxicological evaluation of carbamazepine using six different model systems with eighteen endpoints. *Toxicol in Vitro.* **17**, 525–32.

Kang, I.J., Yokota, H., Oshima, Y., Tsuruda, Y., Shimasaki, Y. and Honjo, T. (2008) The effects of methyltestosterone on the sexual development and reproduction of adult medaka (Oryzias latipes). *Aquatic Toxicology* **87**(1), 37–46.

Kidd, K.A., Blanchfield, P.J., Mills, K.H., Palace, V.P., Rvans, R.E., Lazorchak, J.M. and Flick, R.W. (2007) Collapse of a fish population after exposure to a synthetic estrogen. *PNAS* **104**, 8897–8901.

Kim, Y., Choi, K., Jung, J., Park, S., Kim, P.G. and Jeongim Park, J. (2007) Aquatic toxicity of acetaminophen, carbamazepine, cimetidine, diltiazem and six major sulfonamides, and their potential ecological risks in Korea. *Environment International* **33**(3), 370–375.

Köpf, W. (1995) Effects of endocrine substances in bioassays with aquatic organisms; cited in Webb SF. 2004. A data-based perspective on the environmental risk assessment of human pharmaceuticals II – aquatic risk characterisation. In: Kümmerer K (ed.) Pharmaceuticals in the Environment: Sources, Fate, Effects and Risks, Second Edition, Springer, Berlin, Germany, pp. 345–361.

Kuhn, R., Pattard, M., Pernak, K. and Winter, A. (1989) Results of the harmful effects of selected water pollutants (anilines, phenols, aliphatic compounds) to *Daphnia magna*. *Water Res.* **23**(4), 495–9.

Länge, R., Hutchinson, T.H., Croudace, C.P., Siegmund, F., Schweinfurth, H., Hampe, P., Panter, G.H. and Sumpter, J.P. (2001). Effects of the synthetic oestrogen 17α-ethynylestradiol on the life-cycle of the fathead minnow (*Pimephales promelas*). *Environ Contam Toxicol* **20**, 1216–1227.

Lanzky, P.F. and Halling-Sørensen, B. (1997) The toxic effect of the antibiotic metronidazol on aquatic organisms. *Chemosphere* **35**, 2553–61.

Laville, N., Ait-Aissa, S., Gomez, E., Casellas, C. and Porcher, J.M. (2004) Effects of human pharmaceuticals on cytotoxicity, EROD activity and ROS production in fish hepatocytes. *Toxicology* **196**(1–2), 41–55.

Marking, L.L., Howe, G.E. and Crowther, J.R. (1988) Toxicity of erythromycin, oxytetracycline and tetracycline administered to lake trout in water baths by injection or by feeding. *Prog Fish-Cult.* **50**, 197–201.

Martyniuk, C.J., Gerrie, E.R., Popesku, J.T., Ekker, M. and Trudeau, V.L. (2007) Microarray analysis in the zebrafish (Danio rerio) liver and telencephalon after exposure to low concentration of 17alpha-ethinylestradiol. *Aquatic Toxicology* **84**, 38–49.

Metcalfe, C.D., Metcalfe, T.L., Kiparissis, Y., Koening, B.G., Khan, C., Hughes, R.J., Croley, T.R., March, R.E. and Potter, T. (2001) Estrogenic potency of chemicals detected in sewage treatment plant effluents as determined by invivo assays with Japanese medaka (*Oryzias latipes*). *Environ Toxicol Chem.* **20**, 297–308.

Mimeault, C., Woodhouse, A.J., Miao, X., Metcalfe. C.D., Moon, T.W. and Trudeau. V.L. (2005) The human lipid regulator, gemfibrozil bioconcentrates and reduces testosterone in goldfish *Carassius auratus*. *Aquat Toxicol.* **73**, 44–54.
Monteiro, S. and Boxall, A.B.A. (2009) Fate of human-use pharmaceuticals in the environment. *Reviews in Environmental Contamination and Toxicology* **202**, 53–154.
Nałecz-Jawecki, G. (2007) Evaluation of the in vitro biotransformation of fluoxetine with HPLC, mass spectrometry and ecotoxicological tests. *Chemosphere* **70**(1), 29–35.
Nentwig, G., Oetken, M. and Oehlmann, J. (2004) Effects of pharmaceuticals on aquatic invertebrates – the example of carbamazepine and clofibric acid. In *Pharmaceuticals in the Environment: Sources, Fate, Effects and Risks*, 2nd edn (ed. K. Kümmerer), pp. 195–208, Springer, Berlin, Germany.
Nimrod, A.C. and Benson, W.H. (1998) Reproduction and development of Japanese medaka following an early life stage exposure to xenoestrogens. *Aquat Toxicol* **44**, 141–56.
Nunes, B., Carvalho, F. and Guilhermino, L. (2006) Effects of widely used pharmaceuticals and a detergent on oxidative stress biomarkers of the crustacean Artemia parthenogenetica. *Chemosphere* **62**, 581–594.
OECD-201. (1984). OECD guideline for testing of chemicals: Algae, Growth Inhibition Test.
OECD-202. (1984). OECD guideline for testing of chemicals: Daphnia sp. Acute Immobilisation test and reproduction test.
OECD-203. (1992). OECD guideline for testing of chemicals: Fish, Acute Toxicity Test.
OECD-210. (1992). OECD guideline for testing of chemicals: Fish, Early-life Stage Toxicity Test.
OECD-211. (1998). OECD guideline for testing of chemicals: Daphnia magna Reproduction Test.
Örn, S., Holbech, H., Madsen, T.H., Norrgren, L. and Petersen, G.I. (2003) Gonad development and vitellogenin production in zebrafish (Danio rerio) exposed to ethinylestradiol and methyltestosterone. *Aquatic Toxicology* **65**(4), 397–411.
Owen, S.F., Giltrow, E., Huggett, D.B., Hutchinson, T.H., Saye, J., Winter, M.J. and Sumpter, J.P. (2007) Comparative physiology, pharmacology and toxicology of β-blockers: Mammals versus fish? *Aquatic Toxicology* **82**, 145–162.
Pascoe, D., Karntanut, W. and Muller, C.T. (2003) Do pharmaceuticals affect freshwater invertebrates? A study with the cnidarian Hydra vulgaris. *Chemosphere* **51**(6), 521–528.
Pounds, N.A., MacLean, S., Webley, M., Pascoe, D. and Hutchinson, T.H. (2004) Growth and reproductive effects of ibuprofen in the freshwater ramshorn snail *Planorbis carinatus*. Society of Experimental Biology Annual Meeting, Edinburgh, UK.
Purdom, C.E., Hardiman, P.A., Bye, V.J., Eno, N.C., Tyler, C.R. and Sumpter, J.P. (1994) Estrogenic effects of effluents from sewage treatment works. *Chem Ecol.* **8**, 275–285.
Quinn, B., Gagné, F. and Blaise, C. (2008) An investigation into the acute and chronic toxicity of eleven pharmaceuticals (and their solvents) found in wastewater effluent on the cnidarian, *Hydra attenuate. Science of the Total Environment*, 306–314.
Robinson, C.D., Brown, E., Craft, J.A., Davies, IM., Moffat, C.F., Ririe, D., Robertson, F., Stagg, R.M. and Struthers, S. (2003) Effects of sewage effluent and ethinylestradiol upon molecular markers of estrogenic exposure, maturation and reproductive success in the sand goby (*Pomatoschistus minutes*, Pallas). *Aquat Toxicol.* **62**, 119–134.
Schulte-Oehlmann, U., Oetken, M., Bachmann, J. and Oehlmann, J. (2004) Effects of ethinyloestradiol and methyltestosterone in prosobranch snails. In *Pharmaceuticals in the Environment: Sources, Fate, Effects and Risks*, 2nd edn (ed. K. Kümmerer), pp. 233–247, Springer, Berlin, Germany.
Schweinfurth, H., Länge, R. and Schneider, P.W. (1996) Environmental risk assessment in the pharmaceutical industry. Presentation at the 3rd Eurolab Symposium: Testing and Analysis for Industrial Competitiveness and Sustainability, 5–7 June, Berlin, Germany.
Segner, H., Navas, J.M., Schafers, C. and Wenzel, A. (2003) Potencies of estrogenic compounds in in vitro screening assays and in life cycle tests with zebrafish invivo. *Ecotoxicol Environ Saf.* **54**, 315–22.
Thomas, K.V. and Hilton, M.J. (2004) The occurrence of selected human pharmaceutical compounds in UK estuaries. *Marine Pollution Bulletin* **49**, 436–444.
van der Ven, K., Keil, D., Moens, L.N., Hummelen, P.V., van Remortel, P., Maras, M. and De Coen, W. (2006) Effects of the antidepressant mianserin in zebrafish: Molecular markers of endocrine disruption. *Chemosphere* **65**(10), 1836–1845.
Vandenburgh, G.F., Adriaens, D., Verslycke, T. and Janssen, C.R. (2003) Effects of 17_-ethinyloestradiol on sexual development of the amphipod *Hyalella azteca. Ecotoxicol Environ Saf.* **54**, 216–222.
Webb, S.F. (2004) A data-based perspective on the environmental risk assessment for human pharmaceuticals I – collation of available ecotoxicity data. In *Pharmaceuticals in the Environment: Sources, Fate, Effects and Risks*, 2nd edn (ed. K. Kümmerer), pp. 174–201, Springer, Berlin, Germany.
Wollenberger, L., Halling-Sorensen, B. and Kusk, K.O. (2000) Acute and chronic toxicity of veterinary antibiotics to *Daphnia magna. Chemosphere* **40**, 723–30.
Zerulla, M., Länge, R., Steger-Hartmann, T., Panter, G., Hutchinson, T. and Dietrich, D.R. (2002) Morphological sex reversal upon short-term exposure to endocrine modulators in juvenile fathead minnow (*Pimephales promelas*). *Toxicol Lett.* **131**, 51–63.
Zillioux, E.J., Johnson, I.C., Kiparissis, Y., Metcalfe, C.D., Wheat, J.V., Ward, S.G. and Liu. H. (2001) The sheepshead minnow as an invivo model for endocrine disruption in marine teleosts: a partial life-cycle test with 17 alpha-ethinylestradiol. *Environ Toxicol Chem.* **20**, 1968–1978.
Zurita, J.L., Repetto, G., Jos, A., Salguero, M., Lopez-Artıguez, M. and Camean, A.M. (2007) Toxicological effects of the lipid regulator gemfibrozil in four aquatic systems. *Aquatic Toxicology* **81**, 106–115.

Chapter 7

Environmental risk assessment and prioritization strategies for human pharmaceuticals, review and discussion

Jean Philippe Besse and Jeanne Garric

7.1 INTRODUCTION

It is now recognized that pharmaceutical compounds reach the environment and can be considered as environmental contaminants. As a wide range of drugs including antibiotics, antiphlogistics, blood lipid-lowering agents, antiepileptics, β-blockers, hormones modulators and so on, have been found in the effluents and surface waters of several countries, there is a growing need to assess their environmental risk. Consequently, monitoring pharmaceuticals in the surface water and/or ground water is becoming mandatory.

Prior to implement any monitoring program in the aquatic environment, there is a need to rank pharmaceuticals according to their environmental relevance (i.e., their presence in the environment and their potential for ecotoxicological effects), because of the high number of molecules used as marketed human drugs. Indeed, it is not conceivable to search for all molecules in the environment and test all of them for ecotoxicity. Therefore, methodologies need to be developed to select for priority molecules. Priority molecules can be defined as molecules for which a monitoring strategy and eventually ecotoxicological assays are to be implemented. Similarly to the model used in the existing regulatory approaches for the assessment of the environmental risk of pharmaceuticals, the prioritization of pharmaceuticals can be based on risk quotients, as described in the current European Medicine Evaluation Agency (EMEA, 2006) guideline for environmental risk assessment (ERA) for pharmaceuticals. To build such risk ratios, PEC (predicted environmental concentration) and PNEC (predicted no effect concentration) values are required.

A number of studies have been conducted in the ten past years, with regard to environmental risk assessment (ERA) and/or strategies to prioritize pharmaceuticals. Determination of the exposure is quite similar in all studies but the assessment of the effect differs from a study to another: as a consequence of the lack of the ecotoxicological data, a number of works have focus on alternative ways to assess the biological effects of pharmaceuticals on non-target aquatic organisms; taking into account pharmacological data, quantitative structure relationships (QSARs), or even evolutionary conservation of molecular drug targets.

In this chapter we therefore propose to:

(1) Draw an overview of works conducted in different countries with regard to ERA and/or prioritization of human pharmaceuticals.
(2) Give an example of such a prioritization strategy, which uses ecotoxicological, but also human pharmacological and physico-chemical data.

©2010 IWA Publishing. *Pharmaceuticals in the Environment: Current knowledge and need assessment to reduce presence and impact*. Edited by Benoit Roig. ISBN: 9781843393146. Published by IWA Publishing, London, UK.

(3) Finally, discuss the different criteria used for the exposure and hazard assessment of pharmaceuticals, and notably parameters influencing the quantities of pharmaceuticals reaching the aquatic environment, and parameters used to assess their biological effects on aquatic species.

7.2 REVIEW OF ENVIRONMENTAL RISK ASSESSMENT AND PRIORITIZATION STRATEGIES FOR HUMAN PHARMACEUTICALS IN DIFFERENT COUNTRIES

For the last ten years, a number of risk assessment and/or prioritization strategies have been implemented in Europe and United-States. Here are presented the main works, by country and in chronological order.

7.2.1 Denmark

One of the first environmental risk assessment for human pharmaceuticals has been conducted in 2000 (Stuer-Lauridsen *et al.*, 2000). The methodology was based on the use of risk ratios Eq. (7.1). PEC values were calculated using the EU draft guideline for medicinal products for human use (EU 1994), which was a preliminary version of the EMEA guideline. Risk ratios and PEC for surface water were calculated using Eq. (7.2):

$$RCR = \frac{PECw}{PNEC} \tag{7.1}$$

$$PECw = \frac{A \times (100 - R)}{365 \times P \times V \times D \times 100} \tag{7.2}$$

RCR: *Risk ratio.* ***PECw:*** *Predicted Environmental Concentrations for water.* ***PNEC:*** *Predicted No effect Concentration.* ***A:*** *amount used per year (kg).* ***R:*** *removal in wastewater treatment plants, (in percent, set to 0 as a worst case hypothesis).* ***P:*** *number of inhabitants.* ***V:*** *volume of wastewater per day per capita.* ***D:*** *dilution factor in the environment (a default factor of 10 was used).*

The primary selection of molecules to assess was based on sales figures for human pharmaceuticals in Denmark, published by the Danish Medicine Agency. The 25 most used pharmaceuticals for the year 1997 were selected. Consumption data were given as Defined Daily Dose (DDD) for each active drug; where DDD is "*the assumed average maintenance dose per day for a drug used for its main indication in adults*" (www.whocc.no). The annual consumption for pharmaceuticals in weight units (kg) was then recalculated by multiplying number of DDD with the respective DDD value for each pharmaceutical. Finally, a list of 20 molecules was chosen for the purpose of the risk assessment.

Calculated PECs for surface water were then compared to the threshold value of 1 ng/L. All calculated PECs exceeded this cut-off value, meaning that risk could not be excluded for any of the compounds. Therefore, every of the considered pharmaceuticals had to be submitted to a refined assessment. All PEC values were between 1 and 100 ng/L and were within a factor of 2–5 of measured concentrations.

Risk ratios were calculated to assess the environmental risk, but the authors noticed that there was a lack of ecotoxicity data for pharmaceuticals, which precluded the calculation of risk quotients based on measured ecotoxicological values. Calculated risk ratios exceeded the cut-off value of one (indicating a risk) for ibuprofen, paracetamol and acetylsalicylic acid.

For a few compounds, estimated and experimental Kd values were available, and predicted concentrations in sewage sludge were calculated using Eqs. (7.3), (7.4) and (7.5). Results showed large differences in such predicted concentrations, depending on the mode estimation of Kd: experimental, estimated with Kow, or estimated with Dow.

$$\mathrm{Kd} = \mathrm{f_{oc}} \times 0.41 \times \mathrm{Kow} \tag{7.3}$$

$$\mathrm{Dow} = \frac{\mathrm{Kow}}{1 + 10^{\mathrm{pH-pKa}}} \tag{7.4}$$

$$\mathrm{C_{sludge}} = \frac{\dfrac{\mathrm{M_{act}}}{\mathrm{Vw}}}{\mathrm{Kd} + \mathrm{M_{sludge}}} = \frac{\mathrm{M_{act}} \times (\mathrm{Kd} + \mathrm{M_{sludge}})}{\mathrm{Vw}} \tag{7.5}$$

Kd: *sorption coefficient (L/kg).* $\boldsymbol{f_{oc}}$*: fraction of organic carbon in sludge (default value = 0.35).* ***Dow:*** *Kow corrected for ionisable compounds.* $\boldsymbol{C_{sludge}}$*: concentration of active substances sorbed on sludge.* $\boldsymbol{M_{act}}$*: annual consumption of active compound.* ***Vw:*** *total annual wastewater in the considered country.* $\boldsymbol{M_{sludge}}$*: annual sludge production.*

Main conclusions of this work were that:

- Only partial set of data were available for pharmaceuticals.
- Measured concentrations were within a factor of 2–5 of the PECs.
- Calculation of PNEC was possible only for a few compounds due to limited ecotoxicological data.
- No toxicity data were available for the terrestrial compartment.

7.2.2 United Kingdom

Another aquatic environmental assessment of pharmaceuticals has been realised in England by Jones *et al.* (2002). This assessment was performed on the same basis as the one done by Stuer-Lauridsen *et al.* (2000) and used PEC/PNEC ratios.

PEC values for the 25 most consumed pharmaceuticals, in terms of DDD, were calculated using Eq. (7.1) (EU, 1994). PNEC values were calculated using available ecotoxicological data or modelled using the ECOSAR software (USEPA, 2009). Risk ratios were greater than one (indicating a risk) only for 3 of the 25 selected pharmaceuticals: paracetamol, amoxicillin and mefenamic acid.

In addition to risk assessment, sludge concentrations were calculated as in Stuer-Lauridsen *et al.* (2000) and basic environmental modelling was performed with the determination of the Bioconcentration Factor (BCF) and removal rates and sludge sorption in wastewater treatment plants (STPs). These parameters were modelled using different software from the Estimation Program Interface (EPI) Suite from the USEPA (USEPA, 2009).

Conclusions of this study were similar to those drawn by Stuer-Lauridsen *et al.* (2000).

To assess the environmental risk for human pharmaceuticals, a report was commissioned from the UK Environment Agency. This report aimed at selecting some pharmaceuticals in order to conduct a targeted monitoring programme in a number of UK STPs and rivers. (Hilton *et al.*, 2003).The selection methodology was determined by crossing different approaches: EU technical guidance on risk assessment (EU 1996), OSPAR dynamec system (OSPAR, 2006; 2002) and literature review. This approach is displayed in Figure 7.1 and summarized below:

(1) Calculation of risk ratios using EMEA guideline, i.e., PEC and PNEC determination. Two lists of priority compounds were built using two different ways of calculation for PNEC values (Table 7.1):
- determination of PNEC using therapeutic dose of the active substance (PNECD = therapeutic dose/1000).
- determination of PNEC using a QSAR approach ($PNEC_T$ = QSAR value/100).
- PEC were calculated for the top 500 substances consumed in UK, using Eq. (7.2). Data were taken from the British Pharmaceutical Industry and excluded sales to hospital and sales of over-the-counter (pharmaceuticals delivered without prescription) in outlets other than pharmacies and dispensing doctors.

Table 7.1 Top ten substances identified by the prioritisation approaches following further screening (Hilton *et al.*, 2003).

Therapeutic dose approach	QSAR/experimental approach
Aminophylline	Lofepramine
Beclamethasone	Dextropropoxyphene
Theophylline	Procyclidine
Paracetamol	Tramadol
Norethisterone	Paracetamol
Codeine	Clotrimazole
Furosemide	Thiridazine
Atenolol	Mebeverine
Bendroflumethiazide	Terbinafine
Chlorphenamine	Tamoxifen

(2) *Effect assessment using the OSPAR PBT criterion.* In this system, each single substance is defined by three criteria: Persistence, Bioaccumulation and Toxicity (PBT), defined by physicochemical data. However, required data were available only for a few of the pharmaceuticals investigated.

The PBT index is part of the OSPAR (OSPAR, 2006; 2002) dynamec system to assess the hazard of single substances in the following terms:

- Persistence: ability to resist degradation in the aquatic environment,
- Bioaccumulation: accumulation in adipose tissue of aquatic organisms,
- Toxicity: potential to poison aquatic organisms.

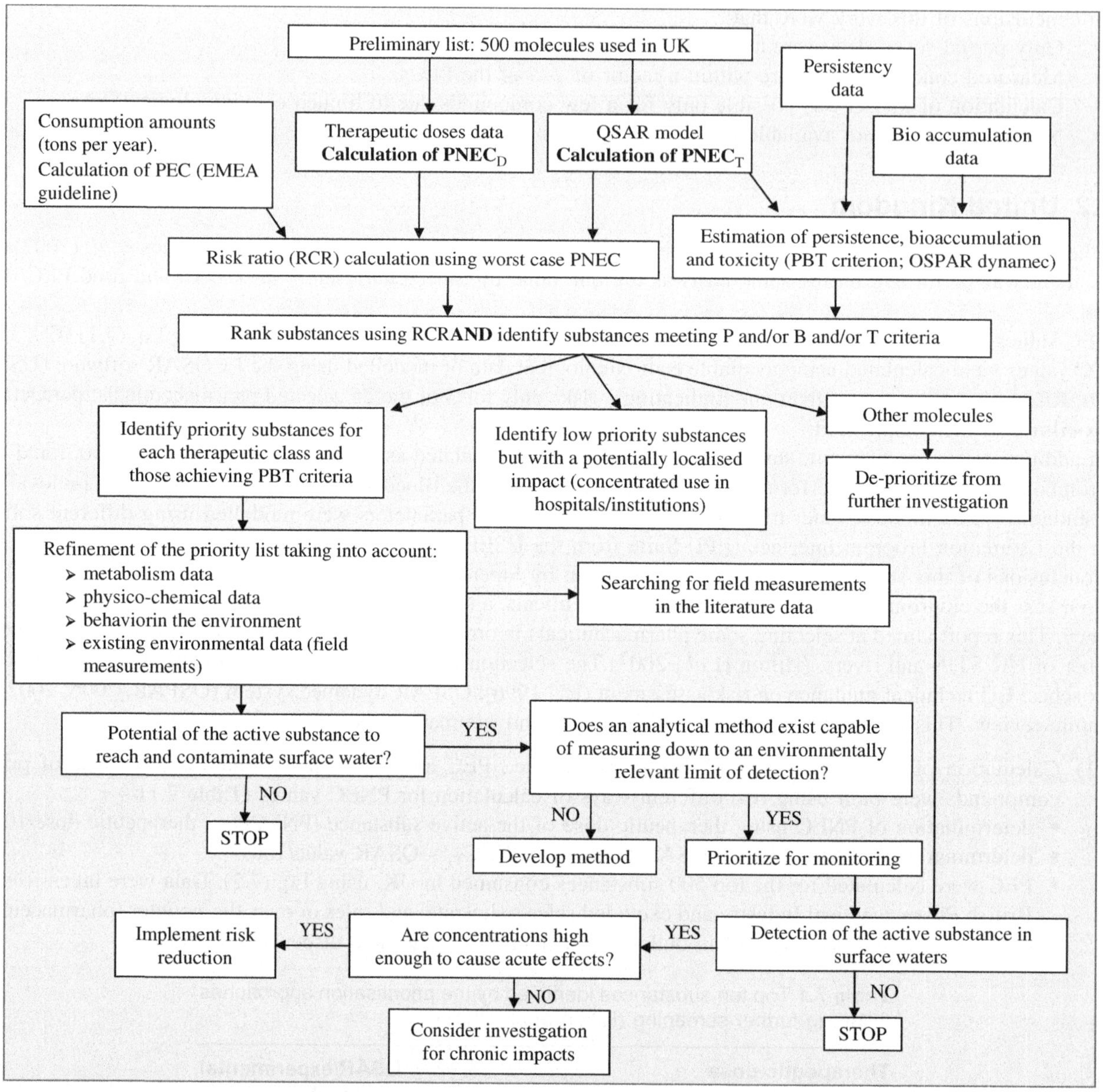

Figure 7.1 Tiered approach for selection of pharmaceuticals (Hilton *et al.*, 2003).

(3) *Review of literature data.* Existing data on occurrence of pharmaceutical substances and existence of analytical method or analytical feasibility were reviewed.

(4) *Derivation of a list of compounds to be included in the monitoring programme according to the following factors:*

- Is the substance near the top of the priority list?
- Is the substance highlighted by PBT criteria?
- Should substances across a range of different therapeutic classes be selected?
- Has the substance been previously detected in either surface waters or STP effluents?
- Is there a reliable analytical method available for the substance?

As this work was directed toward environmental monitoring, it seems that analytical criteria have been determining in the choice of the final priority list: the final list (Table 7.2) only contains 2 of the 19 pharmaceuticals formerly selected on the basis of PEC/PNEC ratios and OSPAR criterion.

Table 7.2 Pharmaceutical compounds selected for targeted monitoring (Hilton *et al.*, 2003).

Pharmaceutical compounds selected for targeted monitoring	
Trimethoprim	Erythromycin
Diclofenac	Dextropropoxyphene
Sulfamethoxazole	Lofepramine
Paracetamol	Tamoxifen
Mefenamic acid	Propranolol
Ibuprofen	

7.2.3 Germany

An extensive environmental risk assessment for the 111 highest-selling human drug substances (annual sales >5000 kg) was conducted in Germany (Huschek *et al.*, 2004). This ERA closely followed the EMEA's stepwise procedure. This procedure (EMEA 2003) is tiered in two main phases: exposure assessment, then fate and effect analysis (see Chapter 1 for details). Results at each step were discussed.

Phase I PECs were calculated using the following equation Eq. (7.6).

$$PEC\ surface = \frac{DOSEai \times Fpen}{WWinhab \times Dilution \times 100} \tag{7.6}$$

DOSEai: *Maximum daily dose consumed per inhabitant (mg/hab/day).* ***Fpen:*** *Percentage of market penetration (default value = 1%).* ***WWinhab:*** *Amount of wastewater per inhabitant per day (l/hab/day; default value = 200 l).* ***Dilution:*** *Dilution factor from STP to surface water (default value = 10).*

Phase 2 PECs were refined accordingly with EMEA requirements, i.e., Fpen refinement (Eq. 7.7), according to the 2003 version of the guideline; but also using actual annual sales of pharmaceuticals, as in previous[1] versions (Eq. 7.2). As PEC calculation has been modified in the 2003 version in comparison with previous versions of the guideline, authors compared PEC values obtained with each way of calculation and differences were noticed.

$$Fpen = \frac{consumption \times 100}{DDD \times hab \times 365} \tag{7.7}$$

consumption: *consumption of active substance (mg/year).* ***DDD:*** *defined daily dose (average daily dose per inhabitant).* ***hab:*** *number of inhabitants.*

As 2003 (and previous) version of the EMEA guideline did not take into account metabolism of pharmaceuticals, Huschek *et al.* choose to consider the fraction of pharmaceuticals excreted. They determined a PT_{index} (which is equivalent to the Fexcreta from EMEA guideline of 2006) to establish refined PEC values. Authors quoted the fact that excretion rates are important to consider for PEC refinement. This study was the first to introduce the excreted fraction of pharmaceuticals for the exposure assessment.

PEC/PNEC ratios were also calculated. Differences were noticed depending on the way of calculation of PECs. Ratios above 1 were achieved for antibiotics ciprofloxacin and clarithromycin in each case, but only with the PEC refinement calculation of the 2003 version of the guideline for acetylsalicylic acid, paracetamol, povidone iodine and ethinylestradiol.

Conclusions of this study were that:

- All of the 111 investigated molecules had PECs above the threshold value of 0.01 μg/L (10 ng/L), meaning that phase 2 environmental effect analysis should be performed for all substances with DDD values greater than 2 mg.
- Information on metabolism data should be considered for an initial risk assessment of potential risks of pharmaceuticals to the environment.
- Results showed that PEC values corrected by the PTindex (human metabolism) were comparable to STP effluent concentrations, except for active substances that metabolize and are biodegradable in STP.
- PNEC calculations were impaired by the lack of ecotoxicological data.

[1] Previous versions are EMEA 2001 and EMEA 2000. EU 1994 can also be considered as a draft version of EMEA guideline for human pharmaceuticals. EMEA 2006 is the version in current use.

7.2.4 Italy

In Italy, an assessment of the contamination of the aquatic compartment by human pharmaceuticals was conducted in 2005 (Zuccato *et al.*, 2005). This assessment, which is not strictly speaking an ERA, was tiered into three phases:

- Pre-selection of pharmaceuticals (prioritization),
- Sampling in STP and surface waters,
- Comparison of results from sampling campaign with theoretical prioritization.

The pre-selection was based on two main parameters: sales figures for pharmaceuticals (consumption amounts) and metabolism rate of the different compounds in the human body. Pharmaceuticals were ranked according to their predicted environmental loads, calculated by multiplying sales figures by the rate of metabolism in man (Eq. 7.8). Only the excretion rates of pharmaceuticals as unchanged molecule were considered and metabolites were not taken into account. According to the authors, this approach gave a good estimation of the quantities of active molecules reaching the environment.

$$\text{Load} = \text{prescription amount} \times \text{excretionrate} \quad (7.8)$$

Load: *predicted environmental load (tonnes).* ***Prescription amount:*** *sale figures (tonnes for year 2001).* ***Excretion rate:*** *excretion rate of pharmaceutical as unchanged molecule (%).*

To this first set of molecules, selected on predicted environmental loads, the authors added molecules known to be persistent and also molecules with high activity and potential toxicity but with low environmental load such as estrogens and anti-cancer drugs.

Field measurements in STP effluents revealed that ofloxacin, furosemide, atenolol, hydrochlorothiazide, carbamazepine, ranitidine, ciprofloxacin, sulfamethoxazole and ibuprofen were found in the range of the hundred ng/L.

Spiramycin, bezafibrate, erythromycin, lincomycin and clarithromycin were found in the ten ng/L range. Despite very high predicted environmental loads, amoxicillin had low concentrations in STP effluents, suggesting that this molecule was rapidly degraded in the environment.

In river waters (Po and Lambro rivers), concentrations were lower than in STP effluents with the exception of spiramycin, bezafibrate and lincomycin, for which concentrations were roughly the same. Table 7.3 displays the results for the molecules found in highest concentrations.

Table 7.3 Comparison of concentrations of pharmaceuticals, detected in STP effluents and surface water (Zuccato *et al.*, 2005).

Concentrations of pharmaceuticals detected in STP effluents and surface water (expressed in ng/L, median value)			
Compounds	**STP effluents**	**Lambro river**	**Po river**
Ofloxacin	600	306	33
Furosemide	585	254	32.5
Atenolol	466	241	23
Hydrochlorothiazide	439	255	17
Carbamazepine	291	175	13
Ranitidine	288	74	10
Ciprofloxacine	251	57	4.5
Sulfamethoxazole	127	24	3.5
Ibuprofen	121	20	3.2

Other selected molecules, salbutamol, cyclophosphamide, omeprazole, methotrexate, ethinylestradiol, enalapril and diazepam were detected at very low concentrations (in the ng/L range) or were below detection limits.

7.2.5 Sweden

In December 2002, as part of the work of achieving national environmental goals, a report was commissioned from the Swedish Medical Products Agency. The report was to include a risk assessment of environmental effects of pharmaceuticals based on their occurrence in the environment in relation to their current sales volumes (Carlsson *et al.*, 2006). The selection of pharmaceuticals was based on the list of the 100 most sold pharmaceuticals, in terms of DDD, for the year 2002. Twenty-seven pharmaceuticals were finally selected using the following criterions:

- molecule occurring on the list of the 100 most sold pharmaceuticals,
- molecule reported as observed in the environment,
- molecule otherwise highlighted in the literature.

The risk assessment was then conducted as described in the EMEA guideline (EMEA, 2003). However, for the calculation of PECs, the sold amount of pharmaceuticals was used as a measure of the factor "DOSEai x Fpen/100" (see Eq. 7.6). It was assumed that the entire amount of each product sold was consumed, and that the predicted consumed amount was evenly distributed over the year and throughout the population, as in Jones *et al.* (2002). Refined PEC were calculated using STP removal rates calculated with the Simple Treat 3.1 modelling software (Struijs *et al.*, 1991). Metabolism data were also took into account, the fraction of parent compound excreted after metabolism was included in the calculation. Metabolites lacking environmental and/or ecotoxicity data were regarded as parent compound and added to this fraction. PNEC values were calculated in accordance with EMEA guideline and the European TGD (EU, 2003) and finally risk ratios were calculated.

All risk quotients were below one, except for ethinylestradiol, estradiol, estriol and paracetamol. It is worth noting that except paracetamol, all the molecules displaying a risk quotient higher than one are steroid estrogens.

The main conclusions of this work were that:

- The acute risk linked to pharmaceuticals is negligible.
- The chronic risk cannot be ruled out.
- Ecotoxicological data are scarce.
- Risk assessment based on acute toxicity test do not adequately reflect the potential for long term exposure to sub-acute levels.
- The ecotoxic potency of a compound is to be taken into account at an early stage of the risk assessment, irrespective of the quantity released into the environment.
- Studies are to be directed toward bioaccumulation potential, indirect effects, and mixture effects.

More recently, a risk assessment has been implemented at a national scale in Sweden (www.janusinfo.se/environment). It aims to provide information for healthcare practitioners in order to reduce the residues of medicinal products in the ground, water and air. Pharmaceuticals are environmentally classified according to two criteria: environmental risk on one hand and environmental hazard, estimated by the PBT index (OSPAR 2006; 2002), on the other hand.

Environmental risk refers to acute toxicity for the aquatic environment. It is based on the ratio between predicted environmental concentration of the substance (PEC) and the highest concentration of the substance that does not have a harmful effect in the environment (PNEC).

Environmental hazard expresses the inherent environmentally damaging characteristics of the substance using the PBT (Persistence, Bioaccumulation, Toxicity) index. Each characteristic is assigned a numeric value (Table 7.4).

- Biodegradability is assessed based on criteria for ready biodegradation according to the OECD's test guidelines (test 301) or another equivalent test of biodegradability.
- Potential bioaccumulation is assessed according to the OECD based on the partition coefficient n-octanol/water, Log Kow, in which substances with log Kow >3 are judged to be potentially bioaccumulating (OECD test 107 or 117). If simulation or test data for bioaccumulation are available, these can be reported as an alternative.
- Toxicity for aquatic organisms is assessed based on the results of toxicity tests including three trophic levels; fish, Daphnia and algae (OECD test guidelines 203, 202 and 201, or equivalent). Data for the most sensitive organisms are used in the assessment, which is divided into four categories (Table 7.4).

In case of lack of data, a worst case scenario is applied, e.g., in the absence of bioaccumulation data the active substance is classified as potentially bioaccumulating.

Detailed assessment methodolgy can be found at www.janusinfo.se/environment and www.fass.se. Below is given an example of the proposed classification (Table 7.5).

Table 7.4 Parameters and cut-off values used in the environmental hazard assessment for pharmaceuticals classification (janusinfo.se/environment).

Parameter		Cut-off value	Ranking Value
Toxicity	very high toxicity	$LC/EC/IC_{50} < 1$ mg/L	3
	high toxicity	$1 < LC/EC/IC_{50} < 10$ mg/L	2
	moderate toxicity	$10 < LC/EC/IC_{50} < 100$ mg/L	1
	low toxicity	$LC/EC/IC_{50} > 100$ mg/L	0
Bioaccumulation (potentially bioaccumulating)	yes	Log Kow > 3	3
	no	Log Kow < 3	0
Persistence (readily biodegradable)	yes	ND	3
	no	ND	0

Table 7.5 Example of pharmaceuticals classification (janusinfo.se/environment).

Substance	Environmental risk	Environmental hazard				Volume in DDD
		PBT	P	B	T	
Amoxicillin	moderate	6	3	0	3	1 187 403
Ceftriaxone	insignificant	3	3	0	0	7 322
Ciprofloxacin	cannot be excluded	5	3	0	2	563 660
Fluoxetine	low	6	3	0	3	3 214 933
Paracetamol	low	5	3	0	2	24 494 039
Propranolol	moderate	3	0	0	3	1 166 665

7.2.6 United States

Jjemba (2006) conducted a study on the excretion and ecotoxicity of pharmaceuticals in the aquatic environment. The author investigated the correlation between several pharmacokinetic parameters and the behaviour of pharmaceuticals in the environment. He also proposed a simple equation (Eq. 7.9) to rank pharmaceuticals according to an ecotoxicity potential:

$$EP = \frac{T}{V \times NOEC} \tag{7.9}$$

EP: *ecotoxic potential.* ***T:*** *overall residence time of the compound in the environment.* ***V:*** *concentration of the compound in the environment.* ***NOEC:*** *No Observable Effect Concentration.*

Author suggests that the quantification of ecotoxicity should reflect an element of the overall residence time of the compound in question, its bioavailability to the organisms, and its concentration in the environment.

Time of residence is a main parameter of this model, but as no data were available at the moment for pharmaceuticals the author assumed a default value of 1 year (i.e., all year continuous discharge). For daphnids, preliminary results of this study indicate a high EP for ethinylestradiol and carbamazepine, and a low EP for diclofenac and ibuprofen.

In order to prioritize human pharmaceuticals, (Cooper *et al.*, 2008) compiled extensive data (Table 7.6) in a free access database (PEIAR database). In this study, pharmaceuticals were ranked using five different combinations (Table 7.6) of physico-chemical, occurence and toxicity data, to emphasize different risks.

Table 7.6 Data categories included in each of the five different rankings (Cooper *et al.*, 2008).

	Ranking				
	ECOSAR	All available data	All available data without ECOSAR	Most available data	Emphasis on aquatic environment
Biologigal half-life		X	X		
Environmental half-life		X	X	X	X
Solubility		X	X		X
Kow		X	X		X
Mammal LD_{50}[1]		X	X		
Crustacean LD_{50} or LC_{50}[1]		X	X		X
Fish EC_{50} or LD_{50}[1]		X	X		X
Surface water concentration		X	X		X
Effluent Concentration		X	X		
2004 prescriptions dispensed		X	X	X	
ECOSAR	X	X		X	

(1) acute data

An additional component of the rankings was an uncertainty value, calculated for each pharmaceutical and taking into account rankings of pharmaceuticals in each individual category.

Ranking results and compiled data are available at www.chbr.noaa.gov/PEIAR/.

Another study (Kostich and Lazorchak, 2008) assessed the environmental exposure and potential biological effects for the 50 most dispensed pharmaceuticals in USA. The consumption masses for pharmaceuticals were indirectly calculated from dollar sales data. Concentrations reaching the environment were preliminary calculated as amount of Active Pharmaceutical Ingredient (API), using Eq. (7.10). This equation takes into account metabolism inactivation (in the human body) and disposal of unused medicine (fraction wasted). For this last parameter, conservative arbitrary values were assumed: 15% for short-term therapy, 5% for long-term therapy and 33% for topical medicines.

$$\text{API activity} = (\text{M} \times (1 - \text{Fi}) \times (1 - \text{Fw})) + (\text{M} \times \text{Fw}) \quad (7.10)$$

M: *mass dispensed.* ***Fi:*** *fraction inactivated (by metabolism).* ***Fw:*** *fraction wasted.*

PEC for wastewaters and biosolids were subsequently calculated with Eqs. (7.11) and (7.12), respectively assuming complete partitioning of raw wastewater API residues into the aqueous phase, and assuming complete partitioning of raw wastewater API residues to biosolids.

$$\text{PEC wastewater} = \frac{\text{API activity introduced annually}}{\text{Annual wastewater volume}} \quad (7.11)$$

$$\text{PEC biosolids} = \frac{\text{API activity introduced annually}}{\text{Annual biosolids volume}} \quad (7.12)$$

Potential biological effects of individual pharmaceuticals, were assessed for human exposure, non human exposure and for antimicrobial effects. The two latter methodologies are summarized below:

(1) For non-human exposure, authors calculated a specific risk quotient (called "exposure multiple") for each pharmaceutical. Such a quotient was expressed as the ratio between wastewater PEC and human peak freely dissolved plasma concentration of the pharmaceutical (Eq. 7.13) and wass based on the following assumptions:

- human plasma concentration can be used as a surrogate measure of cellular sensitivity (i.e., activity on non-target organisms).
- via passive equilibration, the concentration of pharmaceutical dissolved in the modelled organism's extracellular fluid would approach the concentration dissolved in wastewater.

For non human exposure, exposure multiples were above one for 11 of the 50 pharmaceuticals investigated.

$$\text{EM} = \frac{\text{PEC}}{\text{Cmax}} \quad (7.13)$$

EM: *exposure multiple for non-human organisms.* ***PEC:*** *Predicted Environmental Concentration.* ***Cmax:*** *freely dissolved plasma concentration of the pharmaceutical (peak plasma concentration corrected by the fraction bound to proteins).*

(2) For antimicrobial effects, an exposure multiple was expressed for microbial inhibition as the ratio of wastewater PEC and the lowest minimum inhibitory concentration (MIC) found for a specific pharmaceutical. Moreover, potential of antimicrobials residues for selection of resistant bacteria was also evaluated. For microbial inhibition, 4 antibacterial by the 30 investigated had an exposure multiple above 1; the multiple for the mixture effect (assumed as the sum of all individual antibacterial multiples) was about 21.

In addition of such a risk assessment, authors reviewed the molecular targets of the selected studied pharmaceuticals and search for potential similar targets in non-human species. Indeed, one of the main questions addressed in the ERA of pharmaceuticals is the sensitivity of non-target taxa to human pharmaceuticals, due to high differences in biology and physiology. Molecular drug targets for pharmaceuticals were identified from a literature review. Then, human sequences of target proteins were retrieved through a database and non-mamalian sequences similar to putative human drug targets were searched.

Main conclusions of this work were that:

- Significant human, non-human and microbial impacts resulting from exposure to most pharmaceuticals are likely only possible through contact with concentrating sources, unexpected dose responses or in extremely sensitive population.
- For a small subset of pharmaceuticals, effects of direct aquatic exposure to single compounds cannot be ruled out for non target-organisms and microbes.
- Predicted wastewater concentrations of several antibiotics are sufficient to impair growth or select for low-level resistance in microorganisms.
- Analysis of conserved molecular mechanisms, along with the information about the physiological roles of molecular pathways in different organisms, can provide guidance on the range of species and the type of endpoints that should be considered in chronic toxicity studies.

7.2.7 Swiss

To identify pharmaceuticals with high environmental risk, ansd to evaluate if human metabolites' contribution to ecotoxicity needed to be accounted for, Lienert *et al.* (2007) proposed a screening tool mainly based on the use of modelling toxicity from quantitative structure activity relationships (QSAR). Their methodology did not consider chronic toxicity (due to lack of chronic data and appropriate QSAR) or specific toxic mode of action of parent drug or metabolite, but mainly baseline toxicity (i.e., narcosis). Nevertheless, in a more complex approach, the inclusion of known specific toxicity effects is possible and was shown for algal toxicity of β-Blockers. (Escher *et al.*, 2006).

The approach was based on literature data of human metabolism and excretion, pharmaceuticals sales data, physicochemical properties of parent drugs and metabolites, and ecotoxicological data when available. It allowed to estimate a risk quotient of parent drug and metabolite mixture (Eq. 7.14).

$$RQ_{mix} = \frac{PEC_{\text{wastewater}}}{PNEC_{\text{mix}}} \tag{7.14}$$

RQ

***RQ_{mix}:** Risk quotient for parent drug and metabolites.* ***PECwastewater:** Predicted Environmental Concentration for wastewaters.* ***PNECmix:** Predicted No Effect Concentration for parent drug and metabolites.*

When ecotoxicity data were missing, baseline acute toxicity was modelled with lipophilicity (D_{lipw}, pH7) estimates using QSAR. The mixture effect of each parent drug and its metabolites was treated with the model of concentration addition, assuming a similar mode of toxic action of all components. From a literature information, the main phase of their procedure consisted in a tiered estimation of the effect concentration (EC_{50}) of the mixture of the parent drug and metabolites, based on the assessment of the baseline toxicity of parent drug and next the relative potency of metabolites.

The relative toxicity potential of each metabolite was assessed by using the ratio between baseline toxicity of parent drug and modelled baseline toxicity of metabolite. Next it became possible to calculate the toxic potential of the mixture (when no toxicophore is present in metabolites).

This procedure gives help to screen the risk of pharmaceutical including their metabolites.

7.2.8 France

A risk assessment was conducted in France in 2006 (Besse *et al.*, 2008). This assessment was performed on the same basis as the previous ones (Carlsson *et al.*, 2006; Huschek *et al.*, 2004; Jones *et al.*, 2002; Stuer-Lauridsen *et al.*, 2000) and used PEC/PNEC ratios. Using year 2004 French drug consumption data, aquatic PECs for 112 parent molecules and several metabolites were calculated using Eq. (7.15), adapted from the EMEA guideline (EMEA 2006).

$$PEC\ surface\ water = \frac{consumption \times Fexcreta \times Fstp}{WWinhab \times hab \times Dilution \times 365} \tag{7.15}$$

***consumption**: the quantity (mg/year) of an active molecule consumed by the population during 1 year in a defined zone (generally a country).* ***Hab:** number of inhabitants and 100 the correction factor for the percentages.* ***365:** number of days per year (day/year).* ***WWinhab**: volume of wastewater per person per day (default value = 200 L/inhabitants/day).* ***Fexcreta:** excretion fraction of the active molecule.* ***Fstp**: fraction of emission of the drug from wastewater treatment plants (STP) directed to surface water, Fstp can be defined as (1-STP removal fraction).*

A special emphasis was made on human metabolism. Fexcreta values (proportion of drug excreted as unchanged molecule) were determined for almost 80 compounds. Moreover, several metabolites were identified as potentially of concern for the aquatic environment, due to their potential occurrence in the aquatic environment (PEC) and/or pharmacological activity.

PEC/PNEC ratios were calculated but, as ecotoxicological data were lacking, it was only possible to achieve risk quotients for 6 compounds according to the EMEA guideline (i.e., 3 NOEC values from 3 trophic levels). Using limited data (1 or 2 NOEC values) it was possible to determine risk quotients for a further 16 molecules. On these 22 pharmaceuticlas, only the antibiotic amoxicillin showed a ratio above one.

Conclusions of this work were that:

- Considering excretion fractions of pharmaceuticals can lead to drastically reduced predicted concentrations reaching the aquatic environment, and can help to target environmentally relevant pharmaceuticals and metabolites.
- Metabolites should be considered when performing a pharmaceutical ERA.
- Calculated PECs using the described methodology were consistent with French field measurements.
- Due to the lack of ecotoxicological data, the use of PEC/PNEC ratios is not enough informative to prioritize pharmaceuticals likely to pose a risk for surface waters.
- Alternative ways to prioritize risk to pharmaceuticals, combining PEC, pharmacological, and ecotoxicological data available from the literature, should be implemented.

Two prioritization strategies were also conducted in France. The first one, with regard to the use of pharmacological data to assess the biological effects of pharmaceuticals (Besse and Garric, 2008), is fully described in the paragraph 6.3 of the current chapter as it is given as a specific example. The second one, conducted by Jean (2008), with emphasis on bioaccumulation is summarized below.

In year 2008, a prioritization strategy was conducted for hospital effluents and with regard to consumption amounts in hospitals and bioaccumulation potential (Jean, 2008). Bioaccumulation potential was determined by reviewing modelled bioconcentration factor (BCF) from CAS database® values for a high number of pharmaceuticals. To take into account the ionization of pharmaceuticals in the environment, BCF were reviewed for four environmental relevant pH values ranging from 6 to 9. A compound was considered bioccumulating if at least one of the estimated BCF value was higher than the threshold value of 1000. The recommended value of TGD is set at 2000, however, a more conservative value of 1000 was chosen to stand for uncertainties of estimated values (which could differ from real ones).

The main assumption of this work is that, considering the environmental concentrations for pharmaceuticals (ng/L to µg/L range), the main risk is linked to chronic exposure and to bioaccumulation of pharmaceuticals in aquatic organisms. Such a bioaccumulation could lead to critical internal concentration in the organisms and toxicity. Indeed, as pharmaceutical compounds are continuously released in the aquatic environment, the risk of bioaccumulation is to consider.

A final list of 70 substances, potentially bioaccumulating, and with respect to hospital effluents was therefore obtained.

7.2.9 Sum up and compilation of data

In brief, the review provided here allowed to highlight the following facts:

- Most of the studies implemented in Europe use the European Medicines Agency (EMEA) guideline for ERA as a conceptual basis.
- Calculated PEC values are in general in good agreement with measured field values.
- Chronic ecotoxicological data are lacking, which makes hard to implement accurate ERAs.
- Some methodologies have been implemented in order to bypass the lack of ecotoxicological data.

Accuracy and reliability of the different parameters used to assess the exposure and the biological effects of pharmaceuticals are discussed in the paragraph 6.4 of the current chapter.

It is rather difficult to compare the results of risk assessment and prioritization strategies reviewed here; as methodologies, goals, and available information are more than heterogenous.

Table 7.7 is an attempt to compile such information to give an overview of the potentially priority compounds. It gathers available information on pharmaceuticals which have been studied in the framework of ERA or prioritization strategies (Stuer-Lauridsen *et al.*, 2000; Huschek *et al.*, 2004; Jones *et al.*, 2003; Kostich and Lazorchak, 2008; Cooper *et al.*, 2008; Carlsson *et al.*, 2006; Besse and Garric, 2008; Lienert *et al.*, 2007), but does not include studies which have been directed to environmental occurrence (Hilton *et al.*, 2003; Zuccato *et al.*, 2005).

This table does not give information on pharmaceuticals searched and/or found in the aquatic environment. If no definitive conclusion should be drawn from this table, on the priority level of pharmaceuticals for the aquatic environment, such a data compilation allows to point out which pharmaceuticals have been studied and highlights the extent of the remaining gaps.

7.3 EXAMPLE OF PRIORITIZATION STRATEGY

A prioritization strategy, with special emphasis on metabolism data and pharmacological data has been conducted in France in 2008 (Besse and Garric, 2008). This work, commissioned by a French river basin authority, was addressed in the framework of the French Plan National Santé Environnement (PNSE, 2004). In a first step, all ecotoxicological data (acute and chronic) were compiled in a database and a first risk assessment using PEC/PNEC ratios was implemented following the EMEA recommendations. From this approach, only a few pharmaceuticals were classified due to limited ecotoxicological data (Besse *et al.*, 2008). Consequently, a pragmatic prioritization approach was implemented, which aimed to identify priority human pharmaceuticals and metabolites to monitor in surface water. Moreover, this approach allowed to identify some data gaps that should be filled to allow a better assessment of these compounds. The approach used was based both on the calculation of PEC values and on a decision scheme based on selected quantitative and qualitative biological data: ecotoxicological, pharmacological and physicochemical data.

The prioritization scheme (fully detailed and available in Besse and Garric, 2008) consisted of three tiers:

(1) First, a preliminary exposure-based classification was established using PEC values.
(2) Second, this preliminary classification was reviewed using a case-by-case expert approach considering the potential environmental effect using ecotoxicological data, pharmacological data, mammalian toxicological data, physicochemical data, and available data on the environmental behaviour of pharmaceuticals.
(3) Finally, an additional selection was made in order to select priority compounds belonging to the same pharmacological and chemical class (i.e., compounds with the same chemical structure and the same mechanism of action [MoA]).

Table 7.7 Collection of the different established lists of parent pharmaceuticals from eight environmental risk assessment and prioritization strategies.

Compound name	Denmark[1]	Germany[2]	UK[3]	USA[4]	USA[5]	Sweden[6]	France[7]	Swiss[8]
Acetylcysteine							X	
Acetylsalicylic acid	X	X	X		72		X	X
Albuterol				X				
Allopurinol		X	X	X			X	X
Alprazolam				X			X	
Amiodarone							X	
Amitriptyline				X				
Amitriptyline hydrochloride					22			
Amlodipine	X			X				
Amoxicillin			X		95	X	X	X
Amphetamine				X	57			
Amphotericin B							X	
Ampicillin					73			
Atenolol			X	X	33	X	X	X
Atorvastatin				X			X	
Atorvastatin calcium					49			
Azithromycin				X	65		X	
Bacitracin					99			
Baclofen							X	
Bendroflumethiazide	X							
Benfluorex							X	
Benztropine				X				
Betamethasone				X			X	
Bezafibrate		X					X	X
Bisoprolol							X	X
Bromazepam							X	
Budesonide	X				41			
Buflomedil							X	
Buprenorphine							X	
Butalbital					80			
Butoconazole				X				
Caffeine					11			
Capsofungin				X				
Captopril		X			87			
Carbamazepine		X	X	X	23		X	X
Carbocysteine							X	
Carisoprodol					93			
Carvedilol					12		X	
Cefaclor				X	92			
Cefadroxil				X				
Cefdinir				X				
Cefpodoxime							X	
Cefprozil				X				
Ceftazidime							X	
Ceftriaxone				X			X	
Cefuroxime				X				
Cephalexin				X	81			
Cetirizine							X	
Chloramphenicol					26			
Chlortetracycline					43			
Cimetidine			X		35			
Ciprofloxacin		X		X	44		X	X
Citalopram	X						X	X
Clarithromycin		X		X	21		X	
Clavulanic acid							X	
Clindamycin				X				
Clofibrate					6			X
Clonazepam							X	
Clonidine				X				

(*continued*)

Table 7.7 (*continued*)

Compound name	Denmark[1]	Germany[2]	UK[3]	USA[4]	USA[5]	Sweden[6]	France[7]	Swiss[8]
Clonidine hydrochloride					63			
Clorazepate							X	
Clotrimazole				X				
Codeine phosphate					27			
Conjugated estrogens				X				
Cortisone								X
Cyamemazine							X	
Cyclophosphamide						X		
Destrogestrel	X							
Dextropropoxyphene				X		X	X	
Diazepam	X				19	X	X	X
Diclofenac		X	X			X	X	X
Diclofenac sodium					10			
Dicyclomine hydrochloride					34			
Digoxin	X			X	77			
Diltiazem			X	X	50			
Dimethyl sulfoxide		X						
Diosmin							X	
Domperidone							X	
Doxazosin					52			
Doxycycline				X			X	
Econazole				X				
Enalapril	X			X		X		
Enalapril maleate					67			
Enoxacin					89			
Erythromycin			X	X				X
Erythromycin base					1			
Escitalopram							X	
Estradiol	X			X	9	X		
Estriol						X		
Estrogen	X							
Ethinylestradiol		X		X	42	X		X
Felodipine					96			
Fenofibrate		X			30		X	X
Ferrous sulphate			X					
Fluconazole				X			X	
Flucloxacillin sodium			X					
Fluocinonide				X				
Fluoxetine							X	X
Fluoxetine hydrochloride					4			
Fluticasone				X				
Fluvoxamine							X	
Fosfomycin							X	
Fosinopril sodium					29			
Furosemide	X			X	31	X	X	
Gatifloxacin				X				
Gemfibrozil					17			X
Gentamycin				X				
Gestodene	X							X
Glibenclamide							X	
Gliclazide			X					
Glipizide				X				
Glyburide				X	91			
Haloperidol							X	
Heptaminol							X	
Histamine								
Hydrochlorothiazide	X	X		X		X		
Hydrocodone				X	71			

(*continued*)

Table 7.7 (*continued*)

Compound name	Denmark[1]	Germany[2]	UK[3]	USA[4]	USA[5]	Sweden[6]	France[7]	Swiss[8]
Hydrocortisone				X			X	
Hydrogen peroxyde	X							
Ibuprofen	X	X	X	X	7	X	X	X
Ifosfamide		X				X		
Insulin				X				
Irbesartan								
Isosorbide mononitrate				X	82			
Josamycin							X	
Kanamycin A					100			
Ketoconazole	X			X				
Ketoprofen						X	X	
Lactic acid producing organisms	X							
Lactitol		X						
Lansoprazole					88			
Levodopa							X	
Levofloxacin				X	58			
Levomepromazine							X	
Levothyroxine				X			X	
Linezolid				X				
Lincomycin							X	
Liothyronine				X				
Lisinopril				X	36			
Lomefloxacin					70			
Loperamide							X	
Loratadine							X	
Lorazepam							X	
Losartan							X	
Losartan potassium					84			
Loxapin							X	
Mebeverine			X					
Mefenamic acid			X					
Mesalazine			X					
Metamizole sodium		X						
Metformin		X	X	X		X	X	
Methotrexate								X
Methylphenidate HCL					75			
Methylprednisolone				X				
Metoclopramide							X	
Metoprolol				X		X	X	X
Metronidazole				X	13		X	
Methocarbamol					90			
Methotrexate								X
Mianserine							X	
Midazolam							X	
Minocycline hydrochloride					79			
Mitomycin								X
Moxifloxacin				X				
Mupirocin				X				
Nadolol							X	X
Naftidrofuryl							X	
Naproxen			X		18	X	X	
Naproxen sodium					78			
Neomycin				X				
Nicardipine							X	
Nitrafurantoin				X				
Nitrazepam	X							
Nitrofurazone					25			
Nitroglycerin				X	5			
Nordazepam							X	

(*continued*)

Table 7.7 *(continued)*

Compound name	Denmark[1]	Germany[2]	UK[3]	USA[4]	USA[5]	Sweden[6]	France[7]	Swiss[8]
Norethisterone				X	46	X		
Norfloxacin					54			X
Nystatin				X	51			
Ofloxacin							X	
Omeprazole					62		X	
Ondansetron							X	
Oxazepam						X	X	
Oxprenolol							X	
Oxytetracycline			X		2	X		
Pantoprazole							X	
Paracetamol	X	X	X	X	8	X	X	X
Paroxetine					20		X	
Paroxetine hydrochloride					55			
Penicillin v				X				
Perindopril							X	
Phenazone								
Phenobarbital					28		X	
Phenoxymethylpenicillin			X					
Piperacillin				X			X	
Piracetam								
Piroxicam							X	
Potassium chloride	X							
Povidone-iodine		X						
Pravastatin							X	
Prazepam							X	
Prednisolone				X			X	
Prednisone				X	66		X	
Pristinamycin							X	
Procyclidine							X	
Promethazine				X	45			
Propranolol		X		X			X	X
Propranolol hydrochloride					16			
Propyphenazone								
Quinine sulphate			X					
Ramipril							X	
Ranitidine		X	X			X	X	
Rifampicin							X	
Risperidone					24			
Roxytrhromycin		X			69		X	
Salbutamol	X				59	X		
Salicylic acid		X					X	
Sertraline				X			X	
Sertraline hydrochloride					47			
Simvastatin				X	38	X	X	
Streptomycin					37			
Sotalol								X
Spiramycin								
Sulfamethoxazole				X	3		X	X
Sulfadiazine					56			
Sulfafurazole					85			
Sulfamethazine					32			
Sulphasalazine			X					
Tazobactam							X	
Teicoplanine							X	
Terbinafine				X				
Terbutaline	X					X	X	
Tetracycline				X		X		
Tetracycline hydrochloride					15			
Testosterone								X

(continued)

Table 7.7 (*continued*)

Compound name	Denmark[1]	Germany[2]	UK[3]	USA[4]	USA[5]	Sweden[6]	France[7]	Swiss[8]
Theophylline		X		X				
Tianeptine							X	
Tobramycin				X	97			
Tramadol							X	
Tramadol hydrochloride					60			
Triamterene				X				
Trihexyphenidyle							X	
Trimebutine							X	
Trimetazidine							X	
Trimethoprim				X	14		X	
Tropatepine							X	
Troxerutine							X	
Valproic acid			X				X	
Valsartan				X				
Vancomycin					98		X	
Verapamil				X				
Verapamil hydrochloride					39			
Warfarin sodium salt				X	83	X		
Xylometazolin	X							
Zolpidem							X	
Zolpidem tartrate					53			
Zopiclone	X						X	

X or number indicate an investigated compound in the corresponding study; (1) Stuer-Lauridsen *et al.*, 2000. Priority compounds are those for which PEC/PNEC ratios were equal or above one; (2) Huschek *et al.*, 2004. Priority compounds are those for which PEC/PNEC ratios were equal or above one. Not all the 111 compounds investigated are included in this table because complete list was not available through the study of Huschek *et al.*, 2004; (3) Jones *et al.*, 2003. Priority compounds are those for which PEC/PNEC ratios were equal or above one; (4) Kostich and Lazorchak (2008). Priority compounds are those for which an exposure multiple was equal or above one (see Kostich and Lazorchak 2008 for details). Only the 50 first rank compounds and antibacterials compounds, for which an exposure multiple was calculated, are indicated in table 7 (87 coumpounds); (5) Cooper *et al.*, 2008. Numbers indicate the ranking of pharmaceuticals according to the "aquatic-env" (emphasis on aquatic environment) classification (see Cooper *et al.*, 2008 for details). Only the 100 first ranked compounds are indicated. We arbitrary selected the 20 first ranked compounds as priority compounds for table 7; (6) Carlsson *et al.*, 2006. Priority compounds are those for which PEC/PNEC ratios were above one; (7) Besse and Garric 2008. Priority compounds are indicated according to the methodology described in Besse and Garric (2008) and presented in paragraph 6.3. of the current chapter; (8) Lienert *et al.*, 2007. Priority compound is the one for which RQmix is higher than one (*see Lienert et al., 2007 for details*).

The candidate pharmaceuticals list was based on the data provided by the French Health Product Agency (Agence Française de Sécurité Sanitaire des Produits de Santé, AFSSAPS, Paris), which covered exhaustive sales data for drugs delivered in France for hospitals and pharmacies, including over-the-counter drugs, for the year 2004.

The candidate list[2] consisted in 120 parent pharmaceuticals and 30 metabolites, selected metabolites were targeted in a previous work (Besse *et al.*, 2008).

7.3.1 Exposure assessment

The first step of the prioritization strategy classified pharmaceuticals compounds according to their probable exposure concentration in the aquatic environment. This was based upon the premise that the pharmaceuticals used in higher amounts have a potential to reach the aquatic environment in greater quantities and therefore may present a higher risk for the aquatic environment. The exposure-based classification implemented was derived from the EMEA guideline (EMEA, 2006) and used Eqs. (7.16) and (7.17).

$$PECa = \frac{consumption}{WWinhab \times hab \times Dilution \times 365} \tag{7.16}$$

[2] Cytotoxic compounds and sexual steroids (estrogens, progestagens and androgens) were excluded from this strategy. As these molecules display specific MoAs, authors considered that they had to be assessed in a specific procedure; moreover biocides and X-ray contrast media compounds were excluded from this work.

$$PECb = \frac{consumption \times Fexcreta}{WWinhab \times hab \times Dilution \times 365} \quad (7.17)$$

Consumption* (*mg/year*)*: quantity of an active molecule consumed in France (AFSSAPS data).* ***WWinhab: *volume of wastewater per person per day (default of 200 l/inhab/da*[1]*).* ***Fexcreta:*** *excretion fraction of the unchanged active molecule (parent drug or metabolite).* ***Dilution:*** *dilution factor from WWTP effluents to surface water (default value of 10).* ***hab:*** *number of inhabitants in the defined zone (set at 60 million for France).*

Pharmaceuticals were subsequently ranked accordingly and classified in 6 different classes summarized in Table 7.8.

Table 7.8 Ranking of pharmaceuticals according to their exposure assessment (Besse and Garric, 2008).

Priority class	Priority rank according to the exposure criteria	Comments
I A	Highest risk compounds	PECa and PECb higher than 100 ng/l. High consumption and limited metabolism.
I B	Potentially hazardous compounds but limited data	PECa higher than 100 ng/l. High consumption. No data on metabolism
II A	Potentially hazardous compounds	PECa higher than 100 ng/l and PECb higher than 10 ng/l. High consumption and intermediate metabolism.
II B	Unclassified priority risk	PECa lower than 100 ng/l but higher than 10 ng/l. No data on metabolism. No definitive conclusion, need further investigation
III	Very low risk for the environment (extensive metabolism)	PECa higher than 100 ng/l but PECb lower than 10 ng/l. High consumption but extensive metabolism.
IV	Very low risk for the environment (low consumption amount)	PECa lower than 10 ng/l. Low consumption amount.

7.3.2 Effect assessment

Second and third phase were dedicated to the effect assessment. In the second phase, pharmaceuticals were reviewed according to ecotoxicological, physicochemical and pharmacological data with emphasis on human toxicity (mechanism of action, side effects, specific organ toxicity, cytochrome and para-glycoprotein P modulation). The implemented strategy is summarized in Figure 7.2.

This effect assessment was based on the following assumptions:

- For non-mammalian animals with targets similar to those of mammals, biological effects may occur.
- Known adverse human side effects of pharmaceuticals may also be valuable to indicate potential harmful effects on non-target organisms.
- Several APIs are known to interact with Cytochrome P-450, therefore there is a potential risk of disruption in the homeostasis of non-target organisms.
- Several pharmaceuticals are known to interact with Glycoprotein-P (P-gp), a multidrug transporter that actively transports xenobiotics out of the cell, preventing the accumulation of toxic compounds; therefore an inhibition of its expression by a specific drug could result in enhancing the sensitivity of organisms to environmental pollutants.

(a) Yes, if the compound meets at least two of the selected criteria (e.g.,, relevant side effect and cytochrome inhibition). (b) For priority compounds selected based on pharmacological data, the final inclusion on the priority list is to be discussed case-by-case by taking into account PEC values and biological data.

Finally, to select priority pharmaceuticals displaying the same MoA, an additional step was implemented using available data: NOEC, DDD and PEC values for pharmaceuticals.

7.3.3 Results and discussion

A final priority list of 40 compounds and 14 metabolites (Table 7.9) was obtained with several therapeutic and chemical classes represented. 21 parent compounds have been already detected in surface waters (mainly β-blockers, anti-inflammatories and antibiotics). 5 metabolites have also been measured in surface waters, other metabolites have not been searched yet. PEC values refined by STP removal rates were in good agreement with field measurements.

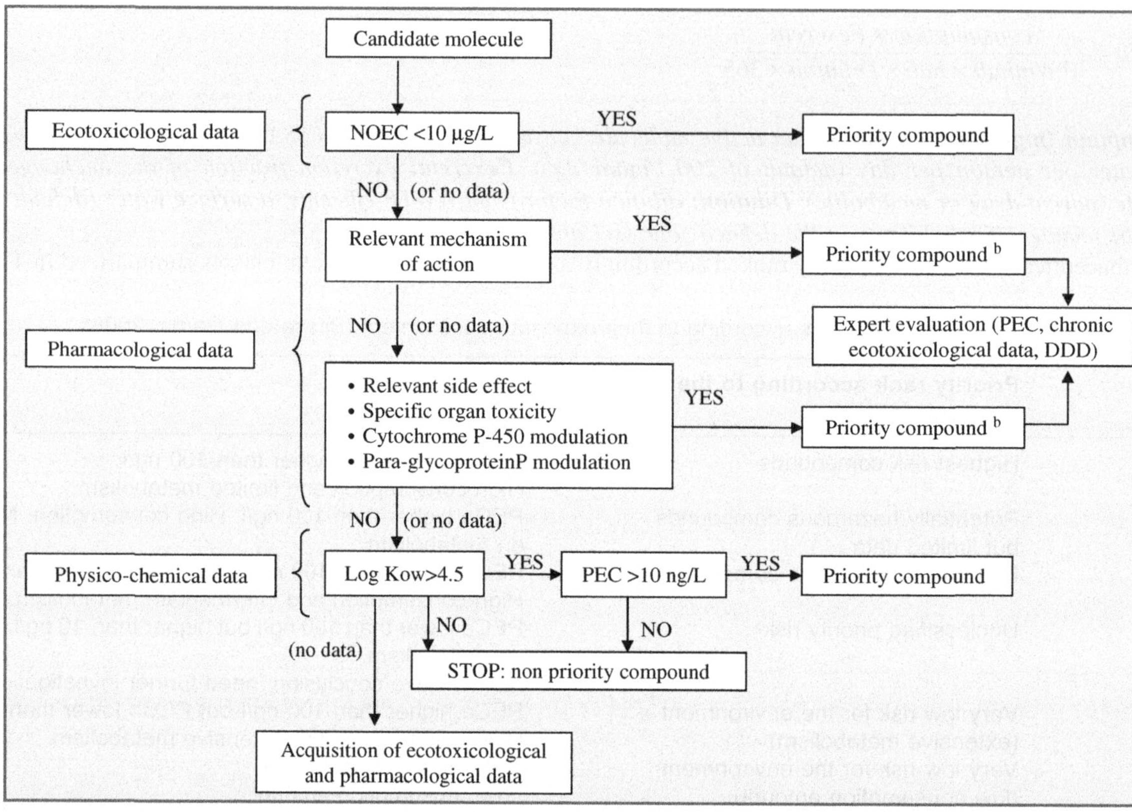

Figure 7.2 Prioritization of pharmaceuticals based on ecotoxicological, pharmacological and physicochemical data.

Table 7.9 List of priority metabolites (Besse and Garric, 2008).

Metabolite	Parent compound	Reason(s) for inclusion on priority list	Pharmacological activity
Salicylic acid	Aspirin	Active metabolite of the prodrug	Responsible for pharmacological activity
Fenofibric acid	Fenofibrate	Active metabolite of the prodrug	Responsible for pharmacological activity
Perindoprilat	Perindopril	Active metabolite of the prodrug	Responsible for pharmacological activity
Ramiprilat	Ramipril	Active metabolite of the prodrug	Responsible for pharmacological activity
Demethyltramadol	Tramadol	Active high excretion rate (60%)	Active analgesic activity
Hydroxy-ibuprofen	Ibuprofen	High excretion rate (25%)	Inactive
Carboxy-ibuprofen	Ibuprofen	High excretion rate (37%)	Inactive
Acetylsulfamethoxazole	Sulfamethoxazole	High excretion rate (60%)	Inactive
14-OH-clarithromycin	Clarithromycin	Active synergy with parent drug	Active
Norfluoxetine	Fluoxetine	Active high excretion rate (28%)	Equipotent to parent drug
OH-metronidazole	Metronidazole	Active high excretion rate (28%)	30% to 65% of metronidazole activity
β-hydroxy-acid metabolite	Simvastatin	Active metabolite of the prodrug	Responsible for pharmacological activity
2-OH-atorvastatin	Atorvastatin	Active metabolite of the prodrug	Responsible for pharmacological activity
4-OH-atorvastatin	Atorvastatin	Active metabolite of the prodrug	Responsible for pharmacological activity

Even if the strategy is widely applicable in different countries, the priority list obtained may not be relevant for others countries than France, mainly due to differences in the consumption profile of pharmaceuticals.

The prioritization strategy was successfully applied to the French situation; however, this methodology can be improved:

- STP removal rates should be included in the PEC calculation, when available.
- Fate data (biodegradation, photodegradation and hydrolysis time) should also be included.
- Degradation by-products of pharmaceuticals should be taken into account.

Finally, there is a need to investigate the relevance of use of pharmacological data to assess the risk of pharmaceuticals for non mammal organisms.

7.4 DISCUSSION

Several risk assessment and prioritization strategies have been conducted for the last ten years, especially in the last two years. Here are discussed the different methodologies and the perspectives given by these different works with regard to exposure and effect assessment.

7.4.1 Exposure assessment

Determining which compounds and in which concentrations they can reach the aquatic environment is obviously the first step of any ERA/prioritization strategy. To do so, a number of parameters are to be considered. Figure 7.3 summarizes the different factors that can modify the quantities of pharmaceuticals reaching wastewaters and surface waters.

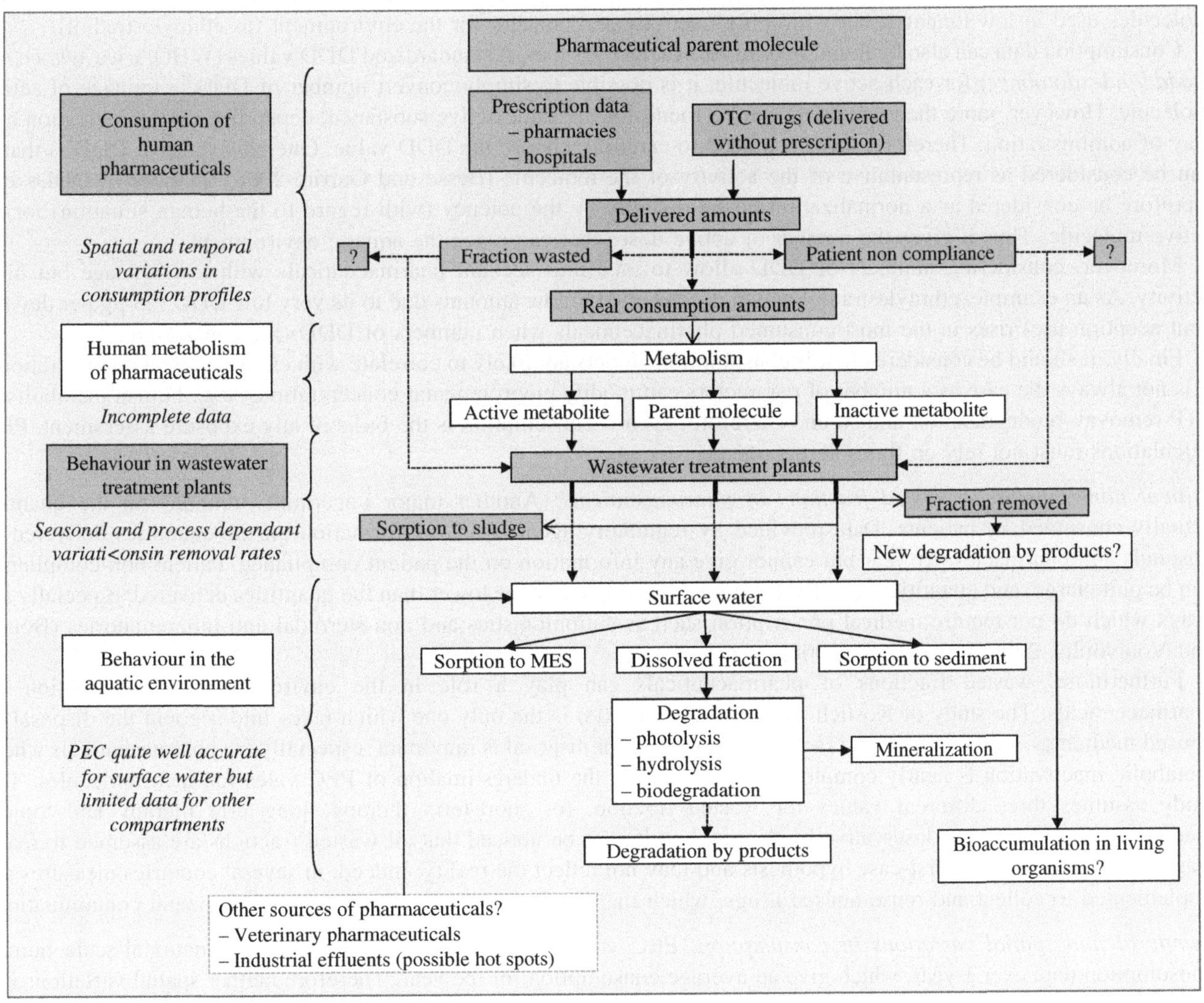

Figure 7.3 Factors that can affect the quantities of pharmaceuticals reaching the aquatic environment.

7.4.1.1 Accuracy of PEC calculation for surface water with field measurements

Most of the conducted studies are based, for the exposure assessment, on the EMEA guideline. The exposure assessment for surface water seems to be quite well estimated. Comparison of PECs with field measurements is generally in good agreement, especially if we consider the simplicity of the models used and the remaining uncertainties. A number of authors agree with this conclusion (Bound and Voulvoulis, 2006; Besse *et al.*, 2008; Castiglioni *et al.*, 2004; Huschek *et al.*, 2004; Carlsson *et al.*, 2006). Conversely, some other studies highlight the limits of such predictions at a local scale (Coetsier *et al.*, 2009) or for specific pharmaceuticals (Siemens *et al.*,

2008). Indeed, some uncertainties are to be taken into consideration in the assessment of the exposure, and are discussed below.

7.4.1.2 Parameters, uncertainties and limitations in the exposure assessment for surface water

Consumption amounts: In every study, the basis of the exposure assessment is the actual amounts of pharmaceuticals dispensed over a year. These actual amounts can be found in tonnage (kg), number of DDDs for every single active molecule, or even in terms of total dollar value of product sold; depending on the availability of the data.

About the consumption data, there is a need to carefully consider whether or not, the consumption amounts are based on prescription data only, or also include over-the-counter (OTC) drugs (drugs which are delivered without prescription) as this can lead to large discrepancies in the estimation of concentrations entering the environment.

Actual amounts in terms of tonnage are the most simple and direct to use, however, the availability of such data depends on the existence of regulatory agencies in a country and the possibility to share such data. It is to be noted that an assessment of pharmaceuticals based on the most consumed pharmaceuticals in terms of tonnage can ignore molecules used in low tonnages but with a high activity and concern for the environment (as ethinylestradiol).

Consumption data can also be found in number of DDD. As it exist standardized DDD values (WHO, *www.whocc.no/atcddd/indexdatabase*) for each active molecule, it is possible to simply convert number of DDD in tonnage of active molecule. However, more than one DDD can be found for the same active substance, depending on the indication and way of administration. Therefore, there is a need to carefully choose the DDD value. One advantage of DDD is that it can be considered as representative of the activity of the molecule (Besse and Garric, 2008); number of DDDs can therefore be considered as a normalization of the quantity by the potency (with regard to the human situation) for an active molecule. Thus it gives the number of active doses that can reach the aquatic environment.

Moreover, considering numbers of DDD allow to take into account pharmaceuticals with low tonnage but high activity. As an example, ethinylestradiol which is consumed in low amounts due to its very low DDD (25 μg per day for contraception use) rises in the most consumed pharmaceuticals when numbers of DDDs are considered.

Finally, it should be considered that if consumption amouts are likely to correlate with environmental concentrations, it is not always the case as a number of parameters can modify environmental concentrations, e.g., human metabolism, STP removal, biodegradation and so on. Therefore, even if consumption is the basis of any exposure assessment, PEC calculations must not rely on this sole parameter.

Patient non-compliance/Wasted fractions of pharmaceuticals: Another major uncertainty remains on the quantity actually consumed by patients. Data provided by regulatory agencies give information on the quantities delivered in hospitals and pharmacies per year but cannot give any information on the patient compliance. Patient non-compliance can be quite large, and quantities actually consumed by people may be lower than the quantities delivered, especially for drugs which do not require medical prescription such as antiphlogistics and non-steroidal anti-inflammatories (Bound and Voulvoulis, 2006; Besse *et al.*, 2008).

Furthermore, wasted fractions of pharmaceuticals can play a role in the environmental contamination by pharmaceuticals. The study of Kostich and Lazorchak (2008) is the only one which takes into account the disposal of unused medicines. Authors quote the fact that accounting for disposal is important, especially for pharmaceuticals where metabolic inactivation is nearly complete, as it can avoid the underestimation of PEC vales for such molecules. The study assumes three different values for wasted fraction, for short-term therapy, long-term therapy and topical medicines, considered as conservative by the authors. It is to be noticed that all wasted fractions are assumed to reach wastewaters, which is a worst-case hypothesis and may not reflect the reality: indeed, in several countries measures are implemented to collect and reuse unused drugs, which therefore contribute to limit the environmental contamination.

Temporal and spatial variations in consumption: PEC values are generally calculated using national-scale human consumption data over 1 year, which give an average consumption for the year. Therefore neither spatial variations nor temporal variations can be taken into account.

Spatial variations can include specific local consumption of pharmaceuticals, which may differ from one region to another. Therefore calculated PECs with national consumption amounts may somewhat vary from local field measurement. Moreover, at a national scale, it should be considered that consumption profiles are likely to differ from a country to another. Therefore, a priority list implemented from a country should not automatically stand for another one.

Temporal variations especially concern pharmaceuticals used in acute treatments, such as antibiotics or antihistaminics: quantities consumed vary over the year and can present seasonal outbreaks. Consequently aquatic concentrations can change throughout the year. This has been recently shown for antihistaminics (Kosonen and Kronberg, 2009). Finally, it should be considered that consumption of pharmaceuticals will also vary over years, as some new marketed drugs take over from more ancient ones; PECs, ERAs and priority lists have therefore to be updated accordingly and regularly.

Metabolism of pharmaceuticals: Human metabolism is an important step which must be taken into account for several reasons:

(1) Metabolism is the primary transformation and inactivation process that undergo pharmaceuticals; therefore, metabolism can lead to drastically reduced amounts of parent pharmaceuticals reaching wastewaters.
(2) Metabolism can give rise to new molecules that have can be of environmental concern if excreted in significant amounts.
(3) Parent compounds and active metabolites with the same MoA are likely to act additively.

Some suggestions have been made in order to include metabolites in ERA procedures (Besse *et al.*, 2008):

- Metabolites should be considered for ERA, unless they are excreted at low rates. Because it is not possible to propose a threshold excretion value using available data (metabolism, occurrence, and ecotoxicity), a threshold value of 10% could be assumed, which is the value proposed by the EMEA (EMEA 2006) to assess the relevance of a metabolic fraction of a pharmaceutical compound.
- Active metabolites with the same mechanism of action than parent compounds should be considered when implementing ERA for pharmaceuticals: PEC of active metabolites should be added to PEC of parent compounds when assessing the environmental risk.
- Metabolites (pharmacologically active or not) with an excretion fraction greater than the parent compound should be considered relevant for the aquatic environment.

Moreover, it should be noted that activity of a metabolite is generally considered according to its activity in mammals and especially humans (i.e., pharmacological activity). However the fact that a metabolite is "pharmacologically inactive" does not necessarily mean that such a compound has no effects on non-target aquatic organisms. Therefore, and because there are no data on the toxicity of pharmacologically inactive metabolites on such organisms, metabolites should be considered on a case-by case approach considering the amount of parent compound and their excretion fraction.

Metabolism data are generally available for pharmaceuticals; however, data can be scarce and incomplete, which sometimes hinder the calculation of reliable excretion rates values (Besse *et al.*, 2008; Huschek *et al.*, 2004).

Industrial sources of pharmaceuticals: The large majority of studies focus on environmental contamination through STP and originating from human consumption. However, other sources of contamination such as industrial effluents (production facilities and conditioning facilities) should be considered: such sources could result in hot spots of contamination and therefore increase environmental concentrations at a local scale. A study on pharmaceuticals occurrence in STP effluent serving about 90 bulk drug manufacturers in India revealed extremely high levels of drugs: from 90 μg/L for ranitidin up to 31 mg/L for ciprofloxacin (Larsson *et al.*, 2007).

Veterinary pharmaceuticals: As some pharmaceutical compounds are used both in human and veterinary medicine, there are still uncertainties about the actual amounts of pharmaceuticals reaching surface waters. This is particularly the case for antibiotic and antiprotozoal compounds. Theoretically, including veterinary consumption is likely to ensure a more comprehensive PEC. However, routes of administration and ways of reaching the aquatic environment differ between veterinary and human pharmaceuticals (Besse *et al.*, 2008); therefore, specific methodologies for assessing exposure for veterinary compounds are needed. Several studies have focused on these latter compounds (Kools *et al.*, 2008; Capleton *et al.*, 2006; Boxall *et al.*, 2003).

Removal in STPs: Uncertainties also lie in the removal rates of pharmaceuticals in STPs. Review of available data shows that existing values are scarce and that there is a high heterogeneity in removal rates. STP efficiency toward pharmaceuticals can be influenced by the season (Castiglioni *et al.*, 2006), and also by the processes used (Miège *et al.*, 2008), therefore leading to varying surface water concentrations among the year. As most of the values reported in the literature are "snapshots" removal rates, i.e., generally based on a grab or 24h composite sample, and with regard to a specific season and STP process, it is difficult to obtain universal values to include in PEC calculations. The extensive review of Onesios *et al.* (2008) on STP removal rates for pharmaceuticals is a good example of the variability of such measured rates.

Finally, results of a study (Yu *et al.*, 2006) on 12 pharmaceuticals belonging to different therapeutical classes suggested that the extents of removals were highly variable and could not be correlated to drug classification or structure.

Degradation of pharmaceuticals in the environment: Abiotic and biotic degradation processes that can occur in surface waters are also important factors to consider. Like human metabolism, such processes are able to modify environmental concentrations and/or lead to the formation of by-products of potential concern. As an example, it has been shown that some photodegradation metabolites could be more toxic than the parent compounds (Isidori *et al.*, 2006; 2005; DellaGreca *et al.*, 2004).

Abiotic processes are reported to be most important ones (Fent *et al.*, 2006). Photolysis and hydrolysis were suggested to be rapid ways of removal of amoxicillin in the environment (Andreozzi *et al.*, 2004). This statement was supported by the fact that amoxicillin was only detected in surface waters at low levels (Zuccato *et al.*, 2005; Paffoni *et al.*, 2006). The β-blocker propranolol was reported to be rapidly photodegraded and therefore was not expected to be persistent in surface waters (Qin-Tao and Williams, 2007).

Conversely, results from a study assessing the biodegradability of pharmaceuticals in batch experiments (Yu *et al.*, 2006) showed that if abiotic degradation was low, biotic degradation for some pharmaceuticals was important (for aerobic biodegradation experiments of selected pharmaceuticals, 8 of the 12 drugs tested exhibited a biotransformation rate greater than 80% after 50 days of incubation).

On the other hand, most of the pharmaceuticals are continuously released in the environment. This fact could balance the environmental degradation processes in the environment, therefore some authors have suggested that pharmaceuticals should be considered as "pseudo-persistent" contaminants, due to this continuous release (Daughton and Ternes, 1999). Moreover, some pharmaceuticals such as carbamazepine are known to persist in surface waters.

Default values used in PEC calculation: Other great uncertainties lie in the default values proposed by the model used to calculate PECs, such as volume of wastewater and dilution factor used in the EMEA guideline. As an example for France, the default value for quantities of effluent is set to 200 L/inhab/day which is a mean value that can be accepted at the national scale. However, for some specific regions, this value may drop to 150 L/inhab/day or even lower; resulting in higher PEC values. Moreover, dilution factor from STP to surface waters is often set at a default value (10) which is not representative for local specifications. To this extent, PEC values calculated for the STP effluents are more reliable than surface water PECs (Besse *et al.*, 2008). This was recently highlighted in the study of Coetsier *et al.* (2009): if the authors found a good agreement between predicted and measured concentrations in STP effluent, not such a good match was observed for surface waters.

7.4.1.3 Calculation of PEC for other compartments (sludge and sediment)

PECs are often calculated for surface waters but more rarely for sewage sludge, suspended matter or sediment. (Stuer-Lauridsen *et al.*, 2000; Jones *et al.*, 2002). In their risk assessment for pharmaceuticals, Stuer-laurdisen *et al.* (2000) calculated predicted concentrations for sludge using estimated and calculated Kd (partition coefficient). Results showed that there were large differences (several orders of magnitude) in the values of PEC depending on the way of calculation and the Kd value (Table 7.10).

Table 7.10 Predicted concentration in sludge based on estimated and experimental Kd (Stuer-lauridsen *et al.*, 2000).

	Predicted concentration in sludge based on estimated and experimental Kd (μg/kg)		
	Kd estimated from Kow	**Kd estimated from Dow**	**Experimental Kd**
Furosemide	150	0.012	1470
Ibuprofen	34210	180	20330
Oxytetracycline	0.026	–	1990
Ciprofloxacin	0.00095	–	130

To our knowledge, no PEC calculation has been conducted for water sediment. This is a gap as some pharmaceuticals can be found in sediment (Feitosa-Felizzola and Chiron, 2009; Petrovic *et al.*, 2002; Daughton and Ternes, 1999).

7.4.1.4 Use of calculated and modelled values to assess the environmental behaviour of pharmaceuticals

Modelled biodegradation of pharmaceuticals: Relative biodegradabilities of chemicals can be predicted using softwares such as the BIOWIN program (USEPA). This program was reported to have shown positive results for predicting biodegradability for some antibiotics and pharmaceuticals (Boethling *et al.*, 2004): readily biodegradable nature of about 60 pharmaceuticals has been shown to be predicted with BIOWIN submodels 5 and 6 with reasonable accuracy (respectively 83 and 87%).

Conversely, another study (Yu *et al.*, 2006) reported inconsistencies between predicted and experimental results (batch experiments). Two submodels of the BIOWIN software (1 and 2) tended to overestimate biodegradability while the remaining two submodels (5 and 6) tended to underestimate it. Author concluded that within the different submodels used by BIOWIN, the non-linear one (submodel 2) seemed to correspond better with the experimental data.

Modelled removal rates in STP: Existing measured removal rates still concern a limited number of pharmaceuticals, therefore modelled values can be used as a surrogate. A number of studies have used modelling software such as STPWIN from the USEPA (Jones *et al.*, 2002) or Simple treat, recommended by EMEA guideline (Carlsson *et al.*, 2006), to predict removal rates in STP for pharmaceuticals.

In their study, Jones *et al.* (2002) used the worst-case STPwin model which assumes no biodegradation in STP. Results from table 11 show that such a worst-case scenario largely underestimates removal rates in STP.

For a few compounds, we run the STPWIN software to compare mean and maximal measured removal rates (data from Miège *et al.*, 2008) observed for some pharmaceuticals with modelled values. Results (Table 7.11) show that there are inconsistencies between modelled and measured values, especially for antibiotics. It is probable that STPWIN can make reasonable predictions for pharmaceuticals within the applicability of the models used, and less reasonable or wrong predictions for compounds which fall outwidth the domain of the models.

Table 7.11 Comparison of measured removal rates and modelled values from STPWIN for some pharmaceuticals.

Name	Measured removal rates[1,2]				Modelled Removal Rates[1,3]			
					Assuming No Biodegradation		Assuming Biodegradation	
	Mean	SD[4]	n	Max	Total[5]	Bio[6]	Total[5]	Bio[6]
17α-ethinylestradiol	68	22.33	46	98	17.51	0.22	31.23	15.24
Aspirin	83	2.12	2	84	1.91	0.09	92.11	91.74
Atenolol	8	4.35	29	10	1.85	0.09	21.99	20.54
Azithromycin	47	0.00	6	47	30.99	0.33	30.99	0.33
Carbamazepine	9	8.91	42	53	2.96	0.1	24.51	22.09
Ciprofloxacin	73	13.98	16	81	1.85	0.09	8.79	7.13
Clarithromycin	45	0.00	6	45	7.3	0.14	7.3	0.14
Diclofenac	32	19.14	79	80	56.55	0.53	86.57	46.94
Ibuprofen	74	28.58	125	100	28.72	0.31	94.93	80.36
Ketoprofen	38	25.67	70	100	6.85	0.13	82.89	79.26
Metoprolol	10	14.42	30	83	2.15	0.09	22.67	20.96
Naproxen	76	16.66	68	98	7.55	0.14	83.68	79.66
Roxithromycin	37	9.03	3	44	4.05	0.11	4.05	3.94
Sulfamethoxazole	59	21.99	6	96	1.88	0.09	22.05	20.55
Trimethoprim	16	20.33	35	83	1.88	0.09	8.83	7.15

(1) All values are given as% of removal rates; (2) All measured data are taken from Miège et al. (*2008*); (3) All calculations were run with default treatment system properties; (4) SD: standard deviation; (5) total modelled removal rate (*biodegradation and sludge adorption*); (6) proportion of the compound's biodegradation in STP.

NB: it is not the aim of the authors to judge or make an appraisal of modelling softwares. However, we consider that modelled results have to be considered carefully for pharmaceuticals and especially that expert judgement is required to ascertain if models are appropriate in this case.

Use of Kow to assess partition to sludge and sediment: Partition is generally assessed by the Koc coefficient, as experimental values of Koc are scarce. However, as Kd (or Koc) modelling is mainly based on the Kow, such estimated values may be inaccurate in the case of pharmaceuticals. as quoted by several authors (Fent *et al.*, 2006; Wells, 2006; Tolls 2001), Kow may not be an accurate descriptor of the environmental behavior (sorption and bioaccumulation) of pharmaceuticals in the environment. As the majority of pharmaceuticals are polar ionisable compounds, the assumption that only the hydrophobic sorption performed by uncharged molecules is significant in the environnement, may be the exception rather than the rule. For ionisable compounds, a correction of the Kow by the pKa (i.e., Dow) is recommended (Stuer-Lauridsen *et al.*, 2000; Wells, 2006), however, even in this case, there is a need to assess logDow accuracy as a descriptor of pharmaceuticals environmental behaviour.

Use of pharmacological data to assess the environmental behaviour of pharmaceuticals: As Kd/Koc values are lacking, some authors have investigated alternative ways to assess the environmental behaviour of pharmaceuticals. As the pharmacokinetic behavior of pharmaceuticals is influenced by the same parameters that can modify environmental behavior such as pH and pKa, it makes sense to draw a parallel between pharmacokinetic and environmental criteria. Williams *et al.* (2006) studied the correlation between the environmental partitioning coefficient Kd and the distribution volume Vd, which measures the distribution of a pharmaceutical within the body. Results from this study suggest that pharmacokinetic parameters could be helpful to estimate environmental behaviour for pharmaceuticals.

7.4.2 Effect assessment

The environmental effects for human pharmaceuticals are much more difficult to assess than the environmental exposure. The different parameters and approaches used to assess the environmental effects of pharmaceuticals are depicted below. Table 7.12 summarizes the main parameters considered in the different assessment strategies.

7.4.2.1 Ecotoxicological data

Experimental data: To assess the risk for the aquatic environment, the most classical approach is to calculate PEC/PNEC risk quotients. The EMEA guideline describes the way to calculate PNEC values; this approach is similar to that proposed by the European TGD (EU 2003). Unlike TGD, the EMEA guideline (EMEA, 2006) enforces the use of chronic toxicity data and requires long-term NOEC for the base set (i.e., three NOEC values from three different trophic levels, applying an assessment factor of 10 to the lowest value). Nevertheless, only very few compounds bring together the conditions required by the EMEA guideline.

Extensive review of available ecotoxicity data already exist (Crane *et al.*, 2003; Fent *et al.*, 2001), but what such reviews show above all is that ecotoxicological data are still limited, therefore precluding the calculation of risk quotients. Indeed the lack of chronic ecotoxicological data has been clearly highlighted by the several risk assessment and prioritization strategies implemented for the last ten years (Cooper *et al.*, 2008; Besse *et al.*, 2008; Jjemba 2006; Carlsson *et al.*, 2006; Jones *et al.*, 2002; Stuer-Lauridsen *et al.*, 2000).

In order to accurately assess the environmental risk for pharmaceuticals, there is a need to acquire ecotoxicological data; however due to the high number of existing pharmaceuticals, it seems rather difficult, indeed impossible to build extensive toxicity data.

Moreover, if data required in ERA procedures concern ecotoxicity of single substances, the main risk for pharmaceuticals (and other chemicals) is probably linked to mixtures but studies carried out on topic are still few (Quinn *et al.*, 2009; Laird *et al.*, 2007; Cleuvers, 2003; Fraysse and Garric, 2005 ; Eguchi *et al.*, 2004).

Modelling of ecotoxicological data with ECOSAR software: ECOSAR is a freely downloadable software from the USEPA website, which has been recognized useful for predicting the environmental effects of some chemicals. As ecotoxicological data are lacking for pharmaceuticals, some authors have investigated the use of the ECOSAR software to predict acute toxicity for pharmaceuticals and to rank pharmaceuticals accordingly (Sanderson *et al.*, 2004a; 2004b; Cooper *et al.*, 2008).

However, there are two limits in the use of ECOSAR software applied to pharmaceuticals. First, QSARs only allow the modelling of acute toxicity data, which are of limited relevance as it is established that the risk linked to pharmaceuticals is chronic (Carlsson *et al.*, 2006; Fent *et al.*, 2006). Second, ECOSAR toxicity assessment is mainly base on Log Kow value meaning that predictions only account for non-specific narcosis endpoints, which may be not relevant for pharmaceuticals.

An assessment of Serotoninergic antidepressant (SSRI) toxicity on algae showed that ECOSAR modelled values were not in accordance with experimental ones and could not account for differences in algae species sensitivity (Johnson *et al.*, 2007). Furthermore, in a recent paper, Madden *et al.* (2009) assessed ECOSAR's accuracy for pharmaceuticals and notably if the assigned chemical classes, wherby models are based, were representative of the pharmaceuticals. Authors conclude that if ECOSAR has been shown to be useful in predicting aquatic ecotoxicity of industrial chemicals, it might be used with caution for pharmaceuticals as many of them fall outside the applicability domain of the models.

In a same way, in the screening tool proposed by Lienert *et al.* (2007), some uncertainties exist due to limitation of QSAR modelling when considering very hydrophilic and ionisable drugs and due to possible specific toxicity (e.g., sulfamethoxazole showed specific toxicity in algae) which is not accounted for when one models the baseline toxicity.

7.4.2.2 PBT Criteria

The OSPAR dynamec (OSPAR 2006, 2002) approach includes a ranking classification for single substances based on three criterions (PBT criterion). This approach has been used to describe the hazard of pharmaceuticals in the environment (www.janusinfo.se). Persistence in the environment (based on degradation time DT_{50}), Bioaccumulation potential (based on BCF or log Kow) and Toxicity (CL_{50} or NOEC).

Table 7.12 Review of existing strategies for assessing the biological effects of pharmaceuticals on aquatic species..

Data	Strategy	Criteria	Comments[1]	Reference
Ecotoxicological data	Experimental data	Acute tests	Some data are available, but the acute risk for pharmaceuticals is negligible. Can be used to derive chronic PNEC values but relevance is questionable.	Besse *et al.*, 2008; Carlsson *et al.*, 2006;
		Chronic tests	The more representative of aquatic hazard for pharmaceuticals but data are lacking.	Jones *et al.*, 2002; Stuer-lauridsen *et al.*, 2000
	QSAR modelling	ECOSAR parameters	Predict acute toxicity and do not take into account specific mechanism of action (only account for baseline toxicity). May not be accurate for pharmaceuticals.	Sanderson *et al.*, 2004a, b; Cooper *et al.*, 2008; Madden *et al.*, 2009
		Lipophilicity	Take into account and add the toxic potential of parent drug and metabolites in the PNEC calculation. Predict acute and baseline toxicity.	Lienert *et al.*, 2007
	OSPAR dynamec	Persistence	Based on experimental data or BIOWIN model. Modelled values remain to be investigated for accuracy.	www.janusinfo.se;
		Bioaccumulation	Only a few measured values of BCF are available.	van wezel and Jager (2002)
		Toxicity	Use of Kow to estimate bioaccumulation may be inaccurate for pharmaceuticals. Based on acute data or chronic ecotoxicological data.	
	Calculated values	Bioaccumulation	Bioaccumulation is a potential risk due to continuous exposure to pharmaceuticals. Only a few measured values of BCF areavailable. Relevance of calculated values based on log Dow needs to be investigated.	Jean, 2008
Pharmacological data	Human and mammalian data[2]	Mechanism of action (MoA)	Can give information on biological effects and toxicity when some similar MoA are identified in human and aquatic species.	Besse and Garric, 2008
		CYP 450 modulation	CYP 450 modulation can interfere with homeostatis in no-target species.	
		P-gp modulation	P-gp modulation can result in enhance sensitivity of organism to environmental pollutants.	
		Human side effects	Can help to understand some observed toxic effect in non-target species.	
		Therapeutic plasma concentration	Proposed to estimate response in fish. Relevance of this criteria needs to be better investigated.	Huggett *et al.*, 2003, Kostich and Lazorchak, 2008
			For antimicrobials, suggested to estimate the response in aquatic organisms (other than microbes).	Kostich and Lazorchak, 2008
	Antimicrobial sensitivity[2]	Inhibitory concentration for antibiotics	Can provide information on environmental effects of antibacterial pharmaceuticals (antimicrobial resistance and growth inhibition of beneficial microbes).	Kostich and Lazorchak 2008
Other data	Evolutionary approach	Conservation of human drug targets in aquatic organisms	Can provide indication on pharmaceuticals molecular targets and taxa of concern. Can be used to assess relevance of ecotoxicological data and to select relevant species for bioassays.	Kostich and Lazorchak 2008; Gunnarsson *et al.*, 2008

(1) Comments reflect the point of view of the authors of the current chapter; (2) Accuracy of pharmacological data for the assessment of pharmaceuticals are to be more investigated.

Estimation of the toxicity is based on available ecotoxicological data, therefore this approach meets the same limitation (data gaps) as the risk quotients approach. Estimation of persistency is based on experimental data, but in the absence of such data, the dynamec guidance document recommends the use of the BIOWIN model to estimate the persistency. Estimation of the bioaccumulation is based on log Kow, which could be inaccurate for pharmaceuticals:

- A study conducted in 2002 (van Wezel and Jager, 2002) suggested that the OSPAR dynamec was not adapted for pharmaceuticals. Log Kow for pharmaceuticals are often lower than 3 (cut-off value selected by the authors for bioaccumulation), which prevented an efficient discrimination between molecules with regard to the estimated bioaccumulation potential.
- As discussed above in the exposure assessment, Log Kow might not be a good descriptor for environmental considerations on pharmaceuticals, this assertion is also valid for the use of Kow to estimate the bioaccumulation potential of pharmaceuticals.

7.4.2.3 Liability to bioaccumulate

For some authors (Jean, 2008), one of the main risk, through continuous exposure to low concentrations of pharmaceuticals, is the risk of bioaccumulation in living organisms which could result in toxic effects. Bioaccumulation is estimated by the Bioaccumulation Factor (BCF) which takes into account the hydrophobicity of the molecule and the living organism considered. If some studies have focused on pharmaceuticals bioaccumulation (Paterson and Metcalfe, 2008; Mimeault *et al.*, 2005), there is still a lack of experimental values for BCF. Therefore, bioaccumulation can only be afforded through calculated values. In his study, Jean (2008) used calculated values from the CAS database®, mainly based on log Dow (log Kow value corrected by pH); environmental relevance of such calculated values must be investigated.

7.4.2.4 Pharmacological data

Several authors consider that the use of existing pharmacological, toxicological and pharmacokinetic data is likely to be helpful in assessing the risk of pharmaceuticals, as they could provide a better understanding of the fate and effect of pharmaceuticals in the aquatic environment (Besse and Garric 2008; Fent *et al.*, 2006; Jjemba, 2006; Lange and Dietrich, 2002; Seiler, 2002). Pharmacological data alone are not sufficient enough to assess the risk for the aquatic environment; however, such data can give information on the mechanism of action (MoA) and the toxicity of the pharmaceuticals.

Mammalian pharmacological and toxicological data: The use of mammalian pharmacological and toxicological data was proposed to help to prioritize the potential impacts of pharmaceuticals to fish (Huggett *et al.*, 2003). One of the key assumption of this model is that many enzyme/receptor systems are conserved between mammalian and teleost fish. The model requires comparison of human plasma concentration, HssPC, (i.e., the highest plasma concentration that corresponds to NOEC) and aquatic vertebrate plasma concentration, FssPC, (derived from environmental exposure). If these data are available, a safety margin can be calculated (Eq. 7.18).

$$Margin\ of\ safety_{PB} = \frac{HssPC\ NOEC}{FssPC} \tag{7.18}$$

Margin of safety$_{PB}$: *plasma besed margin of safety.* ***HssPC NOEC:*** *highest human plasma steady state concentration that corresponds to a NOEC.* ***FssPC:*** *fish plasma concentration derived from environmental exposure.*

As an alternative to this model, if some data are unavailable, a measured human therapeutic plasma concentration (H_TPC) can be used instead of HssPC, and a ER (Effect Ratio) is calculated (Eq. 7.19). The lower the ER, the greater the potential for a biological response in fish is.

$$\mathrm{ER} = \frac{\mathrm{H_TPC}}{\mathrm{FssPC}} \tag{7.19}$$

ER: *Effect Ratio.* ***H_TPC:*** *highest therapeutic plasma concentration.* ***FssPC:*** *fish plasma concentration derived from environmental exposure.*

A major key assumption of this model is that pharmaceutical absorption from water to fish is driven by hydrophobicity, i.e., only passive mechanisms are taken into account. FssPC are therefore calculated according to Eqs. (7.20) and (7.21), based on log Kow once more.

$$\mathrm{FssPC} = EC \times \mathrm{P_{Blood:water}} \tag{7.20}$$

EC: *Predicted/measured environmental concentration of a pharmaceutical.*

$$\mathrm{Log\ P_{Blood:water}} = 0.73 \times \log\ Kow - 0.88 \tag{7.21}$$

Log $P_{Blood:water}$: *Partitioning between aqueous phase and arterial blood.*

This conceptual model is interesting, however, the assumption of passive equilibration may be inaccurate:

- Some active physiogical mechanisms are probably involved in the absorption/concentration/elimination of pharmaceuticals by living organisms.
- Other ways of contamination such as accumulation *via* the food chain or due to contact with sediments are not taken into account.
- Limits of the log Kow as a descriptor of environmental behaviour have been discussed previously.

Use of free plasma concentrations of pharmaceuticals in human have recently been suggested to assess the critical concentrations in aquatic organisms (Kostich and Lazorchak, 2008). As Huggett *et al.* (2003), authors assume that exposure to non-target organisms would occur *via* passive equilibration with wastewater. Authors assumed, as a conservative hypothesis, that the concentration of pharmaceutical dissolved in the modelled organism's extracellular fluid would approach the dissolved concentration in wastewater. Therefore, comparing wastewater concentrations and human plasma concentrations after therapeutic dosing could suggest whether or not a significant effect in an aquatic organism would occur. As said above, such assumption and especially the exposure way (passive equilibration) remains to be confirmed.

Mechanism of action (MoA): Mechanism of action of pharmaceuticals may provide useful information regarding the potential toxic effects on environmental targets. Pharmaceuticals, unlike other pollutants such as polycyclic aromatic hydrocarbons or pesticides, are molecules designed to exert a specific mode of action with a limited toxicity. Extensive metabolic and toxicological studies are central to the development of drugs and can provide valuable information to guide ecotoxicological studies (Owen *et al.*, 2007).

For non-mammalian species with receptors similar to those of mammals, similar biological effects or adverse reactions may occur; it was recently suggested that cardiovascular dysfunction could be one of the consequence of the waterborne exposure of fish to β-blockers (Owen *et al.*, 2007). On the other hand, unexpected chronic effects may occur in lower organisms due to biological differences in pharmacodynamics and physiology (Fent *et al.*, 2006; Seiler, 2002; Lange and Dietrich, 2002).

Side effects: Known side effects of pharmaceuticals may also be valuable to indicate potential harmful effects on non-target organisms as it as already been shown for diclofenac in vultures (Oaks *et al.*, 2004), and in fish (Schwaiger *et al.*, 2004; Triebskorn *et al.*, 2004). Taking into account such effects could make it possible to target the harmful impacts of these compounds, at least on non-target vertebrates. Side effects have been used as an effect criteria in the prioritization strategy from Besse and Garric (2008).

Cytochrome P-450 modulation: Several drugs are known enzymatic inductors or inhibitors of the cytochrome P-450. P-450 isoforms are involved in a number of physiological reactions: transformation of both endogenous compounds and xenobiotics, synthesis and degradation of several steroids, prostaglandins, fatty acids, and other endogenous molecules (Stegeman *et al.*, 1992). Therefore, modulation of the enzymatic response may lead to disruption in the homeostasis of non-target organisms (Besse and Garric, 2008). Interference between pharmaceuticals and the metabolizing enzyme have been shown *in vitro* (Thibaut *et al.*, 2006).

Para glycoprotein-P (P-gp) modulation: Several pharmaceuticals are known to interact with P-glycoprotein (P-gp). P-gp is a protein acting as a multidrug transporter that actively transports xenobiotics out of the cell, preventing the accumulation of toxic compounds (Endicoot and Ling, 1989; Tutundjian and Minier, 2002). Glycoprotein P is involved in the multi-xenobiotic resistance (MXR) system. Increases in P-gp expression have been reported for aquatic organisms from polluted areas (Toomey and Epel, 1993; Britvic and Kurelec, 1999). As P-gp could play an important role in the protection of the organism from toxic effects caused by xenobiotics, a modulation in the expression of the P-gp and particularly an inhibition of its expression by a specific drug could result in enhancing the sensitivity of organisms. P-gp modulation by pharmaceuticals in aquatic species remains to be studied.

Comparative pharmacology: Comparative pharmacology could also be useful to understand toxicity pathways of pharmaceuticals. At the moment, studies have only considered the major MoA of pharmaceuticals in ecotoxicity assays. However, evidence shows that compounds belonging to same pharmacological and chemical classes (i.e., compounds with same mechanisms of action), can display a high variability in toxic values on same species and endpoints (Garric *et al.*, 2006; Dzialowski *et al.*, 2006; Henry *et al.*, 2004). Indeed, pharmaceuticals are not only characterized by their principal MoA but also by some additional pharmacological characteristics that should be taken into account.

In the case of β-blockers, several authors (Fraysse and Garric, 2005; Fent *et al.*, 2006) have suggested that differences in toxicity should be partially explained by pharmacological properties specific to these compounds such as receptor selectivity or membrane-stabilizing activity.

Several studies have shown that serotoninergic antidepressants (SSRIs) effective concentrations ranged over several order of magnitudes, in crustaceans (Henry *et al.*, 2004) and algae (Johnson *et al.*, 2007), despite their same MoA.

Actually, pharmacological data indicate that even if SSRIs have a greater selectivity for blocking serotonin reuptake (their principal MoA), they also have affinities to some other receptors and reuptake inhibitor activities on other systems such as the noradrenergic or dopaminergic systems, (Hyttel 1993; Dulin *et al.*, 2002). Even if designed to exert specific MoAs, pharmaceuticals have generally more than one target. "Secondary" MoAs could help to understand such differences in toxic responses for pharmaceuticals (Besse *et al.*, 2008).

Antimicrobial potency of antibiotic agents: Selection of resistant strains of bacteria due to the release of antibiotics in the environment is a crucial issue. Kostich and Lazorchak (2008) suggested that MIC (Minimum Inhibitory Concentration) of antibacterials could be used as an estimation of the critical concentrations inducing microbial effects in the environment, in particular antimicrobial resistance and growth inhibition of beneficial microbes. By comparing PECs and MICs, authors suggested that wastewater concentrations for pharmaceuticals were sufficient to impair growth or select for low-level resistance in microorganisms. Accuracy of such an approach must be confirmed as mechanisms of selection of resistant bacteria in the environment remain unclear.

7.4.2.5 Evolutionnary approaches

Activity of a pharmaceutical (or its metabolite) is generally considered with respect to its activity in mammals and especially in humans (i.e., pharmacological activity). However the fact that a molecule display biological effects in mammals doesn't mean that it is able to display the same effects in other species (especially invertebrates and plants) indeed any effect. Inversely, a "pharmacologically inactive" compound does not necessarily mean that such a metabolite has no effects on an aquatic organism. There is therefore a need to better understand how drug targets, and physiological functions driven by such targets are conserved or not from a specie to another.

Evolutionary conservation of drug targets have been investigated (Gunnarsson *et al.*, 2008), by prediction of orthologs (proteins derived from a common ancestral protein at the time of speciation). Ortholog were compiled for 1318 human drug targets (see Gunnarsson *et al.*, 2008 for details) and predicted for 16 species, including species used in risk assessment. Prediction results estimated to 86% the homology of drug targets for the zebrafish, to 35% for daphnia and to 35% for green alga. For some pharmaceuticals, individual targets such as enzymes are well conserved, suggesting that tests on evolutionary distant organisms would be highly relevant for these drugs.

Authors suggest that the applied methodology could be useful to (i) guide ERA by identifying sensitive species, and (ii) interpret the relevance of existing ecotoxicological data. The same general conclusions were drawn in the study of Kostich and Lazorchak (2008), in which the sequence conservation in drug target was proposed as a guidance on the range of species and endpoints that should be considered in chronic ecotoxicity studies.

Therefore, assessment of evolutionary conservation of drug targets could be of great significance as it could ensure:

- Identification of drugs with the most potential to elicit adverse effects on non-target organisms.
- Interpretation and assessment of ecotoxicological and pharmacological data.
- Improvement of the possibilities to identify which pharmaceuticals may pose a risk to certain type of species (or inversely identification of specific sensitive species to certain compounds).
- Selection of relevant species and/or endpoints for ecotoxicological studies.

Nevertheless, one must keep in mind that the presence of a similar drug receptor in a specie does not mean that a functional interaction with the drug can occur; moreover, receptors structurally similar may display different biological functions in different species. Inversely a functionnal interaction between a drug and another non-similar receptor cannot be excluded.

7.5 CONCLUSION

Prioritization of human pharmaceuticals is necessary due to the high number of pharmaceuticals used, which hinders the possibility to assess the ecotoxicity of every compound. To implement a relevant prioritization strategy, there is a need to accurately assess the environmental exposure and the environmental effects. From the review of environmental risk assessments (ERA) and prioritization strategies conducted in the last ten years, it is possible to highlight the following parameters and to propose some concluding remarks:

For the exposure assessment:

- The use of simple models, such as the EMEA model, to calculate PECs for surface water is in general in good agreement with field measurements.
- Actual and accurate consumption amounts are essential, but data are sometimes unavailable, depending on the country.
- Metabolism data and excretion rates are essential but data are sometimes incomplete or unavailable.
- STP removal rates are lacking which is a major limitation for the accuracy of ERA.
- Data on environmental behaviour and degradation by-products are scarce.

- PEC for other compartments than water column are not well assessed yet.
- Bioaccumulation of pharmaceuticals remains to be more studied.

For the effect assessment:

- Due to low concentrations in surface waters, acute effects of pharmaceuticals can be ruled out.
- Chronic ecotoxicological data are lacking.
- Due to gaps in ecotoxicological data for pharmaceuticals (in connection with the high number of pharmaceuticals), alternative ways of assessing the effects of pharmaceuticals have to be investigated and validated.
- Pharmacological data can be useful to estimate the biological effects on aquatic organisms however, i) the access to such data is sometimes not possible, ii) the relevance of such data for environmental considerations remains to be confirmed.
- Investigation of the evolutionary conservation of drug targets is important information that can help for a relevant use of pharmacological data, and for targeting sensitive species in bioassays.

To conclude, it becomes necessary to harmonize the different prioritisation strategies and models now available in European countries. As pharmacological and toxicological data are not easily available, the ERA of old pharmaceuticals at least, could greatly benefit from the development of a database with extensive consumption, pharmacological, toxicological and ecotoxicological data.

REFERENCES

Andreozzi, R., Caprio, V., Ciniglia, C., De Champdore, M., Lo Giudice, R., Marotta, R., Zuccato, E. (2004). Antibiotics in the environment: Occurrence in Italian STPs, fate, and preliminary assessment on algal toxicity of amoxicillin. *Environmental Science and Technology* **38**(24), *6832–6838.*

Besse, J.-P., Kausch-Barreto, C., Garric, J. (2008) Exposure assessment of pharmaceuticals and their metabolites in the aquatic environment: Application to the French situation and preliminary prioritization. *Human and Ecological Risk Assessment* **14**(4), 665–695.

Besse, J.-P., Garric, J. (2008) Human pharmaceuticals in surface waters. Implementation of a prioritization methodology and application to the French situation. *Toxicology Letters* **176**(2), 104–123.

Boethling, R. S., Lynch, D. G., Jaworska, J. S., Tunkel, J. L., Thom, G. C., Webb, S. (2004). Using Biowin^(TM), Bayes, and batteries to predict ready biodegradability. *Environmental Toxicology and Chemistry* **23**(4), 911–920.

Bound, J. P., Voulvoulis, N. (2006). Predicted and measured concentrations for selected pharmaceuticals in UK rivers: implications for risk assessment. *Water Research* **40**, 2885–2892.

Boxall, A.B.A., Fogg, L.A., Kay, P., Blackwell, P.A., Pemberton, E.J., Croxford, A. (2003) Prioritisation of veterinary medicines in the UK environment. *Toxicology Letters* **142**(3), 207–218.

Britvic, S., Kurelec, B. (1999) The effect of inhibitor of multixenobiotic resistance mechanism on the production of mutagens by Dreissena polymorpha in waters spiked with premutagens. *Aquatic Toxicology* **47**, 107–116.

Capleton, A.C., Courage, C., Rumsby, P., Holmes, P., Stutt, E., Boxall, A.B.A., Levy, L.S. (2006) Prioritising veterinary medicines according to their potential indirect human exposure and toxicity profile. *Toxicology Letters* **163**(3), 213–223.

Carlsson, C., Johansson, A.K., Alvan, G., Bergman, K., Kuhler, T. (2006) Are pharmaceuticals potent environmental pollutants? Part I: Environmental risk assessments of selected active pharmaceutical ingredients. *Science of The Total Environment* **364**(1–3), 67–87.

Castiglioni, S., Fanelli, R., Calamari, D., Bagnati, R., Zuccato, E. (2004) Methodological approaches for studying pharmaceuticals in the environment by comparing predicted and measured concentrations in River Po, Italy. *Regulatory Toxicology and Pharmacology* **39**(1), 25–32.

Castiglioni, S., Bagnati, R., Fanelli, R., Pomati, F., Calamari, D., Zuccato, E. (2006) Removal of pharmaceuticals in sewage treatment plants in Italy. *Environmental Science and Technology* **40**(1), 357–363.

Cleuvers, M. (2003) Aquatic ecotoxicity of pharmaceuticals including the assessment of combination effects. *Toxicology Letters* **142**(3), 185–194.

Coetsier, C. M., Spinelli, S., Lin, L., Roig, B., Touraud, E. (2009) Discharge of pharmaceutical products (PPs) through a conventional biological sewage treatment plant: MECs vs PECs? *Environment International* **35**, 787–792.

Cooper, E.R., Siewicki, T.C., Phillips, K. (2008) Preliminary risk assessment database and risk ranking of pharmaceuticals in the environment. *Science of the Total Environment* **398**(1–3), 26–33.

Crane, M., Watts, C., Boucard, T. (2006) Chronic aquatic environmental risks from exposure to human pharmaceuticals. *Science of the Total Environment* **367**(1), 23–41.

Daughton, C.G., Ternes, T.A. (1999) Pharmaceuticals and personal care products in the environment: *agents of subtle change? Environmental Health Perspectives* **107** (Supp. 6), 907–938.

DellaGreca, M., Fiorentino, A., Isidori, M., Lavorgna, M., Previtera, L., Rubino, M., Temussi, F. (2004) Toxicity of prednisolone, dexamethasone and their photochemical derivatives on aquatic organisms. *Chemosphere* **54**(5), 629–637.

Dulin, R., Silberstein, N., Bonnin, M. Saux, M. C. (2002). Comparison and practical guidelines of selective serotonin reuptake inhibitors [Comparaison et critères de choix des inhibiteurs sélectifs de la recapture de la sérotonine]. *Journal de Pharmacie Clinique* **21**, 39–46.

Dzialowski, E. M, Turner, P. K., Brooks, B. W. (2006) Physiological and reproductive effects of beta adrenergic receptor antagonists in Daphnia magna. *Archives of Environmental Contamination and Toxicology* **50**(4), 503–510.

Eguchi, K., Nagase, H., Ozawa, M., Endoh, Y. S., Goto, K., Hirata, K., Miyamoto, K., Yoshimura, H. (2004) Evaluation of antimicrobial agents for veterinary use in the ecotoxicity test using microalgae. *Chemosphere* **57**(11), 1733–1738.

EMEA (2000) European Agency for the Evaluation of Medicinal Products. Guideline on environmental impact assessment (EIAS) for veterinary medicinal products—phase I. CVMP/VICH/592/98–Final. London: EMEA.

EMEA (2001) European Agency for the Evaluation of Medicinal Products. Discussion paper on environmental risk assessment of non-genetically modified organism (non-GMO) containing medical products for human use. CPMP/SWP/4447/00 draft corr. London: EMEA.

EMEA (2003) European Agency for the Evaluation of Medicinal Products Note for guidance on environmental risk assessment of medical products for human use. CPMP/SWP/4447/00 draft. London: EMEA.

EMEA 2006. Note for guidance on environmental risk assessment of medicinal products for human use. Doc. Ref. EMEA/CHMP/SWP/4447/00. Committee for proprietary medicinal products. European Agency for the Evaluation of Medicinal Products, London, UK. http://www.emea.eu.int/pdfs/human/swp/444700en.pdf.

Endicoot, J.A., Ling, V. (1989) The biochemistry of P-glycoprotein-mediated multidrug resistance. Annual review of biochemistry **58**, 137–171.

Escher, B.I., Bramaz, N., Richter, M., Lienert, J. (2006) Comparative ecotoxicological hazard assessment of beta-blockers and their human metabolites using a mode-of-action-based test battery and a QSAR approach. *Environmental Science and Technology* **40**(23), 7402–7408.

EU (1994). Assessment of potential risks to the environment posed by medicinal products for human use, excluding products containing live genetically modified organisms. EU. Ad Hoc Working Party, III/5504/94 Draft 4.

EU (1996). Technical guidance document in support of commission directive 93/67/EEC on risk assessment for new notified substances and commission regulation (EC) No 1488/94 on risk assessment for existing substances, Part II.

EU (2003). TGD Technical Guidance Document. Technical Guidance Document in support of Council Directive 93/67/EEC on risk assessment for new notified substances and Commission.

Regulation (EC) 1488/94 on risk assessment for existing substances. Office for Official Publications of the European Communities, Luxembourg.

Feitosa-Felizzola, J., Chiron, S. (2009) Occurrence and distribution of selected antibiotics in a small Mediterranean stream (Arc River, Southern France). *Journal of Hydrology* **364**(1–2), 50–57.

Fent K, Weston, A. A, Caminada, D. (2006) Ecotoxicology of human pharmaceuticals. *Aquatic Toxicology* **76**(2), 122–159.

Fraysse, B, Garric, J. (2005) Prediction and experimental validation of acute toxicity of β-blockers in Ceriodaphnia dubia. *Environmental Toxicology and Chemistry* **24**(10), 2470–2476.

Garric, J., Ferrari, B., Fraysse, B., Mons, R., Vollat, B. (2006) Effects of some human pharmaceutical on freshwater organisms | [Impact de médicaments à usage humain sur les organismes aquatiques d'eau douce]. *Environnement, Risques et Sante* **5**(4), 290–295.

Gunnarsson, L., Jauhiainen, A., Kristiansson, E., Nerman, O., Larsson, D. G. J. (2008) Evolutionary conservation of human drug targets in organisms used for environmental risk assessments. *Environmental Science and Technology* **42**(15), 5807–5813

Henry, T.B, Kwon, J.W, Armbrust, K.L, Black, M.C. (2004) Acute and chronic toxicity of five selective serotonin reuptake inhibitors in Ceriodaphnia dubia. *Environmental Toxicology and Chemistry* **23**(9), 2229–2233.

Hilton, M.J., Thomas, K.V., Ashton, D. (2003) Targeted monitoring programme for pharmaceuticals in the aquatic environment. R&D Technical report P6-012/06/TR UK Environment Agency.

Huggett, D.B., Cook, J.C., Ericson, J.F., Williams, R.T. (2003) A theoretical model for utilizing mammalian pharmacology and safety data to prioritize potential impacts of human pharmaceuticals to fish. *Human and Ecological Risk Assessment* **9**(7), 1789–1799.

Huschek, G., Hansen, P.D., Maurer, H.H., Krengel, D., Kayser, A. (2004) Environmental risk assessment of medicinal products for human use according to European Commission recommendations. *Environmental Toxicology* **19**, 226–240.

Hyttel, J. (1993) Comparative pharmacology of selective serotonin re-uptake inhibitors (SSRIs). *Nordisk Journal of Psychiatry* **47**(30), 5–12.

Isidori, M., Lavorgna, M., Nardelli, A., Parrella, A., Previtera, L., Rubino, M. (2005) Ecotoxicity of naproxen and its phototransformation products. *Science of the Total Environment* **348**(1–3), 93–101.

Isidori, M., Nardelli, A., Parrella, A., Pascarella, L., Previtera, L. (2006) A multispecies study to assess the toxic and genotoxic effect of pharmaceuticals: Furosemide and its photoproduct. *Chemosphere* **63**(5), 785–793.

Jean J. (2008) Identification et hiérarchisation des substances médicamenteuses bioaccumulables rejetées dans les effluents hospitaliers. Thèse pour le diplôme d'état de Docteur en Pharmacie. Thèse n° 48–2008. Université Claude Bernard-Lyon 1. Faculté de pharmacie. LYON.

Jjemba, P.K. (2006) Excretion and ecotoxicity of pharmaceutical and personal care products in the environment. *Ecotoxicology and Environmental Safety* **63**(1), 113–130.

Jones, O.A.H, Voulvoulis, N., Lester, J.N. (2002) Aquatic environmental assessment of the top 25 English prescription pharmaceuticals. *Water Research* **36**, 5013–5022.

Jones, O.A.H., Voulvoulis, N., Lester, J.N. (2003) Pharmaceuticals – their potential impact on environmental health. *Bulletin of the World Health Organisation* **81**, 768–769.

Kools, S.A., Boxall, A., Moltmann, J.F., Bryning, G., Koschorreck, J., Knacker, T. (2008) A ranking of European veterinary medicines based on environmental risks. *Integrated Environmental Assessment and Management* **4**(4), 399–408.

Kosonen, J., Kronberg, L. (2009). The occurrence of antihistamines in sewage waters and in recipient rivers. *Environmental Science and Pollution Research* **16**(5), 555–564.

Kostich, M. S., Lazorchak, J. M. (2008) Risks to aquatic organisms posed by human pharmaceutical use. *Science of the Total Environment* **38**, 329–339.

Laird, B.D., Brain, R.A., Johnson, D.J., Wilson, C.J., Sanderson, H., Solomon, K.R. (2007) Toxicity and hazard of a mixture of SSRIs to zooplankton communities evaluated in aquatic microcosms. *Chemosphere* **69**(6), 949–954.

Lange, R., Dietrich, D. (2002) Environmental risk assessment of pharmaceutical drug substances–conceptual considerations. *Toxicology Letters* **131**(1–2), 97–104.

Larsson, D.G.J., de Pedro, C., Paxeus, N. (2007) Effluent from drug manufactures contains extremely high levels of pharmaceuticals. *Journal of Hazardous Materials* **148**(3), 751–755.

Lienert, J., Güdel, K., Escher, B.I. (2007) Screening method for ecotoxicological hazard assessment of 42 pharmaceuticals considering human metabolism and excretory routes. *Environmental Science and Technology* **41**(12), 4471–4478.

Madden, J.C., Enoch, S.J., Hewitt, M., Cronin, M.T.D. (2009) Pharmaceuticals in the environment: Good practice in predicting acute ecotoxicological effects. *Toxicology Letters* **185**(2), 85–101.

Miège, C., Choubert, J.M., Ribeiro, L., Eusèbe, M., Coquery, M. (2009) Fate of pharmaceuticals and personal care products in wastewater treatment plants - Conception of a database and first results. *Environmental Pollution* **157**(5), 1721–1726.

Mimeault, C., Woodhouse, A.J., Miao, X.-S., Metcalfe, C.D., Moon, T.W., Trudeau, V.L. (2005) The human lipid regulator, gemfibrozil bioconcentrates and reduces testosterone in the goldfish, Carassius auratus. Aquatic *Toxicology* **73**(1), 44–54.

Oaks, J.L., Gilbert, M., Virani, M.Z., Watson, R.T., Meteyer, C.U., Rideout, B.A., Shivaprasad, H.L., Ahmed, S, Chaudhry, M.J.I., Arshad, M., Mahmood, S., Ali, A., Khan, A. A. (2004) Diclofenac residues as the cause of vulture population decline in Pakistan. *Nature* **427**(6975), 630–633.

Onesios, K.M., Yu, J.T., Bouwer, E.J. (2008) Biodegradation and removal of pharmaceuticals and personal care products in treatment systems: a review. *Biodegradation* **20**(4), 441-66.

OSPAR (2002). Dynamic Selection and Prioritisation Mechanism for Hazardous Substances (DYNAMEC). http://www.ospar.org/documents/dbase/publications/p00146_DYNAMEC%20Manual.pdf

OSPAR (2006) Dynamic Selection and Prioritisation Mechanism for Hazardous Substances (New DYNAMEC Manual). http://www.ospar.org/documents%5Cdbase%5Cpublications%5Cp00256_New%20DYNAMEC%20Manual.pdf

Owen, S.F., Giltrow, E., Huggett, D.B., Hutchinson, T.H., Saye, J., Winter, M.J., Sumpter, J.P. (2007) Comparative physiology, pharmacology and toxicology of β-blockers: mammals versus fish. *Aquatic Toxicology* **82**, 145–162.

Paffoni, C., Welte, B., Gousailles, M., Montiel, A. (2006) Nouvelles molécules mises en cause par les directives Européennes: de la station d'épuration à l'usine de traitement d'eau potable. *Journal Européen d'Hydrologie* **37**(1), 21–38

Paterson, G., Metcalfe, C. D. (2008) Uptake and depuration of the anti-depressant fluoxetine by the Japanese medaka (Oryzias latipes). *Chemosphere* **74**(1), 125–130.

Petrovic, M., Solé, M., Lopez de Alda, M. J., Barcelo, D. (2002) Endocrine disruptors in sewage treatment plants, receiving water and sediments, integration of chemical analysis and biological effects on feral carp. *Environmental Toxicology and Chemistry* **21**, 2146–2156.

PNSE (2004). Plan National Santé Environnement. Ministère de la Santé et de la Protection sociale. Ministère de l'Ecologie et du Développement durable. Ministère de l'Emploi, du Travail et de la Cohésion sociale. Ministère délégué à la Recherche. Available at http://www.sante.gouv.fr/htm/dossiers/pnse/rapport.pdf

Qin-Tao, L, Williams, H.E. (2007) Kinetics and degradation products for direct photolysis of β- blockers in water. *Environmental Science and Technology* **41**, 803–810.

Quinn, B., Gagné, F., Blaise, C. (2009) Evaluation of the acute, chronic and teratogenic effects of a mixture of eleven pharmaceuticals on the cnidarian, Hydra attenuata. *Science of the Total Environment* **407**(3), 1072–1079.

Sanderson, H., Brain, R.A., Johnson, D.J., Wilson, C.J., Solomon, K.R. (2004a) Toxicity classification and evaluation of four pharmaceuticals classes: antibiotics, antineoplastics, cardiovascular, and sex hormones. *Toxicology* **203**(1–3), 27–40.

Sanderson, H., Johnson, D.J., Reitsma, T., Brain, R.A., Wilson, C.J., Solomon, K.R. (2004b) Ranking and prioritization of environmental risks of pharmaceuticals in surface waters. *Regulatory Toxicology and Pharmacology* **39**(2), 158–183.

Schwaiger, J., Ferling, H., Mallow, U., Wintermayr, H., Negele, R.D. (2004) Toxic effects of the non-steroidal anti-inflammatory drug diclofenac: Part I: histopathological alterations and bioaccumulation in rainbow trout. *Aquatic Toxicology* **68**(2), 141–150.

Seiler, J.P. (2002). Pharmacodynamic activity of drugs and ecotoxicology–can the two be connected? *Toxicology Letters* **131**(1–2), 105–115.

Siemens, J., Huschek, G., Siebe, C., Kaupenjohann, M. (2008) Concentrations and mobility of human pharmaceuticals in the world's largest wastewater irrigation system, Mexico City-Mezquital Valley. *Water Research* **42**(8–9), 2124–2134.

Stegeman, J.J., Brouwer, M., Richard, T.D.G., Förlin, L., Fowler, B.A., Sanders, B.M., van Veld, P.A. (1992). Molecular responses to environmental contamination: enzyme and protein systems as indicators of chemical exposure and effect. In Biomarkers: biochemical, physiological and histological markers of anthropogenic stress, (eds. R.J. Hugget, R.A. Kimerly), pp. 235–335, Lewis publishers, Chelsea, MIn USA.

Struijs, J., Stoltenkamp, J., van de Meent, D. (1991) A spreadsheet based box model to predict the fate of xenobiotics in a municipal wastewater treatment plant. *Water Research* **25**, 891–900.

Stuer-Lauridsen, F., Birkved, M., Hansen, L. P., Holten Lutzhoft, H. C., Halling-Sorensen, B. (2000) Environmental risk assessment of human pharmaceuticals in Denmark after normal therapeutic use. *Chemosphere* **40**(7), 783–793.

Thibaut, R., Schnell, S., Porte, C., (2006) The interference of pharmaceuticals with endogenous and xenobiotic metabolizing enzymes in carp liver: an in vitro study. *Environmental Science and Technology* **40**, 5154–5460.

Tolls, J. (2001) Sorption of veterinary pharmaceuticals in soils: A review. *Environmental Science and Technology* **35**(17), 3397–3406.

Toomey, B.H., Epel, D. (1993) Multixenobiotic resistance in Urechis caupo embryos: protection from environmental toxins. *Biology Bulletin* **185**, 355–364

Triebskorn, R., Casper, H., Heyd, A., Eikemper, R., Köhler, H. R., Schwaiger, J. (2004) Toxic effects of the non-steroidal anti-inflammatory drug diclofenac: Part II. Cytological effects in liver, kidney, gills and intestine of rainbow trout (Oncorhynchus mykiss). *Aquatic Toxicology* **68**(2), 151–166.

Tutundjian, R., Minier, C. (2002) Les protéines de résistance multiple et leur exploitation pour la biosurveillance chez les organismes aquatiques. *Regard sur la biochimie* **4**, 37–50.

USEPA (2009). Estimation Programs Interface Suite™ for Microsoft® Windows. United States Environmental Protection Agency, Washington, DC, USA.

van Wezel, A.P, Jager, T. (2002) Comparison of two screening level risk assessment approaches for six disinfectants and pharmaceuticals. *Chemosphere* **47**(10), 1113–1128.

Wells, M.J.M. (2006) Log Dow: Key to understanding and regulating waste-water-derived contaminants. *Environmental Chemistry* **3**, 439–449.
Williams, M., Saison, C. L. A., Williams, D. B., Kookana, R. S. (2006) Can aquatic distribution of human pharmaceuticals be related to pharmacological data? *Chemosphere* **65**, 2253–2259.
Yu, J.T., Bouwer, E., Coelhan, M. (2006) Occurrence and biodegradability studies of selected pharmaceuticals and personal care products in sewage effluent. *Agricultural Water Management* **86**, 72–80.
Zuccato, E, Castiglioni, S, Fanelli, R. (2005) Identification of the pharmaceuticals for human use contaminating the Italian aquatic environment. *Journal of Hazardous Materials* **122**(3), 205–209.

Chapter 8

Recommendations on research and development

Richard Greenwood, James Clark, Graham Mills, Fabien Deswarte, Louise Summerton, Evelyne Touraud and Benoit Roig

8.1 INTRODUCTION

The KNAPPE project promoted discussion of the areas the areas of ecopharmacostewardship, and ecopharmacovigilance. The former aims encourage the development of compounds and practices that will reduce environmental impacts, whilst the latter involves the monitoring of the sources, distribution, fate and biological impact (on ecosystems and human health) of pharmaceutical products (PPs). Overall the aims were to identify stewardship and vigilance approaches that were available or might be developed to minimise the environmental impacts of PPs throughout their life cycle, including design, manufacture, use, disposal, emission to the environment, and transfer between environmental compartments. Methods investigated involved life cycle assessment to evaluate the sustainability of new and existing drugs. This should aid comparisons between different products, and cost benefit analyses. An objective of this area of the project was to identify examples of greener drugs, and to identify eco-compatibility criteria that could be used to assess the overall environmental footprint of PPs. The area of the project dealing with aspects of environmental vigilance aimed to find what was needed to improve monitoring of the environmental impact of PPs after registration, so that reliable information is available for managing risks, and identifying means of reducing impacts. It was recognised that feedback from a range of those involved at various stages of the life cycle of PPs was essential for the achievement of this aim. In order to obtain this, a range of players was consulted, including those involved in the design, manufacture, prescribing, use and disposal of pharmaceutical products; monitoring the treatment of sewage, waste water, and the production of potable waters; and monitoring the chemical quality of surface and ground waters. Development of discussion documents involved the use of the current scientific literature, national and international reports, and consultations with representatives from the various areas listed above.

In order to obtain a complete picture of the life cycle of a medicine it is necessary to obtain reliable and representative information on levels of medicines in various divisions of the pathways to and within the aquatic environment, including manufacture and distribution, end use, sewage treatment plants (STPs), waste water, surface and ground waters, sludge, soil, soil and aquatic biota, drinking water. It is not feasible to obtain detailed information on all of the medicines in use, or to monitor them, even only in surface waters, and it will therefore be necessary to assess potential risk on the basis of available information in order to prioritise research and monitoring efforts. Compounds could be categorised as of high or of medium to low risk. For newly registered medicines monitoring is not necessary until the product has been in use for sufficient time to allow it to reach the treatment and environmental distribution phases. For veterinary medicines

 Pharmaceuticals in the Environment: Current knowledge and need assessment to reduce presence and impact. Edited by Benoit Roig. ISBN: 9781843393146. Published by IWA Publishing, London, UK.

guidelines for environmental impact assessment are already in place for new compounds, but it will be necessary to assess existing compounds and medicines for human use in the same way.

Monitoring needs to establish locations and times where and when environmental concentrations are likely to be highest to quantify the worst case scenario. There is also a need to obtain a reliable, representative picture of the average overall concentrations in various divisions of the aquatic environment. This may identify where local remedial measures could be useful, and avoid unnecessary large scale operations. For high risk compounds detailed investigations would be appropriate in order to underpin potentially expensive remedial actions and monitoring regimes. For lower risk compounds it may be possible to use modelling, and to identify indicator compounds that can be used to represent a set of similar medicines. Where regulatory monitoring is required, then work will be necessary to develop validated extraction and instrumental methods that can be used by routine analytical laboratories. This is particularly problematic for some difficult, complex matrices such as sludge, and sediments. In order to keep the costs of monitoring within reasonable bounds, it may be necessary to adopt different strategies for newly approved medicines, existing medicines, and those identified during authorisation as potentially posing a high risk, to target monitoring activities, and to consider possibly more cost effective alternative approaches.

There is a need for an overall evaluation of the current situation and for a consensus to be reached between the pharmaceutical industry and environmental legislators to produce objective guidelines that will afford protection to environment and drinking water but that will not prevent valuable medicines that will alleviate human or animal suffering from reaching the market place.

8.2 DESIGNING 'GREENER' PHARMACEUTICAL PRODUCTS

Green chemistry is recognised internationally as the term to describe the development of more environmentally acceptable and sustainable chemical processes and products. Application of green chemistry in the chemical industry as a whole is fragmented. However, an increasing number of drivers exist that will strengthen the need for a greater adoption of green chemistry clean synthesis methods and technologies. These include increasing process costs (waste treatment and disposal, energy costs), increasing costs of and demand for diminishing traditional petroleum-derived raw materials and the increasingly strict legislation governing the production, storage, use and disposal of chemicals (Clark *et al.,* 2005). Some representatives of the pharmaceutical industry also believe that the new European chemicals legislation, REACH (Registration, Evaluation and Authorisation of Chemical Substances), which entered into force on 1st June 2007, may pose a serious threat to the supplies of a large number of pharmaceutical intermediates (O'Driscoll, 2006). Figure 8.1 shows some of the methods that can be applied in the design of 'greener' pharmaceutical products. Review of available literature has revealed that benign by design clean synthesis methods and green production technologies are being adopted by the pharmaceutical industry in the syntheses of pharmaceutical products to a certain extent. Some companies, including Merck and GlaxoSmithKline, have also established in-house best practice awards to reward their staff for embracing these initiatives[1,2].

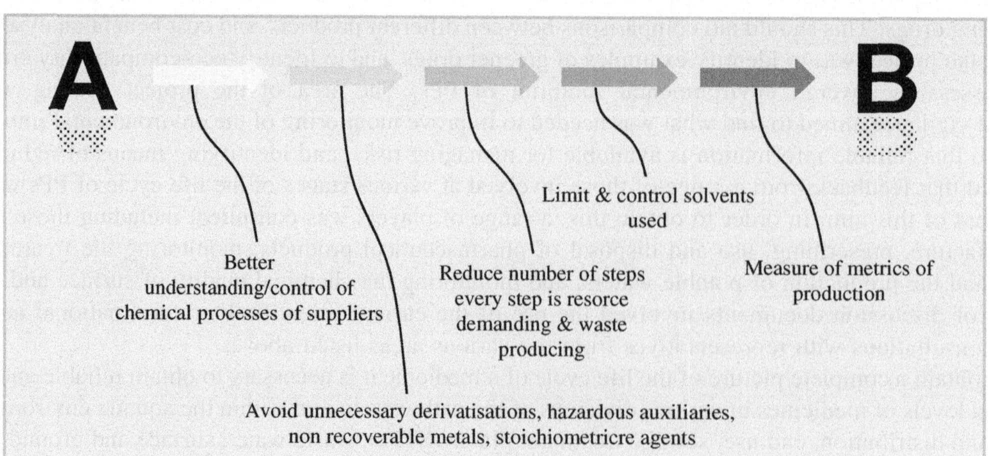

Figure 8.1 Clean synthesis of pharmaceutical products.

[1]http://www.merck.com/cr/environment_health_and_safety/

[2]http://www.copdtreatment.co.uk/investors/reps03/EHS03/GSKehs-32.htm

8.2.1 Towards greener drugs

8.2.1.1 Existing "greener" drugs

Since the greening of Ibuprofen synthesis in 1992 by the BHC Company, there has been a steadily increasing drive within the pharmaceutical industry towards the synthesis of 'greener' pharmaceutical products. This is highlighted through the US Environmental Protection Agency's prestigious Presidential Green Chemistry Challenges Awards, which annually receive entries from pharmaceutical companies on the successful adoption of benign-by-design clean synthesis methods and green technology to produce pharmaceutical products. Table 8.1 highlights some of the most widely publicised improvements in the environmental impact of pharmaceutical product synthesis. These range from novel green catalytic methods, to reductions in solvent use, waste minimisation and elimination of hazardous reagents. The majority of these improvements were made after the drug was already on the market, providing an opportunity to address these issues at an earlier stage in the lifecycle. Until recently improvements to the processing of pharmaceutical ingredients was not permitted after the new drug application (NDA) had been filed with the FDA. This is a major step towards recognising the need to improve the environmental profile of drug manufacturing processes. The sooner environmental impact is considered during the development of pharmaceutical products, the greater the chance of adopting prevention and minimisation techniques (Butters *et al.,* 2006).

8.2.1.2 Improving the eco-compatibility of pharmaceutical products

As highlighted in previous section, the pharmaceutical industry has recently achieved numerous successes in greening its synthetic processes (gate to gate approach). However, little consideration has, so far, been given to the environmental impact of pharmaceuticals once they have left the industry (gate-to-grave approach) (Larsen *et al.,* 2004). In fact, most drugs are not fully metabolised in the body of animals or humans and are subsequently introduced into the environment. Since these chemicals are biologically very active, their presence in the environment is of growing concern. It is therefore crucial that we develop strategies to prevent pollution and limit the harm of pharmaceuticals on the environment. Such strategies could include:

(1) *Reducing burden of pharmaceutical products at source*: This can be done by reducing the amount of product ending up in the environment after use. Drugs should therefore be designed to minimise therapeutic doses. This approach is already employed by the pharmaceutical industry, which is dedicating lots of time and effort to produce enantiomerically pure drugs (Daughton *et al.,* 2003). This method presents the added benefit of eliminating exposure to the non-therapeutic isomers, which are often responsible for unwanted side effects. However, more research is required to improve the economics of racemic separation.

(2) *End-of-pipe technologies*: Most pharmaceuticals that originate from human and animal excretion are collected by wastewater treatment plants. A number of promising end-of-pipe technologies (including biological treatment and ozonation), which will allow efficient treatment of these micropollutants, are under development (Khetan and Collins, 2007). However, it is recognised that large parts of these products will always be lost to the environment (Butters *et al.,* 2006).

(3) *Benign by design or design for biodegradability*: Pharmaceutical products should be designed so that at the end of their function they break down into innocuous products and do not persist in the environment (Jorgensen and Halling-Sørensen, 2000). In fact, an environmental evaluation of a drug could potentially take place as an integrated part of the already long list of examinations required for approval of a new drug (Rieger *et al.,* 2002; Jorgensen and Halling-Sørensen, 2000).

There is broad knowledge that both therapeutic properties and biodegradability of pharmaceutical products can be improved by systematic variation of chemical structure (Kümmerer, 2007). This approach, however, is very ambitious since active pharmaceutical ingredients rely upon their precise chemical structure to exhibit the desired biological activity and must demonstrate adequate stability and acceptable shelf life (Tucker, 2006). The challenge for the pharmaceutical industry over the next decades will be, as highlighted by Kümmerer, to design new greener drugs in such a way that their lifetime is sufficient for their use but short enough under environmental conditions (i.e. a suitable balance between stability and biodegradability) (Rieger *et al.,* 2002). Nevertheless, it is important to note that, since the potential for degradation of pharmaceuticals greatly varies across their life cycle (i.e. different pH, temperature, redox conditions, bacteria and enzymes present), stability during the application phase and degradability thereafter are not necessarily mutually exclusive (Figure 8.2) (Kümmerer, 2007).

To the best of our knowledge, very little work has been done, so far, on the development of pharmaceuticals with improved biodegradability (benign by design strategy). Only two publications related to this subject could be found in the literature. Wenlin and co-workers introduced in 2000 the concept of a built-in chemical switch that once activated would lead to rapid decomposition and reported a new type of β-lactam antibiotic presenting in its structure a functionality that unmasks over several hours of exposure to light (Wenlin *et al.,* 2000). This light-induced unmasking generates a reactive functionality, which destroys the β-lactam moiety and hence eliminates its antibiotic properties.

Table 8.1 Examples of 'greener' drugs.

Drug name/active ingredient	Company	Indication	Improvement	Stage in lifecycle of PP
Januvia™ (Sitagliptin)	Merck & Co., Inc	Type 2 diabetes	Developed a novel catalytic method for asymmetric hydrogenation. The new route has fewer steps and >50% increase in yield, providing a significant reduction in raw materials, processing time, waste and energy. Strategy is broadly applicable to other pharmaceutical syntheses[1].	New process developed after clinical trials
Lipitor® (Atorvastatin)	Codexis, Inc	High cholesterol	Greener reaction conditions developed which employ the directed evolution of three biocatalysts to produce the key chiral building blocks for the active ingredient in Pfizer's Lipitor®. Resulted in increased yield, improved worker safety, reduced waste and use of solvents, avoids use of hydrogen gas and reduces the use of purification equipment[2].	?
Celebrex® (Celecoxib)	Pfizer Global R&D	Arthritis	Eliminated need for final re-crystallisation step in manufacturing process – final isolation made direct from reaction mixture. Process efficiency increased in terms of raw materials, solvents, energy and waste[3]	New process approved after drug was already on the market
Lyrica® (Pregabalin)	Pfizer Global R&D	Neuropathic pain	Adoption of biocatalytic route, which eliminates need for organic solvents. Also allows undesired enantiomer to be recycled (previously could not be resulting in vast quantities of waste) (Boethling *et al.,* 2007)	New process approved shortly after drug was placed on the market
Irbesartan® (Avapro)	Bristol-Myers Squibb Co.	Hypertension & renal disease	Modification of synthesis to avoid production of non-biodegradable by-product requiring disposal by incineration. Reduction in use of hazardous materials, improved worker safety and energy savings through solvent recycling (Wackett and Ellis, 1999)	New process developed after clinical trials
Emend® (Aprepitant)	Merck & Co., Inc	Chemo-therapy induced emesis	New shorter and more efficient synthesis, which uses milder reaction conditions, doubles overall yield and reduces use of raw materials by 80%[4].	New process approved shortly after drug was placed on the market
Zoloft® (Sertraline)	Pfizer Global R&D	Depression	Manufacturing process streamlined from 3 steps to one, using a more selective catalyst and more benign solvent. This achieved improved safety, reduced consumption of resources (energy and raw materials) and doubled the overall product yield[5].	Improved commercial manufacturing process
Radafaxine	GlaxoSmithKline	Depression	Discovery and development of a green process for separating the two enantiomers of Radafaxine. Produces the desired enantiomer without expensive and environmentally unacceptable chiral catalysts or templates. It also replaces the environmentally undesirable solvents, allows recycling of the undesired enantiomer and significantly reduces waste[6].	?

(*continued*)

Table 8.1 (*continued*)

Drug name/active ingredient	Company	Indication	Improvement	Stage in lifecycle of PP
Accupril™ (Quinapril hydrochloride)	Pfizer Global R&D	Hypertension and heart failure	The redesigned process reduces raw material, water, and energy use significantly. It also has improved yields and a reduced number of steps. The new process avoids the use of a problematic solvent and solvent waste as a whole was reduced by 90%[6]	?
Cymbalta®, (Duloxetine) & Strattera® (Atomoxetine)	Eli Lilly & Company	Depression & ADHD	An improved synthetic route avoiding the use of traditional N-methyl protecting group, which reduces the combined environmental footprint by an average of 44%, as measured by the weight of materials used to produce one kilogram of product (E-factor)[6].	?
Taxol® (Paclitaxel)	Bristol-Myers Squibb Company	Cancer	Green synthesis via plant cell fermentation and extraction. Replaces previous complex semi-synthetic method, requires no chemical transformations eliminating six intermediates and vastly reduces the amount of solvent, energy and chemicals required[7].	New process approved after drug was already on the market
Cytovene® (Ganciclovir)	Roche Colorado Corporation	Viral disease	More efficient process developed which halves the number of synthetic steps, reduces the number of hazardous waste streams and recycles/reuses a number of ingredients not incorporated in the final product[8].	New process approved after drug was already on the market
LY300164	Lilly Research Laboratories	CNS disorders	New synthetic pathway using a biocatalyst and greener oxidation method. Fewer and inherently safer process steps, elimination of hazardous chemical and waste (chromium oxide & chromium waste), and reduction in use of solvent[9].	Compound in early stages of development
Ibuprofen	BHC Company	Pain-inflammatory	New manufacturing process using three catalytic steps vs. old route of six stoichiometric steps. Virtually all starting materials are either converted to product or reclaimed by-product, or are recovered or recycled in the process, virtually eliminating all waste[10].	New process approved after drug was already on the market

(1) http://www.epa.gov/greenchemistry/pubs/pgcc/winners/gspa06.html; (2)http://www.epa.gov/greenchemistry/pubs/pgcc/winners/grca06.html; (3)http://www.epa.gov/greenchemistry/pubs/docs/award_entries_and_recipients2006.pdf; (4)http://www.greenchemex.org/?module=resources.edit&id=4&fs=1; (5) http://www.greenchemex.org/?module=resources.edit&id=19&fs=1; (6) http://www.epa.gov/greenchemistry/pubs/docs/award_entries_and_recipients2005.pdf; (7) http://www.greenchemex.org/?module=resources.edit&id=9&fs=1; (8) http://www.greenchemex.org/?module=resources.edit&id=29&fs=1; (9)http://www.greenchemex.org/?module=resources.edit&id=34&fs=1; (10) http://www.greenchemex.org/?module=resources.edit&id=45&fs=1

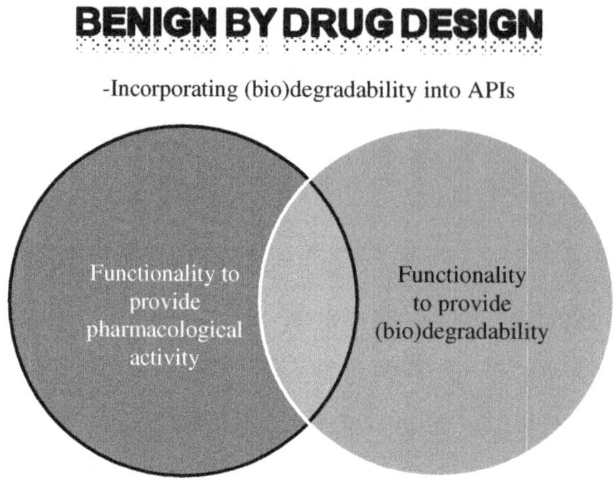

Figure 8.2 Benign by design or design for biodegradability.

Kümmerer also demonstrated that changing the chemical structure of an anti-tumour drug could dramatically enhance biodegradability with no negative effects on therapeutic activity (Kümmerer *et al.,* 2000).

In the meantime, several computer-based programs have been developed over the last few decades to predict the biodegradability of chemicals (including pharmaceutical products) in the environment (Boethling *et al.,* 2007; Wackett and Ellis, 1999). These can be summarised under the generic term Quantitative Structure Activity Relationship (QSAR). The aim is to use descriptors that influence biodegradability and describe the chemical and physical properties of a new substance (Rieger *et al.,* 2002). In principle, this method may avoid both expensive and time-consuming field degradation and animal tests as well as complex chemical syntheses (Kümmerer, 2007). However, predicting biodegradation pathways using QSAR systems is, and will probably remain, an inexact science, since it does not allow biodegradability to be predicted under real conditions (Rieger *et al.,* 2002). Still, this can provide a convenient way for the pharmaceutical industry to make a quick comparison of alternatives during both research and development.

Finally, although recent interest in pharmaceuticals in the environment focuses on the active ingredient, consideration should also be given to packaging and excipients, which are inherent components of pharmaceutical products. Packaging size and package materials could indeed be improved to minimise waste (e.g. drugs passing their expiration date before use and use of non reusable or non recyclable packaging materials) and thus reduce the impact of pharmaceutical products on the environment (Daughton, 2003). Similarly, it would be important to assess the environmental profile of some of the excipients, including preservatives and surfactants, which may exhibit significant biological activity.

8.3 ECOLOGICAL PHARMACOVIGILANCE

For many years regulatory authorities have dealt with a wide range of industrial compounds, concentrating on non-polar organics and heavy metals. In contrast with industrial chemicals that enter the environment either accidentally or through regulated discharges, pesticides are designed to be applied in the environment. For the latter, there are special requirements, including environmental fate studies, for the registration of these compounds. Since they are designed to be biologically active/toxic, post-registration monitoring of these compounds is required. Chemicals of particular concern appear on priority lists in many sets of national and international environmental legislation, and there are well established monitoring requirements, and defined environmental quality standards for these pollutants. In the European Union the recent REACH legislation has widened the range of substances for which environmental risk assessments are required, and will eventually be extended to compounds produced or used in Europe in relatively small quantities (greater than one tonne per year). This will involve large numbers (an estimated 30, 000) of compounds and will have significant impacts on many categories of products (including household cleaning and personal care products). More recently pharmaceuticals have been detected widely in the aquatic environment, and in drinking water. Although these compounds are usually present at only low concentrations in surface waters, they have attracted the attention of the public and regulators because they are designed to be biologically active at very low concentrations. Concern has been increased by finding that removal of some pharmaceuticals from waste by STPs is variable and in some cases inefficient. Further relatively little is known of their fate and behaviour in the environment, and their effects on aquatic biota are poorly defined. There is a need for more work in this area, and experience gained in monitoring priority pollutants, pesticides under established regulations, and now a widening range of compounds under REACH, will be useful in guiding the development work. The guidelines for assessing the environmental impacts of medicinal products for veterinary use may not be directly applicable to products for human use since many of these are discharged more directly into the environment, and not into waste that is treated. There is a need for research to inform the regulatory processes (discussed in chapter seven of this book) for medicines for human use.

Work is needed in a number of areas to supply information needed for use in the development of realistic and reliable risk assessments. Some of these areas (exposure routes, measured environmental concentrations, performance of STPs, ecotoxicological assessments) are addressed in various chapters of this book. This section of this chapter aims to identify key areas where further research and development are required.

8.3.1 Monitoring strategies and scenarios for newly registered, existing and high risk medicines

Clear objectives need to be defined in order to ensure both the representativeness and usefulness of information from monitoring programs. Selection of matrices to be monitored, monitoring points, sampling frequencies and indicators to be used to fulfill monitoring objectives will be based on the pattern of use of the pharmaceutical of interest, its life cycle assessment, whether it has recently been approved, has been on the market for some time or was identified as a high risk compound during registration (e.g. based on volumes or toxicity). For example, the monitoring of newly approved human medicines in surface or drinking waters may not be relevant initially but, if the level of use of the medicine is sufficiently high, measurements in waste water treatment plant influents and effluents may provide relevant information on transformation and removal efficiencies. In addition, such matrices are likely to present highest concentrations thereby the highest chance of detection. Monitoring strategies for pharmaceuticals that have been longer in the market place may focus on providing data and information that can help with mitigating and controlling concentrations of the compounds in various environmental compartments with the aim of reducing exposure of sensitive species or of minimising the threat to drinking water resources.

Monitoring strategies may be developed to:

- Refine estimates for the calculation of predicted environmental concentrations (PECs) or for comparison of measured environmental concentrations (MECs) with PEC/NOEC values (for influent and effluent waste waters and surface waters), where the NOEC is the no observed effect concentration.
- Estimate the proportion of medicines reaching waste waters (prior to STPs) based on usage (consumption and market penetration data).
- Determine (or refine knowledge of) the efficiency of removal and/or transformation of medicines in various types of waste water treatment plants.
- Estimate total load to the aquatic environment from local STP or at the river basin level.
- Understand/assess the fate and distribution of medicines in river, estuarine and marine environments, e.g., transfer to and accumulation in sediments.
- Refine estimates of releases of veterinary product to soils, movement to water, and direct introduction into and dissipation from aqua-cultural installations.
- Identify hotspots of contamination with pharmaceuticals and possible exposure of biota.
- Identify sites and periods where and when risk from the presence of pharmaceuticals is highest (low flow/high proportion of waste water in surface water).
- Identify high risk situations based on the use of indicator substances for the assessment of the proportion of waste water in surface waters and the efficiency of STPs.
- Determine the fraction associated with suspended matter/sludge removed during the STPs and assess the fraction disposed of as manure.
- Assess threat and possible leaching to ground water or run-off to surface waters following manure application or excretion (for veterinary pharmaceutical products).
- Assess threat and possible leaching to ground water and run-off to surface waters from pharmaceutical disposal through land filling.
- Assess threat to sensitive sites such as drinking water intakes, zones with surface/ground water interactions.
- Assess the exposure, effects and impacts of pharmaceutical products on biota (at various trophic levels) in water, sediment and soil.

Other possible monitoring tasks or challenges may be to:

- Assess the effects and impacts of combinations of pharmaceutical products present in sewage effluents or pharmaceutical products in combination with other commonly found contaminants (taking into account possible interactions, including synergism).
- Assess the effects and impacts of pharmaceutical presence in soil/manure combined with pesticide exposure on terrestrial organisms.

- Link monitoring of pharmaceuticals to current legislation; e.g., the WFD, Marine Framework Directive, Environmental Impact Assessment (EIA) Directive, Environmental Liability Directive.
- Optimise monitoring in order to reduce costs by, for example, use of indicator substances (investigate suites of pharmaceuticals for analysis; replacement of suites of analysis by use of indicator substances).
- Trace sewer leakage, and possible ground water contamination in the vicinity of urban areas.
- Trace various important sources of pharmaceuticals such as hospitals, residential care homes, or industrial effluents by using indicator substances.
- Consider alternative tracers, other than pharmaceuticals.
- Investigate transfer from surface water/sediment/sediment/soil to ground water.
- Assess the impact from waste water re-use.
- Assess the impact of combining storm water and raw sewage during high/flash floods.

The overall aim of the implementation of all/some of the monitoring strategies listed above whether they are for newly approved, high risk or existing medicines is the same: to provide realistic assessments of the risk and threat of pharmaceuticals released into aquatic environments. However, because of differences in the pattern of use both within and between nation states, and changes in the local/global distribution of the various medicines, it is necessary to identify the scenario that is relevant in each individual case. It is also essential to identify at as early a stage as possible those compounds that are of highest priority in each situation so that monitoring resources can be focused where they are needed and are not wasted on products that do not pose a risk, and that only relevant parameters are measured.

Monitoring every site for all possible pharmaceutical products would not be sensible, and a screening approach might provide protection at an acceptable cost. One approach would be to assess the likelihood of particular compounds appearing in emissions, and to use a crude measure of presence or absence to focus more detailed investigations of the pharmaceutical and/or its metabolites (in STPs, STP effluent, and adjacent surface waters) where high risk medicines are detected. Other approaches to reducing the monitoring burden would be to identify indicators for specific compounds, or to treat similar compounds as a group. The selection of matrices for these monitoring scenarios is dependent on the available extraction efficiency for the matrix, and the availability of suitable analytical methods.

8.3.2 Removal and/or transformation of medicines in sewage treatment plants

If measurable concentrations of a pharmaceutical considered to be a potential risk are observed in surface waters then it would be worthwhile evaluating the removal efficiency of the various types of treatment plants involved. Here the collection of influent and effluent samples needs to be coordinated and based on the analyte travel/residence time in the STP. Modelling would be necessary to determine mass balances based on measured concentrations in these media and on information on rates of usage, and excretion. However, concentrations of individual pharmaceuticals in environmental waters may be highly variable and may increase as the use of the product increases, and this has an impact on the frequency of sampling and the limits of detection required for monitoring them.

A number of factors can affect the concentrations in the environment, and include distance from the source, and the efficiency of STPs making discharges to surface waters. For instance, Vieno *et al.* (2005) found seasonal variations of up to a factor of four in the concentrations of certain pharmaceuticals in a Finnish river, and this was in addition to large variation associated with the distance from the point of effluent release. These seasonal variations could be explained by variations in the efficiency of the STP and river discharge. Appropriate sampling frequency, sampling period and pattern are prerequisites for representative sampling to be obtained. Castiglioni *et al.* (2006) found a difference of a factor of two between maximum day time and minimum night time influent loads. Such a variation may introduce bias when using time-proportional sampling methods, in this case with an estimated underestimation of influent load of 5–15%. This can, in turn, directly affect removal rate calculations. A seasonal variability of removal rates was also observed by the authors and at one treatment plant lower removal rates were found for some pharmaceuticals during winter when microbial activity is reduced due to low temperatures. Pharmaceuticals were classified into three groups depending on their removal rate (RR). The first group had RRs higher in summer than in winter and included amoxicillin, bezafibrate and ibuprofen. The second group had RRs similar in summer and winter and included ciprofloxacin, hydrochlorothiazide and ofloxacin, whilst the last group was classified as not removed (i.e. erythromycin, salbutamol, carbamazepine). At a wider level, differences may also exist between countries in Europe depending on geographical situation and climate. Vieno *et al.* (2005) observed a reduction of 25% in the elimination efficiency of a STP in Finland in winter compared with values obtained in spring and summer for the drugs benzafibrate, ketoprofen, ibuprofen, diclofenac and naproxen.

Removal of compounds in an STP may be through sorption to sludge with subsequent removal or degradation by biological or non-biological routes. For instance, Golet *et al.* (2003) observed a significant association of the fluoroquinolones, the antibacterials ciprofloxacin and norfloxacin with sewage sludge in an STP in Switzerland. The proportion of antibacterial agents (based on mass flows) in filtered tertiary effluent relative to initial input in raw sewage were in the range 8–11%, while the fraction observed in digested sludge was between 73–78%. Approximately 10% losses were observed during anaerobic treatment of the sludge. Losses of clofibric acid, diclofenac, ibuprofen, ketoprofen, mefenamic acid, and naproxen in activated sludge and a membrane bioreactor were investigated by Kimura *et al.* (2007). Higher solid retention times in membrane bioreactors resulted in higher elimination, especially for ketoprofen and diclofenac. For the latter biodegradation was slow, but important when long retention times were used. Overall in this study while sorption to sludges was identified as an important factor in batch experiments, the primary mode of elimination was found to be biodegradation.

Losses of any compound may differ widely depending on the capability of the STP and the type of treatments and other factors such as residence times (sludge retention time, hydraulic retention time) or reactor configuration. In order to identify the factors or processes that contribute to the elimination of pharmaceuticals from waste waters it is necessary to use mass balance calculations. In order to be able to evaluate fluxes from this modelling it is essential to know the values of a number of parameters or factors including potential for volatilisation, sorption to sludge and degradation in the various stages of the treatment process. In some cases it may be possible to significantly reduce costs by using measured and accurate sludge/water partition coefficients rather than measure pharmaceuticals in both phases. However, accurate values are needed. Goebel *et al.* (2005) measured concentrations of antimicrobial agents at various stages of two STPs in Switzerland to perform mass balances at the STP level. While a reduction was apparent for sulfamethoxazole (and its metabolite N4-acetylsulfamethoxazole), trimethoprim and clarithromycin, significant proportions were still present following the tertiary treatment (38, 36 and 79%, respectively). These data also suggested that re-transformation of sulfamethoxazole occurred during treatment processes. In addition, wide daily and seasonal temporal variations in concentrations were observed but were in agreement with patterns of use of these antimicrobials. For macrolides used to treat respiratory tract infections, loads measured in February-March were approximately twice those measured in September, corresponding to highest sales from February to April. There are, however, differences between different treatment works, and a similar study of antibiotics in Swedish STPs demonstrated minimal removal of sulfamethoxazole and trimethoprim during treatment while doxicycline and fluoroquinolones (e.g. norfloxacin) appeared to be removed reasonably efficiently with a high content found in sludge. The authors used the final effluent and sludge concentrations to estimate the environmental loads per inhabitant and found good correlations with those predicted from consumption data. The data collected from these studies may then be used to refine PEC values for surface waters and STP effluents, and provide checks on the representativeness of available MEC values.

Many of the studies have focused on antibiotics because many of these medicines (for human or veterinary use) and antibacterial agents are released into waste waters, and they have the potential to produce antibiotic resistance in some populations of microorganisms. There is a further problem with these compounds that needs to be taken into account when measuring the efficiency of STPs. Depending on the concentration, constancy of the input, and the impact of mixtures of antibiotics, these compounds may affect the efficiency of sludge or of specific microbial populations in treating waste water. Amin *et al.* (2006) found a 5% reduction in biogas production when the antibiotic erythromycin was added to a pilot scale anaerobic sequencing batch reactor at a concentration of 1 mg/L. However, no further decrease was observed when the concentration of antibiotic was increased to 200 mg/L, though there was a slight increase in the effects on methogenic activity and conversion of butyric acid. Exposure time also appeared to affect butyric acid inhibition.

Where degradation is found to be a significant source of losses within an STP, then it may be necessary to identify, detect and quantify the principal metabolites/breakdown products in the system. In some cases this may be important where breakdown products may have high intrinsic toxicity. This is illustrated by the study of Bedner and Maccrehan (2006) who followed the fate of acetaminophen in a range of treatment processes. They found that some waste water treatment processes such as chlorination could result in the production of toxic derivatives of the pharmaceutical that could result in toxicological effects in the vicinity of sewage effluent outfalls. Where major degradation products are shown to be toxic in their own right then it is important to consider ways of reducing their production, and to carry out a risk assessment to ensure that final environmental concentrations fall below the predicted no effect concentration (PNEC).

In some cases investigations of the transformation and breakdown products of PPs can yield useful information on the factors affecting removal from waste in an STP. An example of this sort of study is provided by the investigation of the fate of carbamazepine in an STP in Peterborough (Canada). In this trial the parent compound and a number of metabolites were measured both in treated/untreated dissolved phases and in solids. The findings were consistent with those of previous studies where carbamazepine was found to be relatively persistent in this system that incorporated secondary treatment technology. Losses throughout the treatment plant were in the region of 29%. Concentrations of certain metabolites were shown to increase during treatment. This may have been due to the presence of conjugated forms of hydroxylated metabolites that were produced in treated patients, excreted in the urine or faeces, and then transformed back into the free form during the sewage treatment process, possibly by microbial metabolism. A mass balance, though based on only single 24 h composite grab samples of waste water and biosolids, showed that carbamazepine and 10,11-dihydroxy-10,11-dihydrocarbamazepine were the main analytes present in the water fraction of the influent and effluent, while less than 1% of the total was associated with bio-solids.

When no data are available on removal efficiency during sewage treatment, it is possible to use models to predict the fate of pharmaceuticals in standard STPs. A good example of this approach is provided by the study of Jones *et al.* (2002) who carried out a risk analysis and used the STP fugacity model STPWIN^{TM} model to estimate the fate of 25 pharmaceuticals used in highest amounts in the United Kingdom. This model was used to estimate values for the total removal, as well as the contributing processes (biodegradation, sorption to sludge, and stripping to air) for a standard system and set of operating conditions. When experimental data are not available the model can be used to provide an indication of how a compound might react within such an environment. Most pharmaceutical products were expected to be removed from waste water by up to 97–98%. Sorption to sludge was predicted for ibuprofen, sulphasalazine and mebeverine hydrochloride, quinine sulphate and mefenamic acid. This means that should the sewage sludge containing any of these compounds be used on land many terrestrial ecosystems could be affected. Models such as STP, SimpleTreat 3.0 or STPWIN^{TM} have been used by the US Environmental Protection Agency and Environment Canada to predict the fate of new chemicals in STPs through mass balance calculations. Recently, Seth *et al.* (2008) attempted to address some of the deficiencies of the current models for use at the screening level. These are (i) inability to handle ionised compounds (ii) the need to characterize biodegradation rates adequately and accurately, (iii) the flexibility to account for the various configurations of STPs, (iv) the need to account for surface volatilisation and non-equilibrium between air/solution and (v) increase flexibility for the input or prediction of sludge-water partition coefficients. There is a continuing need to improve and validate these models since they offer a cost effective way of assessing the behaviour and fate of novel and existing compounds in STPs, and thus underpinning risk assessments without the need for expensive monitoring campaigns.

Biodegradability for environmental risk assessment purposes (e.g. EMEA guidelines, Phase II) may be undertaken following the OECD 303A test. These tests often fail to provide an adequate characterisation of the proportional contributions of the biotic/abiotic degradation and the sorption processes to the removal of the medicines from waste water. Another approach is to use radio-labelled chemicals at environmentally relevant concentrations to evaluate the fate of pharmaceuticals in laboratory/pilot scale waste water treatment plants (Junker *et al.*, 2006). This is particularly useful for antibiotics that may be resistant to biodegradation or may affect sludge processes. The lossless system developed by Junker *et al.* (2006) yields more accurate information on the fate of antibiotics in STPs but there are problems associated with using radio-labeled compounds: one of the most important being the inability to distinguish the parent compounds and metabolites or adducts.

Building on the experience gained from studies such as those outlined above, work has been carried out to monitor the efficiency of novel technology for removing more polar compounds such as pharmaceuticals from waste water. The efficiency of nano-filtration with two membranes (NF90 and NF270) to remove three pharmaceuticals (sulfamethoxazole, carbamazepine and ibuprofen) from water was investigated by Nghiem *et al.* (2005). Retention by the NF90 membrane is primarily the result of steric exclusion, while effects from electrostatic repulsion can be observed for ionised chemicals. The functioning of the system is pH dependent, and at a higher pH a higher retention for negatively charged compounds was observed. Other types of waste water treatments that have been considered include constructed wetlands that are often designed to treat effluents from small scale housing communities. Matamoros *et al.* (2005) investigated the fate of clofibric acid, ibuprofen and carbamazepine in a sub-surface constructed wetland in Barcelona (Spain) by spiking waste water with a solution of pharmaceuticals and taking measurements of concentrations in the wetland. A bromide injection served as a tracer. The behaviour of clofibric acid was similar to that of bromide and could itself serve as a tracer in such environments to evaluate hydraulic conditions. Losses of carbamazepine in this system were attributed to sorption to the gravel bed, and the very large losses of ibuprofen in this shallow sub-surface constructed wetland were attributed to degradation, probably due to the prevailing oxidising conditions. Pierini *et al.* (2004) while developing a simple liquid-liquid extraction procedure for enrofloxcin in swine sewage observed a significant reduction of enrofloxacin concentration when comparing samples from raw sewage with those from the lagoon. Raw sewage treatment included mixing of pig faeces with 10% (v/v) olive oil/ground water, storage in aerated pits and finally retention in an open-air lagoon for a further eight months. This treatment noticeably reduced the amount of enrofloxacin to levels below the detection limit of 0.6 μg/L, and all samples that were treated using this method were found to be free of enrofloxacin after eight months. This treatment also resulted in a reduction in dry weight, pH value (to ~6.5), and bad odours. It was concluded that the aerobic conditions in the storage pits combined with the long retention in open-air lagoons encouraged bacterial growth and chemical reactions that assisted in the decomposition of organic compounds, such as enrofloxacin. Despite this being a relatively cheap and simple method to implement the time frame and space required for the process could be a limiting factor, and it may not be easy to adapt it for other medicines or other types of sewage. However, it does indicate that local in-farm sludge treatment may be able to significantly reduce the inputs and impacts of medicines for veterinary use. It may be possible to develop a system for optimising treatment plant efficiency for removing a particular pharmaceutical product of concern providing there is sufficient knowledge and understanding of the important factors that determine this. However, care must be taken to ensure that toxic products do not result from the treatment, and it will be necessary to balance any increased efficiency against increased costs, and carbon consumption and emissions. The information gained in developing monitoring methodologies would support such efforts to improve the removal of high risk medicines.

The chief conclusion of this discussion was that there was a need to determine (or refine knowledge of) the efficiency of removal and/or transformation of medicines in various types of waste water treatment plants.

8.3.3 Hospital wastewater

In hospitals a large variety of substances (e.g. pharmaceuticals, radionuclides, solvents and disinfectants) are used for medical treatment, and other purposes such as diagnosis and research. Pharmaceuticals and their metabolites excreted by patients, primarily in the urine, along with used diagnostic agents and disinfectants are disposed of in the wastewater. The disposal causes significant problems for waste treatment because hospitals consume large volumes of water per day. Whereas domestic water consumption is generally around 100 L/person/d, hospitals use far more, typically between 400 and 1200 L/bed/d (Gautam *et al.,* 2007). This varies between countries (e.g. United States of America: 968 L/bed/d; France: 750 L/bed/d; developing countries: 500 L/bed/d). These high water consumptions generate large volumes of waste water that need to be treated. The most frequent contaminants in hospital wastewater (apart from pathogens) are pharmaceuticals and their metabolites, significant concentrations (up to 10 mg/L) of organohalogen compounds (measured as those that adsorb to activated carbon, and including iodized X-ray contrast media, and chlorine containing molecules such as solvents, components of cleaning materials and disinfectants, and products of reactions of a range of compounds with hypochlorite bleach), and radioisotopes (Emmanuel *et al.,* 2005). It has been suggested that hospital waste waters contribute to the spread of antibiotic resistance in the environment. Ternes and Joss (2007) reported concentrations of antibiotics in hospital waste water that were 4 to 100 times higher than those found in municipal wastewater. This is not surprising, given that some 25% of antibiotics for human use are applied in hospitals (Kümmerer, 2001).

Since waste water from hospitals contains such a large variety of potentially hazardous pollutants at high concentrations, it would be better to treat this separately, possibly on-site, in order to reduce the load of contamination in municipal sewage systems. Hospital effluents generally contain only low levels of faecal micro-organisms because of the routine use of disinfectants, and despite significant dilution in the sewer system they may impact negatively on the biological processes of STPs (Gautam *et al.,* 2007). There are several processes that could be used to treat hospital waste water (Kajitvichyanuku *et al.,* 2006), and these include the photo-Fenton process, reverse osmosis, activated carbon, ozonation, submerged hollow fibre membrane bioreactor (MBR) (Mahnik *et al.,* 2007), coagulation/flocculation combined with filtration and disinfection, or a combination of some of these. These have been demonstrated to have the potential to remove efficiently the hazardous components of hospital waste waters. Whatever the treatment regime, it would be necessary to apply the same precautions to the disposal of the sludge from hospital waste as those used in the disposal of municipal sludge. It should not be used as manure unless treated to make it safe (Gautam *et al.,* 2007).

8.3.4 Urine separation

Technologies for the separation of urine from other waste, and the recycling of anthropogenic nutrients (nitrogen, phosphorous, and potassium) as fertilizer in agriculture (Larsen *et al.,* 2001) have been demonstrated at local household scale, and this provides a possible model for use by hospitals. Urine separation has a number of benefits, including improvement and simplification of the control of water pollution (most nutrients and some important micropollutants are discharged in urine), the possibility of closing the nutrient cycle, and reductions in the size of municipal STPs. Further, it is generally assumed that the majority of pharmaceutical products are excreted via urine, either un-metabolized, or as detoxification products (usually conjugation products produced in secondary detoxification).

Separation of urine means that there will be only a small dilution of the waste to be treated, and PPs and their metabolites will be present at concentrations up to 100 times higher than those found in typical wastewater. This means that it becomes feasible, and cost-effective to use treatments, such as AOPs or nano-filtration, that are not useful for treating large volumes of dilute waste water in municipal STPs. It has been mooted in Sweden that technologies for separating urine from other waste should be installed in public buildings, hospitals and retirement homes where the input of PPs in urine is particularly high. An alternative way of using the separation of urine is to manage the operation of existing waste water treatment systems to achieve much higher efficiencies (Larsen *et al.,* 2001; Rauch *et al.,* 2003). It would be possible to feed the separated urine into the sewer system in a controlled manner so as to ensure the stability and reliability of the nitrification process. A significant advantage would be the reduction in size of municipal STPs since these are designed to cope with peak loads of ammonia where there is a marked diurnal rhythm in inputs to the system. Inputs of waste are very much reduced during the night. It would also be possible to stop discharges of urine to the sewers during storm events, and this would be particularly important where sewage and storm water are mixed. Such systems cause pollution of environmental waters with untreated sewage, and hence increase concentrations of a wide range of pollutants (including pharmaceuticals) that would normally be reduced in STPs. Environmental concentrations of PPs and metabolites would be reduced by the reduction in quantities of urine in mixed storm and waste waters. These approaches show great potential, and there is a need for further research and development work in the areas of urine separation, separate treatment of urine, and the control of inputs of urine to sewage systems.

8.3.5 Spatial and temporal variations in concentrations of pharmaceutical products in the aquatic environment

8.3.5.1 Factors affecting spatial and temporal variations in concentrations of pharmaceuticals

It has been established that concentrations of PPs in the aquatic environment vary spatially and temporally depending on a number of factors including volume of use of individual medicines, rate of excretion, and inputs to waste water from other sources, and also on the efficiency and distribution of STPs. Following entry to the aquatic environment other factors affect the final concentration. Flow determines the dilution factor, and hence the proportion of waste water in surface waters. The properties of the aquatic system that determine the rate of degradation of any particular PP, and the quantity that becomes associated with sediments. The latter is particularly important in the case of more hydrophobic PPs.

Risk assessments must use concentrations based on the worst case scenarios, and this section of the chapter explores monitoring strategies that could be useful in identifying sites, environmental matrices, and periods when a water body is at highest risk. It is necessary to determine the mixing pattern for sewage effluent and surface waters, any seasonality of use of pharmaceuticals, pattern of use (e.g. in home, hospitals, residential/care homes), pattern of effluent discharge, and range of variation of environmental conditions of the receiving waters throughout the seasons. The highest concentrations of pharmaceutical products in environmental waters may be observed when the ratio of waste to receiving water is highest (e.g. in dry periods when the flow of receiving water is much reduced), and during storm events when waste is mixed with storm water, and released without treatment. In order to identify sites and periods where there will be a high risk it is necessary to monitor to obtain representative measures of concentrations of PPs in surface waters and sediments, to track plumes of waste waters entering surface waters. It is particularly important to monitor intensively during periods (e.g., of low flows, dry weather, and floods) when high concentrations of pharmaceuticals would be expected.

8.3.5.2 Use of indicators of input of pharmaceutical products into surface and ground waters

In order to identify high risk situations, zones or periods when the risk is highest, without having to make full analysis for a range of potential contaminants it may be possible to use indicators to estimate the proportion of waste water in surface waters, and to differentiate between veterinary and human medicinal sources of pharmaceuticals that are used in both areas. It may be possible to use indicators to assess the proportion of input to STPs coming from hospital-related sources. They may also be used to assess the efficiency and changes in efficiency of STPs.

For a substance to be useful as an indicator of the proportion of waste water in surface waters, the following criteria should be met:

- waste water effluents should be the sole source of these substances (therefore substances such as antibiotics for human and veterinary use may not be considered as indicators)
- they should be present at relatively high concentrations allowing reliable measurements
- losses (either through sorption to sludge, volatilization or biodegradation) through STPs should be minimal
- mass flow through the STP should be proportional to the population/community that it serves
- the tracer should have an adequate market penetration factor
- it should have minimal sorption and degradation upon release into surface waters (or other media that are being monitored).

Indicator substances may be pharmaceutical compounds or other organic or inorganic compounds that are present in waste water effluent. They may help to identify sources of pharmaceutical contamination such as untreated effluent, inefficient treatment plants, or release of untreated sewage together with storm water following heavy rainfall. For example Chang *et al.* (2007) measured concentrations of glucocorticoids in STP influents and effluents, and river water. The removal in STPs was in the range 98–100% except for prednisolone for which only 78% removal was achieved. The authors investigated the ratio of combined cortisol/cortisone to prednisolone concentrations. Ratios decreased from 10–35 to 1–4 following STP treatment. Discrepancies between effluent ratios and higher ratios measured in surface water at certain sites led the author to conclude these were the result of releases of untreated sewage effluents. This work illustrates the potential of this approach, but further research is required to determine whether these particular analytes are good candidates for use as indicators.

Another example is provided by a study of the rare earth element gadolinium (Gd) in waste water, and it has been suggested that it could serve as a potential tracer of waste water input into surface waters on the basis of the use of Gd-organic compounds as X-ray contrasting media (Bau and Dulski, 1996). Verplanck *et al.* (2005) showed that minimal losses of these compounds to sludge occur during treatment processes and where Gd-organic compounds break down then the Gd remains in the system mainly in the form of carbonate and phosphate complexes, and these are likely to remain in the dissolved fraction in water. Despite the persistence of Gd compounds, surface water concentrations decreased with increasing distance from effluent output points, and this was attributed to significant inputs to the river from ground water. Despite these compounds being mainly dispensed in hospital for use as X-ray contrasting media (e.g. GdDTPA), excretion can take place both at home and in hospital. This makes Gd unsuitable as a specific indicator substance for hospital-related waste water. However, it is potentially very useful in detecting sources and points of effluent dilution. In another study this element was used to measure the effects of waste water input/recycling on concentrations of pharmaceuticals. This was undertaken in the Hérault watershed (France) where recycled waste water is expected to be close to 20–30% of supplied drinking water (Rabiet *et al.*, 2006). They measured the concentrations of nitrate, chloride and boron (since these are found for instance in household detergents) and evaluated the Gd as a means of estimating the input of waste water into surface/ground waters. High concentrations of boron and Gd were observed at a few monitoring sites across the watershed, in one case associated with relatively high pharmaceutical concentrations. The study confirmed that several wells supplying drinking water to inhabitants of the Herault valley were contaminated with waste water, and that pharmaceutical substances were present in some of the wells. However, Gd was not always present in wells and at sites where pharmaceuticals were found, and this may be due to fluctuating inputs to the system. In this study Gd was not a universally useful indicator of the presence of hospital waste. There is still much work to be done to evaluate the potential utility of indicator substances in estimating waste water inputs to surface and ground waters, and as indicators of concentrations of various classes of organic anthropogenic contaminants including pharmaceuticals.

A study that illustrates a number of the potential uses of markers was undertaken by Kim and Carlson (2007) who monitored antibiotic concentrations along the Cache la Poudre River (Colorado, USA) both in the water and sediment phases. Seasonal variations in concentrations in human and veterinary antibiotics were observed with the highest concentrations generally being found in the winter, probably a result of low flow conditions and possibly minimal biodegradation due to low temperature. Concentrations of chlortetracycline (used to treat cattle) appeared to be an adequate indicator of the area of the catchment dominated by agriculture. The highest concentrations of sulfamethoxazole were associated with two waste water effluents and this compound may be a useful marker of such input into surface waters. While most studies tend to focus on measurement in the water phase, Kim and Carlson (2002) also undertook measurements of antibiotic concentrations in the sediment phase, and found global sediment/water pseudo-partitioning coefficients of 305–1267, 20–517 and 91–402 L/kg for tetracyclines, sulphonamides and macrolides, respectively. These values indicate the importance of measuring the concentrations of this compound in the sediments when assessing its global distribution.

It has been suggested that biological monitoring may be useful as a screening tool to map locations at risk and to identify where it would be worthwhile carrying out intensive chemical monitoring. However, with current direct toxicological methods it would not be possible to differentiate between toxicity caused by different groups of contaminants, and this poses a challenge to environmental toxicologists. There is one exception, however, and that is the compounds that affect reproductive endocrine function, and this illustrates the potential utility of this approach. For instance, Kidd *et al.* (2007) noted that fish populations inhabiting waters close to STPs could be exposed to chemicals affecting reproductive endocrine function, and these are known to cause male fish to produce vitellogenin (VTG), mRNA and proteins that are associated with oocyte maturation in female fish. It is thought that high levels of estrogenic substances such as the synthetic estrogen 17α-ethynylestradiol (EE2) in waste water cause these abnormalities. This study found that neither natural nor synthetic estrogens were fully removed during waste water treatment, and discharges of effluent into surface waters produced concentrations of around 5 ng/L in the aquatic environment. These levels are sufficient to induce the biochemical changes described above, and such levels of EE2 in breeding waters will reduce the reproductive success of specific species and therefore decrease the population size. It is therefore feasible to use this biological system to indicate the presence of waste water contamination, and to indicate where chemical monitoring might be necessary.

There is a need for further work to identify and develop methods of monitoring and distinguishing between the impacts of various inputs to an aquatic system (including surface water and sediment). This would be aided by the availability of reliable indicator compounds since this would significantly reduce the cost of monitoring sets of individual compounds where pollutants are present in complex and varying mixtures. Such efforts would also help to simplify and to improve the reliability of monitoring to determine the locations and times when the risks associated with a wide range of pharmaceuticals would be highest.

8.3.6 Assessment of possible movement of pharmaceutical products from landfill and following agricultural applications of sludge and manure

The guidance provided by the Committee for Medicinal Products for Veterinary Use (CVMP) (2004) on phase II environmental impact assessment for veterinary products provides clear recommendations for the ways in which PECs for surface waters, sediments, ground water, soil, and dung should be calculated for medicinal products for use in aquaculture, intensively reared animals, and pasture animals. These are used in combination with eco-toxicological data to estimate risk as part of the product registration package. The risk assessment process is tiered and the calculations are guided by decision trees. Initially a minimal amount of data is supplied by the manufacturer but if a significant risk is identified in preliminary steps, then the amount of data required increases. If no risk is identified at an early stage, then the assessment process stops at an early stage. A similar process could be used for all medicinal products, but it does rely on predictive modelling and needs accurate values for variables such as mass balances, market penetration, usage, waste disposal pattern, and prediction of disposal through landfills. This could allow monitoring to be focused on those products where a possible problem is identified.

The modelling process is not straight forward or easy. The interactions of veterinary medicines with soil components, organic materials such as found in manure, compost, and soil organic matter are all important in determining the potential for these drugs to move into ground water, run-off to surface waters, or to impact on biota. This is complicated by the fact that many of these compounds are ionized, and this makes it difficult to predict their sorption characteristics under various conditions. Work by Kahle and Stamm (2007) showed that pH and contact time were important factors in the sorption of the veterinary antimicrobial sulfathiazole to compost, manure and humic acid. Highest KOC values were observed for the cationic species followed by the neutral and anionic species. An increase by a factor of 5–20 is observed with a drop in pH from 9 to 4 with similar effects by contact time. Sorbent composition accounted for only a factor of 2–3 difference in measured KOC values. A pH drop for the manure following rainfall and mixing with soil is likely to increase the sorption of sulfathiazole and related ionized compounds.

Similar problems are encountered when dealing with medicinal products for human use; it is just the routes of transfer into the environment that are different. Oppel *et al.* (2004) investigated the leaching behaviour of a number of pharmaceutical compounds in soils with different properties. Relatively low mobility was found for carbamazepine, diazepam, ivermectin and ibuprofen resulting in a low potential for reaching and contaminating ground waters. On the other hand the mobility of clofibric acid and iopromide was high and could result in contamination of ground water. This is consistent with detection of these pharmaceuticals in ground water. The low mobility of carbamazepine was surprising, however, since it is frequently detected in ground water. It may be judicious to undertake mass balance studies, possibly using radio-labelled chemicals to understand the fate of carbamazepine in top soil or when in river sediments. Waste water re-use, such as for agricultural applications, may result in the release of waste water-associated pharmaceuticals during run-off. Pedersen *et al.* (2005) measured concentrations of carbamazepine, carisoprodol (muscle relaxant), gemfibrozil and ibuprofen of 320–440, 680, 190–790 and 11 ng/L, respectively, in run-off from fields irrigated using waste water.

Whilst this area can be complex, much can be learned about modelling to evaluate leaching potential from soils to water from the recent work in the area of veterinary medicines, and from the long experience in modelling as part of the registration process for pesticides.

8.3.7 Fate and distribution of medicines in river, estuarine and marine environments

This involves following the distribution of parent compounds and major transformation products in a range of environmental compartments. This is complex and involves many different matrices, and for some of these extraction and analysis of compounds of interest is challenging. Estimates of partition between the various compartments, degradation in sediments, soil, and water, and bioavailability are associated with large uncertainties, and this makes reliable risk assessments based on these processes difficult to achieve. Often it is necessary to carry out small scale trials (e.g. laboratory or mesocosm scale) in order to extrapolate to environmental conditions, and to identify and understand the key processes.

Once released into surface waters, further losses can occur through processes such as microbial metabolism, abiotic degradation such as hydrolysis or photolysis, and sorption to bed-sediments. Different classes of pharmaceuticals are likely to behave differently and it may be necessary to investigate the fate of some medicines in water in detail. For instance, Edhlund *et al.* (2006) conducted a number of tests to evaluate the effects of water physicochemical parameters on the photolysis of nitrofuran in water. This provides not only necessary information on their fate in surface water following release, but also indicates how treatment processes might be improved. Fastest transformation occurred by direct photolysis, and the presence of reactive oxygen species at environmentally relevant concentrations yielded significantly lower rates of degradation. Indirect photolysis did not appear to promote further degradation while acid conditions acted as a catalyst. These results provide guidance on the conditions and treatments that will degrade this class of chemical, and on the identity of the most significant degradation products likely to be found in surface waters.

Small mesocosm experiments (e.g. following OECD 308 Guidelines) may be undertaken to assess the effect of sorption of pharmaceuticals to bed-sediment and losses in the system due to abiotic/biotic degradation. It is possible to use ^{14}C-labelled analytes to obtain accurate mass balances providing that appropriate traps are used to retain volatile analytes, and to absorb any $^{14}CO_2$ produced if mineralisation occurs. Loeffler *et al.* (2005) used such experiments to evaluate the fate of a series of medicines selected on the basis of their high consumption and occurrence in the environment. Ibuprofen, 2-hydroxyibuprofen and paracetamol dissipated rapidly from the system with half-lives of 7 days or less. Radio-labelled ibuprofen and paracetamol were used, and this allowed the degree of mineralisation (77 and 19% respectively of the initial concentration in the system) to be estimated. Oxazepam, iopromide and ivermectin were shown to have moderate persistence, whilst carbamazepine and its degradation product (carbamazepine-diol), clofibric acid and diazepam were observed to be persistent in this system. Even though the sediment selected for this study contained only a small fraction of organic carbon, significant accumulation of paracetamol, diazepam and carbamazepine was observed in it. Degradation products of carbamazepine, diazepam and ibuprofen were found to have lower half-lives than parent compounds.

It is not possible to use radiolabeled compounds in natural surface waters, but the results of studies of persistence and fate of pollutants in these are more directly relevant. However, interpretation of the findings is not always easy, as exemplified by the study of Peschka *et al.* (2006) who investigated the occurrence of barbiturates in waste water, surface and ground water at a number of sites in Germany. Concentrations of pentobarbital, butalbital, secobarbital and phenobarbital were in the range 0.09–5.4 μg/L in the River Mulde, a tributary of the River Elbe. Samples collected at an upstream location as well as those from waste water treatment plants did not show the presence of any of these compounds. Based on the current human consumption of barbiturates in the catchment area of the STP they estimated that concentrations in the waste water treatment plant were likely to be below the level of detection. These results are difficult to interpret, and it may possible that concentrations observed in the River Mulde are the result of veterinary use or of past contamination. The latter possibility is plausible since these compounds are shown to be recalcitrant in aquatic systems; however, more work is necessary to investigate another possibility, leaching from landfill sites.

Since these systems are complex, modelling is difficult, and a range of models may be necessary to deal with the wide range of classes of pharmaceuticals. There is more experience with modelling the fate of non-polar organic pollutants than is available for polar compounds, and several multimedia fate screening models have been applied to predict the fate of the former. Fewer models are available for polar compounds such as pharmaceuticals, but some have been developed, for example the linear solvation energy relationship of Sprunger *et al.* (2007), and the poly-parameter linear free energy relationship-based model of Zukowska *et al.* (2006). Much more work is underway to develop predictive models for a wide range of chemical types in order to support the recently enacted REACH legislation. Once validated, these modelling approaches could be acceptable within a regulatory context for either veterinary medicines and/or those for human use.

There is a need for further research into the behaviour of pharmaceuticals in aquatic environments. In particular factors affecting transfer to, and accumulation in various compartments such as sediments, biofilms, and biota is poorly understood for many compounds. However, it will not be possible to study these for all of the compounds in use, and it will be essential to identify those of greatest concern, and if possible to use compounds whose behaviour is representative of that of a set of other similar compounds.

8.3.8 Protection of drinking water

Some sensitive sites, such as intakes for drinking water treatment plants, need protection. Since some pharmaceuticals have been detected in water sources exploited for the production of potable water, it would be advisable to assess of risk from the presence of pharmaceutical products in drinking water. Depending on the location of the site relative to points of entry of pharmaceuticals into the water body, and on the properties of the individual medicinal products it may be necessary to screen drinking water, and to remove some contaminants during its production. Significant concern has been expressed about the presence of pharmaceuticals in drinking water but this has been based on perceptions concerning the high potency of these compounds rather than an on a proper evaluation of the potential risk to human health or environmental health. For instance, concern about two oxazaphosphorine cytostatic drugs, prompted a screening of waste water and surface water in Switzerland. Measurements of these compounds in waste water influent and effluent indicated that both cyclophosphamide and ifosfamide were poorly removed by STPs (Buerge *et al.*, 1996). Batch experiments revealed some slow chemical degradation of these compounds in dark conditions, but half-lives in these conditions were long. While these pharmaceuticals do not appear to be lost appreciably during waste water treatment or when released into the environment, the small amounts dispensed result in extremely low concentrations (less than 50 to 170 pg/L) in surface waters, and these are orders of magnitude lower than levels that cause acute ecotoxicological effects. A human health risk assessment would require knowledge of pharmaceutical losses during various types of drinking water treatment processes, and a consideration of the daily and life time intakes (de Voogt *et al.*, 2008).

One approach based on the precautionary principle has been to introduce measures to reduce the levels of pharmaceuticals in potable water. Much work has been done on various scales to assess the efficacy of various treatments for removing medicinal compounds during the production of drinking water. For example, bench-scale experiments were conducted by Westerhoff *et al.* (2005) to evaluate the removal of various pharmaceuticals in the drinking water treatment plants of three different water supplies. The processes investigated included the use of coagulants to reduce turbidity, lime for softening the water, powder activated carbon to remove organic pollutants, and oxidative processes (chlorine and ozone) for disinfection. All of the trials were conducted under relevant drinking water treatment plant conditions and used appropriate, realistic concentrations of pharmaceuticals of interest. While only minimal losses of pharmaceuticals were observed in aluminium sulphate and ferric chloride coagulant treatments, and with lime softening, the addition of activated carbon (5 mg/L for 4 h) resulted in losses ranging from 14% for iopromide to 73% for diazepam or 93% for trimethoprim. The extent of removal appeared to be related to log Kow for most analytes, except those that are ionised under the conditions tested in this study. Ozonation generally increased losses of some pharmaceuticals (e.g. ibuprofen and gemfibrozil) over those observed with chlorination. Other pharmaceuticals such as iopromide and meprobamate showed low removal rates in most treatments and may need new procedures for their removal from drinking water. The removal of some compounds by carbon treatment is consistent with the proposed role of sorption to sediment, and suspended particles in reducing the availability and activity of pharmaceuticals released into surface waters. However, the processes involved are not simple, and Figueroa *et al.* (2004) showed that sorption of antibiotics such as tetracyclines onto clay in soils and sediments is dependent on the pH, ionic strength of the solution (soil or sediment pore water) and on the presence of competitor ions since sorption was best modelled as ion/cation exchange and surface complexation with zwitterion forms of these analytes.

Vieno *et al.* (2007) studied the removal of beta-blockers, an antiepileptic drug, one lipid regulator, four anti-inflammatory and three fluoroquinolones during their passage through a pilot-scale drinking water treatment plant with the following series of processes: ferric salt coagulation, sedimentation, rapid sand filtration, ozonation, filtration with granular activated carbon and finally UV disinfection. On average just over 10% of the pharmaceuticals were removed by the first three steps. A dose of ozone (1 mg/L) was found to be very efficient for the removal of all pharmaceuticals. Only ciprofloxacin remained above limits of quantification following ozone treatment. The high hydrophilicity of atenolol, ciprofloxacin and sotalol resulted in breakthrough in the activated carbon step. Only ciprofloxacin was observed to pass through all treatments unaffected. However, these results apply only to the parent compound, and more information on potential degradation products may be required. For instance, Bedner and Maccrehan (2006) recorded as many as eleven degradation products for acetaminophen during chlorination studies both in pure water at pH 7 and in waste water. A significant proportion of the parent compound was found to be transformed within an hour of treatment with an equivalent half-life of 26 minutes and resulted in the production of two toxic products, 1, 4-benzoquinone and N-acetyl-p-benzoquinone (NAPQI). The latter causes hepatic necrosis at high doses in humans. Sulphite treatment for dechlorination was shown to transform NAPQI back to acetaminophen and 1, 4-benzoquinone into 1, 4-hydroquinone. This has implications for waste water treatment plants employing chlorination processes and for the aquatic environment in the vicinity of the sewage effluent outfall.

UV disinfection of drinking and waste water not only removes bacteria but also has the potential to transform analytes present in the media. This step is often undertaken using a low pressure mercury lamp that emits a monochromatic light with a wavelength of 254 nm. Pereira *et al.* (2007) investigated the potential for UV degradation of certain pharmaceuticals, namely, carbamazepine, iohexol, ciprofloxacin, ketoprofen, clofibric acid and naproxen in laboratory-grade water and surface water using a bench-scale UV quasi-collimated beam. Photolysis appeared to be related to parameters such as fundamental photolysis characteristics, photon irradiance, solution depth or solution absorbance. Iohexol and clofibric acid were transformed more rapidly than carbamazepine or naproxen. Hydrogen peroxide was observed to increase transformation rates especially for carbamazepine and naproxen in surface water. This was due to hydroxide (-OH) formation. High UV irradiation combined with H_2O_2 removed all pharmaceuticals to below limits of detection. However, the issue of cost/benefit of more complex/advanced waste water or drinking water treatments needs to be addressed. While in some cases pharmaceuticals may be removed by processes such as ozone, hydrogen peroxide treatments sometimes combined with UV disinfection, reverse osmosis and nanofiltration, not all pharmaceuticals will be removed and the cost as well as energy required may be questioned.

Transfer of research work in bench scale and small pilot scale experiments to the development of full scale systems requires a sustained investment, and this is not always available. There are financial risks attached, and there is a need for further work to decrease these. Further, there is a need for a cost-benefit analysis of removal of all contaminants from drinking water, not just pharmaceuticals.

8.3.9 Assessment of the impact of pharmaceuticals on biota

The assessment of the impact of pharmaceuticals on aquatic ecosystems needs to be based on an adequate understanding of fate of these compounds in the aquatic environment. This is necessary to determine which organisms are at risk either due to low tolerance (high sensitivity) and/or to exposure to high concentrations (e.g. those in close contact with fish farming or populations living close to sewage outfalls). There is therefore a need to identify relevant test organisms that are representative of the most susceptible species (worst case scenario), and to develop relevant tests and models that provide results that correlate well with toxicity in the natural environment. In some cases it might be appropriate to use indigenous organisms in-situ, or caged organisms, and to link observed effects on health with those found under relevant controlled conditions. The available methods and biological endpoints available for assessing acute and chronic toxicity are extensively reviewed in chapter 6 of this book. These measure the potential hazard that is needed to feed into risk assessment models. The factors affecting exposure are discussed in the preceding sections. The combination of these in screening models is associated with high levels of uncertainty. Where there is doubt about the validity of a prediction (because of lack of appropriate toxicological data on the most sensitive species, or exposure is predicted to be high) for veterinary medicines, then an assessment factor (AF) of between 10 and 1000 is applied to provide a margin of safety (CVMP, 2004). Montforts (2006) has published a discussion of the validity of some of the exposure assessment models available for veterinary medicinal products. Here global national assumptions used in models were not necessarily appropriate under the local conditions. Local inhomogeneities add to the uncertainties due to the variation in soil types, and properties between areas. Montforts (2006) found that it was difficult to obtain the representative field data (where there can be a thirty fold difference between individual samples from the same field) necessary to validate models. However, he found differences between observed and predicted levels (in soils) of antibiotics used in intensive pig and poultry rearing in a study in the Netherlands fell within of an order of magnitude, and concluded that regional scale modelling may provide useful predictions of concentrations for use in risk assessments. In this context the work of Schneider *et al.* (2007) to develop a model that is applicable across various sets of environmental and soil conditions across Europe shows great potential. Whilst much more data is necessary to validate these models, and obtaining this would entail significant work, this would be worthwhile if it gave confidence in the models, and made them transferable between regions with similar conditions (analogous to the combination of water bodies for monitoring purposes under the EU Water Framework Directive).

8.3.10 Estimation of total load of pharmaceutical products in the aquatic environment at local, river basin levels

In order to estimate any environmental risk associated with individual pharmaceuticals, it is necessary to estimate the movement of compounds of interest to the aquatic environment. This must take into account the various possible sources, and in particular inputs from waste water effluent. This presents a number of challenges because of the variable nature of the inputs over diurnal and seasonal periods, and the changes in volumes of drug used during its lifetime. In addition, it is important to take account of the spatial variation in concentrations due to inhomogeneities in the mixing zone below a discharge point into a river, and in large rivers due to inputs from multiple sites in the river basin system. It is not straightforward to obtain a representative picture of concentrations above and below a discharge point to estimate input from individual STPs.

The fate of a compound in surface waters will be affected by a number of factors including susceptibility to biodegradation, abiotic transformation (e.g. hydrolysis or photolysis), binding to suspended matter, and retention by sorption to bed sediments. Most studies of the potential for attenuation of pharmaceuticals once released into surface waters are conducted in microcosm systems or in laboratory-based batch experiments. Another approach involves taking multiple samples along the path of a river in combination with measurements of removal rates in STPs, and concentrations in solids (sludge or sediments). This can provide clues as to the fate of pharmaceuticals and potential removal processes for surface waters. Castiglioni *et al.* (2006) undertook such a study that provided estimates of the probabilities of particular processes being involved in the removal of some pharmaceuticals during transport in surface waters. For instance, the significant attenuation of hydrochlorothiazide over a 1 km stretch of river was attributed to sorption to sediments, and samples of the particulate phase of the river confirmed this. Where compounds (e.g. erythromycin) were found to be resistant to degradation during sewage treatment, attenuation in the river was suggested to be primarily mediated by sorption. However, this must be viewed with caution since the environment in the sewage treatment processes may not represent the most favourable for degradation of pharmaceuticals. Coetsier *et al.* (2009) compared PECs and MECs for a set of eleven pharmaceuticals (representing the main therapeutic classes) in discharges from a conventional STP in the Languedoc Roussillon region of France, and in the receiving river (the Gardon) 10 m downstream of the discharge. The predicted and measured concentrations in the waste discharges were in reasonable agreement for seven of the eleven compounds, but concentrations of the other four were overestimated. A similar comparison of PEC and MEC in the river water showed that the predicted and measured concentrations were in reasonable agreement for six compounds, and strongly overestimated (PEC/MEC >8) for the remaining five. This group concluded that many of the variables used in calculating the PEC were associated with potentially large and undefined uncertainties. The use of national average values can be inappropriate in some locations where local conditions deviate significantly from those assumed in the model, and can vary markedly between seasons.

Some large scale environmental investigations have been carried out, for instance Fono *et al.* (2006) conducted measurements of the decrease in concentration of several pharmaceuticals in the Trinity River (a large river in the USA) and compared the results with those from laboratory-based microcosm experiments. Adsorbable organic iodide, an X-ray contrast medium, was used as a tracer to determine whether fresh input of river water (or dilution of river/waste water) occurred. Since the concentration remained essentially constant, the authors concluded that if decreases in pharmaceutical concentrations were observed downstream, they would not be the result of dilution. Significant reductions in concentration of gemfibrozil, ibuprofen, metoprolol and naproxen were observed over the course of the 500 km in which measurements were made. In addition, the comparison with microcosm experiments showed that both biodegradation and photolysis are important factors in the attenuation of these drugs. While in certain cases such as fast flowing rivers, there is insufficient time for significant biodegradation to occur, the hydraulic retention time of two weeks for this 500 km stretch of the Trinity River and strong attenuation of light in the water increased the relative significance of biodegradation in the attenuation of these drugs.

Another example of a large scale survey is provided by the work of Wiegel *et al.* (2004) who conducted a number of surveys along the River Elbe for a range of pharmaceuticals and detected diclofenac, ibuprofen, carbamazepine, some antibiotics and lipid regulators at levels in the range 20–140 ng/L. Phenazone, isopropyl-phenazone, paracetamol were also detected during a more thorough investigation in 2000. Two metamizole metabolites were measured at higher concentration levels than those for pharmaceuticals under investigation. Multivariate analyses (factor and cluster analysis) were used to understand the pattern of concentrations along the longitudinal path of the river, and integrated contaminant input, dilution effects, and contaminant elimination (through sorption to bed-sediments and degradation). Three factors were found to provide an adequate description of the distribution of six compounds. Calculation of factors for each monitoring site allowed an understanding and identification of hot spots of contamination by certain substances and changes in concentration along the length of the river. Cluster analysis of all monitored sites resulted in three major groups: low, increasing, and high concentrations of drugs in water. This study provided a wealth of information that underpinned a working model of the distribution of pharmaceutical contamination along this river, and enabled the researchers to gain a good understanding of the system. However, it should be noted that in some of the sampling campaigns it was necessary to use a helicopter to transport the sampling team in order to cover a large area of the river system, and this would not be feasible for routine monitoring.

Since water management is required to be at the level of the river basin, it is important to develop affordable methods for wide scale monitoring that takes into account widely varying conditions in the river system, and inputs from multiple STPs that may differ greatly in efficiency of removal of individual pharmaceuticals. One approach is to use extrapolation to estimate effects of multiple STP outputs on overall concentrations in surface waters, and to combine this information with modelling using for example GIS. Reemtsma *et al.* (2006) proposed a Water Cycle Spreading Index (WCSI) to estimate the potential for a polar analyte present in waste water to spread in the aquatic environment. This index has units of concentration, and is calculated as the ratio of the effluent concentration and the normalized removal in the sewage treatment plant:

$$\mathrm{WCSI} = \frac{\mathrm{CEffluent}}{\mathrm{Normalised\ removal\ in\ STP}} = \frac{\mathrm{CInfluent} \times \mathrm{CEffluent}}{\mathrm{CInfluent} - \mathrm{CEffluent}} \tag{8.1}$$

where, CInfluent and CEffluent are the concentrations in the influent and effluent respectively, and the normalised removal in the STP is calculated as ((CInfluent-CEffluent)/CInfluent).

A high WCSI value indicates a high potential for spreading into surface waters based on removal rate (in the treatment plant(s)) and concentration entering the treatment plant. This model can be applied to a single STP or to multiple plants when an estimate of the potential for spreading at a larger scale is required. The authors found a good and positive correlation between WCSI values for three European waste water treatment plants and surface water concentrations for a number of waste water-associated contaminants. This did not, however, take into account removal of analytes following entry into the aquatic environment. More generally the use of geo-referenced exposure assessment tools for the prediction of environmental and more specifically surface water concentrations at the catchment or river basin level, taking into account multiple treated/untreated waste water effluents and surface run-off can provide valuable support for environmental risk assessments. Schowanek and Webb (2002) used the Geo-referenced Regional Exposure Assessment Tool for European Rivers (GREAT-ER) model to estimate PEC values for a number of pharmaceuticals in three European river catchments (Aire, UK; Lambro, Italy and Rhur, Germany). In this system modules are designed to incorporate hydrological information (e.g., river characteristics, flow distribution or velocity) and information on waste pathways (source emissions, metabolism, removal expressed as % removed or predicted using the SimpleTreat modelling software). Initial PEC values can be estimated for points downstream of emission points or representative values for the entire catchment can be calculated. PEC values estimated for the Aire catchment were 0.89–1.61, 0.03–0.06, 0.01–0.02, 0.05–0.09 and 4.3–6.8 μg/L for oxytetracycline, clofibrate, dextropropoxyphene, and paracetamol, respectively. Based on these predictions risk quotients (RQ) measured as PEC/PNEC ratios were found to be 0.3, 0.74 and 7 for clofibrate, paracetamol and oxytetracycline, where a RQ less than one implies that no adverse effects are anticipated.

In this section the emphasis has been on inputs from STPs; however, other important routes can include leaching landfill sites and from sludges and slurries applied to farmland, and in the case of aquaculture through direct input to the environment. The main pressure points will vary depending on the product and its pattern of use.

8.3.11 Estimation of fraction of pharmaceutical products associated with suspended matter/sludge and risk posed following land application

When significant removal of substances is observed during the STP process, and substances are known to associate strongly with particulate matter, or with sewage sludge, then it will be necessary to measure the concentrations of those substances in the sludge in order to evaluate the risk of disposal and application to fields. There is some experience in this area with non-polar industrial chemicals, but much less with pharmaceuticals. Schoof and Houkal (2005) reviewed the current state of risk assessment for pollutants (using dioxins as an example) in biosolids applied to agricultural land and potential risk to human health. The main pathways for risk of biosolid-associated compounds following agricultural applications include run-off and erosion with potential for transfer to surface and ground water resources.

– Pathway 1: biosolid –> soil –> surface water –> drinking/ground water, biota and humans
– Pathway 2: biosolid –> soil –> ground water –> drinking water/humans

In the USA regulation of application of biosolid to land uses a hazard quotient to evaluate the risk of compounds present in biosolids. This is the ratio of a measure of the exposure dose (maximum exposures or worse case scenarios: 50th, 90th percentiles, and maximum biosolid concentration) to a toxicological benchmark (no-observed-adverse effect level). There are two phases in the assessment. In this example based on monitoring levels of dioxins the phase 1 target organisms (receptors) used to represent the fresh water margin (impact through run-off) were the osprey (*Pandion haliaens*), kingfisher (*Ceryle alcyon*) and mink (*Mustela vison*). In phase 2 many more receptors were used, and the hazard quotient was based on combined lowest-observed-adverse-effect levels and 90th percentile concentrations in soil, sediment, plants and surface water derived from fate and transport modelling in these media. In most cases there is a requirement for this approach to monitoring to be complemented by a combination of experimental work (including mesocosm studies) and modelling of leaching through soil following field application of solid waste from STPs. This approach provides a potential model that could be applied to pharmaceuticals that accumulate in sludge.

Much experience is available from the work required for the registration of pesticides, and more recently detailed studies have been undertaken for some medicines for veterinary use. This could be helpful where pharmaceuticals are present in high concentrations in sludge. The first step would be to determine thresholds for concentrations in sludge that may lead to leaching to ground water or run-off into surface waters. These levels may be determined by application of models such as those (e.g. FOCUS) used for the prediction of pesticide leaching following field application. However, it may not be necessary to work through all possible scenarios of application, and the full range of soil types required for pesticides. The approach developed by Schneider *et al.* (2007) is based on a number of parameters (including manure load (estimated on the basis of phosphorous excretion), topsoil type, sand and organic carbon content, distance to surface water bodies, rainfall), and identifies geographical areas and exposure scenarios at a European level that represent the highest risk of contamination of surface waters by veterinary medicine products. Where available, literature data were used to validate the models. One area where there was a notable lack of information was for the transfer of agricultural inputs to sea water, and more work is needed to validate models such as this.

The potential for ground water contamination following land-application of sludge will be determined to an extent by partitioning and sorption of pharmaceuticals to aquifer material. This will also contribute to the potential for ground water contamination insituations where ground water/surface water interactions occur. Some work in this area has been carried out for pharmaceuticals. Hari *et al.* (2005) assessed the sorption of acetomiphen, carbamazepine, nalidixic acid and norfloxacin to an aquifer material collected from the alluvium channel of a Canadian river under a range of pH values, and in the presence or absence of surfactants. The presence of surfactants and pH value can modify the adsorption of these compounds with maximum adsorption occurring around pH 7, close to the pK_a values of the drugs. Further work (Kahle and Stamm, 2007) with propranolol, metoprolol and nadolol indicated that beta-blocker hydrophobicity is an important factor in its adsorption to aquifer mineral surfaces. Authors suggested that differences in adsorption to minerals could be due to interactions with transition metals such as iron. For this class of pharmaceuticals adsorption appeared to produce a significant reduction in the in leaching rate. Similar conclusions were reached by Drillia *et al.* (2005) in a lysimeter study of the sorption and mobility of six pharmaceuticals (carbamazepine, sulfamehtoxazole, ofloxacin, diclofenac, clofibric acid, and propanolol), where leaching was around 100% for the weakly adsorbed clofibric acid and not detectable for the strongly adsorbed ofloxacin.

Concentrations of compounds in sludge may be determined directly or indirectly following determination in waste water depending on the availability of reliable values of sludge/water partition coefficients for these pharmaceuticals. For example, fluoroquinolone antibacterial agents were shown to be removed efficiently (88–92%) from waste waters mainly through sorption and removal with activated sludge in an STP in Zurich. A sludge/water partition coefficient ($logK_d$) of approximately 4 was found for an activated sludge with a hydraulic retention time of 8 h. While biodegradation of ciprofloxacin and norfloxacin once applied to soil appeared limited, the threat to ground water may be reduced by strong sorption to the topsoil investigated, where there was limited movement. However, with repeated sludge applications, concentrations are likely to rise, and there could be breakthrough as maximum capacity is approached. It is important to assess the potential for transfer from sludge to soils, and from there to ground water by leaching or to surface waters by leaching, run-off, or soil erosion.

Partition and adsorption are very important in determining the rate and extent of movement from sludge to soil, and to surface and ground waters. Where significant quantities of pollutants are found in sludge, then estimates of sludge/soil partition coefficients would be useful in predicting the potential contamination of soil, and ground water, and thus in making decisions as to whether to permit the use of contaminated sludge as manure. There is a need for more research in this area.

8.4 CONCLUSIONS

There have been significant developments in the area of ecopharmacostewardship in recent years. There is much evidence of cleaner synthetic methods, and of more sustainable processes. However, this has been fragmented and there is a need for stimulation of the movement to produce new PPs with lower environmental impacts. These could include extensions of patent life for such compounds, and introduction of environmental risk assessment into the drug registration process.

Much work has been undertaken to improve understanding of the factors that affect the impact of PPs for human and veterinary use on the quality of environmental waters. However, in some areas this has indicated a need for further research to provide the information needed for the development of reliable risk assessments for a large number of compounds in this complex system. The potential costs of detailed investigations of the full set of PPs in current and recent use would be unsustainable, and representative information is needed to focus efforts where they are needed most; for compounds that pose a high risk through a combination of high use, toxicity, persistence, and accumulation. The identification of markers that can be used as surrogates for a large number of other compounds would be helpful. In a number of areas there is large uncertainty associated with predictions of environmental behaviour and risks to biota and human health. In particular there is an urgent need for improved information on the factors affecting the removal of PPs in waste water treatment and in the production of potable water so that improvements can be made in the operation of existing installations, and in the development of new technologies. It would be helpful for the purposes of planning at a river basin scale if STPs with similar operational properties could be grouped and treated as a set, but currently this is difficult to achieve because of poorly understood large variations between the performances of apparently similar STPs. Other areas where information is lacking, or where there are large uncertainties associated with conclusions based on available information include the factors determining the extent and properties of binding of PPs to solids in STPs, sediments, and phases used in water treatment. This information is also needed for the assessments of risks associated with sludge that is applied to land as manure.

Overall there is a need to improve monitoring strategies to ensure that the data obtained are reliable and representative. One improvement to the quality of monitoring data would be to replace spot sampling with integrative methods such as passive sampling. There is a need for closer connections between biological and chemical monitoring information. In order to maximise the utility of monitoring data, that are expensive to obtain, it would be useful to introduce screening criteria to ensure that only data that are fit for purpose are used. Further the introduction of standard formats for storing the screened monitoring data, and reliable and accessible archiving of the data so that it can be recovered by those who need it would bring large savings in the costs of model development, and would reduce unnecessary duplication of work. Reliable, validated predictive models offer a way of protecting the environment and human health without the need for large expenditure on comprehensive packages of measurements for each new product. This may help to prevent babies being thrown out with the bath water at an early stage.

REFERENCES

Amin, M.M., Zilles, J.L., Greiner, J., Charbonneau, S., Raskin L. and Morgenroth, E. (2006) Influence of the antibiotic Erythromycin on anaerobic treatment of a pharmaceutical wastewater. *Environmental Science and Technology* **40**(12), 3971–3977.

Bau, M. and Dulski, P. (1996) Anthropogenic origin of positive anomalies in river waters. *Earth Planet. Sci. Lett.* **143**, 245–255.

Bedner, M. and Maccrehan, W.A. (2006) Transformation of acetaminophen by chlorination produces the toxicants 1,4-Benzoquinone and *N*-Acetyl-*p*-benzoquinone Imine. *Environmental Science and Technology* **40**(2), 516–522.

Boethling, R.S., Sommer, E. and DiFiore, D. (2007) Designing small molecules for biodegradability. *Chem. Rev.* **107**, 2207–2227.

Buerge, I.J., Buser, H.-R., Poiger, T. and Mueller, M.D. (2006) Occurrence and fate of the cytostatic drugs Cyclophosphamide and Ifosfamide in wastewater and surface waters. *Environmental Science and Technology* **40**(23), 7242–7250.

Butters, M., Catterick, D., Craig, A., Curzons, A., Dale, D., Gillmore, A., Green, S. P., Marziano, I., Sherlock, J.-P. and White, W. (2006) Critical Assessment of pharmaceutical processes – A rationale for changing synthetic route. *Chem. Rev.* **106**, 3002–3027.

Castiglioni, S., Bagnati, R., Fanelli, R., Pomati, F., Calamari, D. and Zuccato, E. (2006) Removal of pharmaceuticals in sewage treatment plants in Italy. *Environmental Science and Technology* **40**(1), 357–363.

Chang, H., Hu, J. and Shao, B. (2007) Occurrence of natural and synthetic glucocorticoids in sewage treatment plants and receiving river waters. *Environmental Science and Technology* **41**(10), 3462–3468.

Clark, J.H. and Smith, P. (2005) Basic principles of green chemistry. *Innovations in Pharmaceutical Technology*, 94–97.

Coetsier, C.M., Spinelli, S., Lin, L., Roig, B. and Touraud, E. (2009) Discharge of pharmaceutical products (PPs) through a conventional biological sewage treatment plant: MECs vs PECs? *Environment International* **35**, 787–792.

Committee for Medicinal Products for Veterinary Use (CVMP). (2004) VICH-GL38 Environmental impact assessment for veterinary medicinal products phase II guidance. European Medicines Agency (EMEA) veterinary medicines and inspections, *London*. 1–39.

Daughton, C.G. (2003) Cradle-to-Cradle stewardship of drugs for minimizing their environmental disposition while promoting human health. I. Rationale for and avenues toward a green pharmacy. *Environ. Health Perspect.* **111**(5), 757–774.

de Voogt, P., Janex-Habibi, M.-L., Sacher, F., Puijker, L. and Mons, M.N. (2008) Development of an international priority list of pharmaceuticals relevant for the water cycle. Oral presentation, KNAPPE Conference, February, 19–20, Nimes, France.

Drillia, P., Stamatelatou, K. and Lyberatos, G. (2005) Fate and mobility of pharmaceuticals in solid matrices. *Chemosphere* **60**, 1034–1044.

Edhlund, B., Arnold, W.A. and McNeill, K. (2006) Aquatic photochemistry of nitrofuran antibiotics. *Environmental Science and Technology* **40**(17), 5422–5427.

Emmanuel, E., Perrodin, Y., Keck, G., Blanchard, J.-M. and Vermande, P. (2005) Ecotoxicological risk assessment of hospital wastewater: a proposed framework for raw effluents discharging into urban sewer network. *Journal of Hazardous Materials* **117**, 1–11.

Figueroa, R.A., Leonard, A. and Mackay, A.A. (2004) Modelling tetracycline antibiotic sorption to clays. *Environmental Science and Technology* **38**(2), 476–483.

Fono, L.J., Kolodziej E.P. and Sedlak, D.L. (2006) Attenuation of wastewater-derived contaminants in an effluent-dominated river. *Environmental Science and Technology* **40**(23), 7257–7262.

Gautam, A.K., Kumar, S. and Sabumon, P.C. (2007) Preliminary study of physicochemical treatment options for hospital wastewater. *Journal of Environmental Management* **83**(3), 298–306.

Goebel, A., Thomsen, A., Mcardell, C.S., Joss, A. and Giger, W. (2005) Occurrence and sorption behaviour of sulfonamides, macrolides, and trimethoprim in activated sludge treatment. *Environmental Science and Technology* **39**(11), 3981–3989.

Golet, E.M., Xifra, I., Siegrist, H., Alder, A.C. and Giger, W. (2003) Environmental exposure assessment of fluoroquinolone antibacterial agents from sewage to soil. *Environmental Science and Technology* **37**(15), 3243–3249.

Hari, A.C., Paruchuri, R.A., Sabatini, D.A. and Kibbey, T.C.G. (2005) Adsorption of beta blockers to environmental surfaces. *Environmental Science and Technology* **39**(8), 2592–2598.

Jones, O.A.H., Voulvoulis, N. and Lester, J.N. (2002) Aquatic environmental assessment of the top 25 English prescription pharmaceuticals. *Water Research* **36**, 5013–5022.

Jorgensen, S.E. and Halling-Sørensen, B. (2000) Drugs in the environment. *Chemosphere* **40**, 691–699.

Junker, T. Alexy, R., Knacker, T. and Kümmerer, K. (2006) Biodegradability of 14C-labeled antibiotics in a modified laboratory scale sewage treatment plant at environmentally relevant concentrations. *Environmental Science and Technology* **40**(1), 318–324.

Kahle, M. and Stamm, C. (2007) Sorption of the veterinary antimicrobial sulfathiazole to organic materials of different origin. *Environmental Science and Technology* **41**(1), 132–138.

Khetan, S.K. and Collins, T.J. (2007) Human pharmaceuticals in the aquatic environment: a challenge to green chemistry. *Chem. Rev.* **107**, 2319–2364.

Kajitvichyanukul, P. and Suntronvipart, N. (2006) Evaluation of biodegradability and oxidation degree of hospital wastewater using photo-Fenton process as the pretreatment method. *Journal of Hazardous Materials* **138**(2), 384–391.

Kidd, K.A., Blanchfield, P.J., Mills, K.H., Palace, V.P., Evans, R.E., Lazorchak, J.M. and Flick, R.W. (2007) Collapse of a fish population after exposure to a synthetic estrogen. *PNAS.* **104**(21), 8897–8901.

Kim, S-C. and Carlson, K. (2007) Temporal and spatial trends in the occurrence of human and veterinary antibiotics in aqueous and river sediment matrices. *Environmental Science and Technology* **41**(1), 51–57.

Kimura, K., Hara, H. and Watanabe, Y. (2007) Elimination of selected acidic pharmaceuticals from municipal wastewater by an activated sludge system and membrane bioreactors. *Environmental Science and Technology* **41**(10), 3708–3714.

Kümmerer K. (2007) Sustainable from the very beginning: rational design of molecules by life cycle engineering as an important approach for green pharmacy and green chemistry, *Green Chem.* **9**, 899–907.

Kümmerer, K., Al-Ahmad, A., Bertram, B. and Wiessler, M. (2000) Biodegradability of antineoplastic compounds in screening tests: influence of glucosidation and of stereochemistry. *Chemosphere* **40**, 767–773.

Kümmerer, K. (2001). Drugs in the environment: emission of drugs, diagnostic aids and disinfectants into wastewater by hospitals in relation to other sources – a review. *Chemosphere* **45**, 957–69.

Larsen, T., Lienert, J., Joss, A. and Siegrest, H. (2004) How to avoid pharmaceuticals in the aquatic environment. *Journal of Biotechnology* **113**, *295–304*.

Loeffler, D., Rombke, J., Meller, M. and Ternes, T.A. (2005) Environmental fate of pharmaceuticals in water/sediment systems. *Environmental Science and Technology* **39**(14), 5209–5218.

Mahnik, S.N., Lenz, K., Weissenbacher, N., Mader, R.M. and Fuerhacker, M. (2007) Fate of 5-fluorouracil, doxorubicin, epirubicin, and daunorubicin in hospital wastewater and their elimination by activated sludge and treatment in a membrane-bio-reactor system, *Chemosphere* **66**, 30–37.

Matamoros, V., Garcia, J. and Bayona, J.M. (2005) Behaviour of selected pharmaceuticals in subsurface flow constructed Wetlands: a pilot-scale study. *Environmental Science and Technology* **39**(14), 5449–5454.

Montforts, H.M.M. (2006) Validation of the exposure assessment for veterinary medicinal products. *Science of the Total Environment* **358**, 121–136.

Nghiem, L.D., Schaefer, A.I. and Elimelech, M. (2005) Pharmaceutical retention mechanisms by nanofiltration membranes. *Environmental Science and Technology* **39**(19), 7698–7705.

O'Driscoll, C., (2006) Medicines threatened. *Chem. Ind.*, 5–6.

Oppel, J., Broll, G., Loeffler, D., Meller, M., Roembke, J. and Ternes, T. (2004) Leaching behaviour of pharmaceuticals in soil-testing-systems: a part of an environmental risk assessment for groundwater protection. *The Science of the Total Environment* **328**, 265–273.

Pedersen, J.A., Soliman, M. and Suffet, I.H. (2005) Human pharmaceuticals, hormones, and personal care product ingredients in runoff from agricultural fields irrigated with treated wastewater. *Journal of Agriculture, Food & Chemistry* **53**, 1625–1632.

Pereira, V.J., Weinberg, H.S., Linden, K.G. and Singer, P.C. (2007) UV degradation kinetics and modelling of pharmaceutical compounds in laboratory grade and surface water via direct and indirect photolysis at 254 nm. *Environmental Science and Technology* **41**(5), 1682–1688.

Peschka, M., Eubeler, J.P. and Knepper, T.P. (2006) Occurrence and fate of barbiturates in the aquatic environment. *Environmental Science and Technology* **40**(23), 7200–7206.

Pierini, E., Famiglini, G., Mangani, F. and Capiello, A. (2004) Fate of Enrofloxacin in Swine sewage. *Journal of Agriculture, Food & Chemistry* **52**(11), 3473–3477.

Rabiet, M., Togola, A., Brissaud, F., Seidel, J-L., Budzinski, H. and Elbaz-Poulichet, F. (2006) Consequences of treated water recycling as regards pharmaceuticals and drugs in surface and ground waters of a medium-sized Mediterranean catchment. *Environmental Science and Technology* **40**(17), 5282–5288.

Rauch, W., Brockmann, D., Peters, I., Larsen, T.A. and Gujer, W. (2003) Combining urine separation with waste design: an analysis using a stochastic model for urine production. *Water Research* **37**(3), 681–689.

Reemtsma, T., Weiss, S., Mueller, J., Petrovic, M., Gonzaalez, S., Barcelo, D., Ventura, F. and Knepper, T.P. (2006) Polar pollutants entry into the water cycle by municipal wastewater: A European perspective. *Environmental Science and Technology* **40**(17), 5451–5458.

Rieger, P.G., Meier, H.-M., Gerle, M., Vogt, U., Groth, T. and Knackmuss, H.-J. (2002) Xenobiotics in the environment: present and future strategies to obviate the problem of biological persistence. *J. Biotechnol.* **94**, 101–123.

Schneider, M.K., Stamm, C. and Fenner, K. (2007) Selecting scenarios to assess exposure of surface waters to veterinary medicines in Europe. *Environmental Science and Technology* **41**(13), 4669–4676.

Schoof, R.A. and Houkal, D. (2005) The evolving science of chemical risk assessment for land-applied biosolids. *Journal of Environmental Quality* **34**, 114–121.

Schowanek, D. and Webb, S. (2002) Exposure simulation for pharmaceuticals in European surface waters with GREAT-ER. *Toxicology Letters* **131**, 39–50.

Seth, R., Webster, E. and Mackay, D. (2008) Continued development of a mass balance model of chemical fate in a sewage treatment plant. *Water Research* **42**(3), 595–604.

Sprunger, L., Blake-Taylor, B.H., Wairegi, A., Acree W.E. Jr. and Abraham, M.H. (2007) Characterization of the retention behavior of organic and pharmaceutical drug molecules on an immobilized artificial membrane column with the Abraham model. *Journal of Chromatography A* **1160**, 235–245.

Tucker, J.L. (2006) Green Chemistry, a Pharmaceutical Perspective, *Org. Process Res. Dev.* **10**, 315–319.

Ternes, T. and Joss, A. (2006) Human Pharmaceuticals, Hormones and Fragrances – The Challenge of Micropollutants in Urban Water Management. *Water Intelligence Online* © IWA Publishing. 1765–1774.

Verplanck, P.L., Taylor, H.E., Nordstrom, D.K. and Barber, L.B. (2005) Aqueous stability of Gadolinium in surface waters receiving sewage treatment plant effluent, Boulder Creek, Colorado. *Environmental Science and Technology* **39**(18), 6923–6929.

Vieno, M., Tuhkanen, T. and Kronberg, L. (2005) Seasonal variation in the occurrence of pharmaceuticals in effluents from a sewage treatment plant and in the recipient water. *Environmental Science and Technology* **39**, 8220–8226

Vieno, N.M., Harkki, H., Tuhkanen, T. and Kronberg, L. (2007) Occurrence of pharmaceuticals in river water and their elimination in a pilot-scale drinking water treatment plant. *Environmental Science and Technology* **41**(14), 5077–5084.

Wackett, L.P. and Ellis, L.B.M. (1999) Predicting biodegradation. *Environ. Microbiol.* **1**, 119–124

Wenlin, L., Li, Z.-H., Vakulenko, S. and Mobashery, S. (2000) A light-inactivated antibiotic. *J. Med. Chem.* **43**, 128–132.

Westerhoff, P., Yoon, Y., Snyder, S. and Wert, E. (2005) Fate of endocrine-disruptor, pharmaceutical, and personal care product chemicals during simulated drinking water treatment processes. *Environmental Science and Technology* **39**(17), 6949–6663.

Wiegel, S., Aulinger, A., Brockmeyer, R., Harms, H., Loeffler, J. Reincke, H., Schmidt, R., Stachel, B., von Tuempling, W. and Wanke, A. (2004) Pharmaceuticals in the River Elbe and its tributaries. *Chemosphere* **57**, 107–126.

Zukowska, B., Breivik, K. and Wania, F. (2006) Evaluating the environmental fate of pharmaceuticals using a level III model based on poly-parameter linear free energy relationships. *Science of the Total Environment* **359**, 177–187.

Chapter 9

Recommendations on communication and education

Rodrigo Vidaurre, Evelyne Touraud, Benoit Roig and Eleftheria Kampa

9.1 INTRODUCTION

Not only scientific factors, such as the physico-chemical nature of the drugs, or technical aspects, such as the kind of treatment processes that sewage is submitted to, exert an influence on the concentration of pharmaceutical compounds in the environment. The behaviour of the actors involved in the prescription and consumption process is of crucial importance for the actual amounts of pharmaceutical compounds that make their way into the environment.

Whereas consumers or prescribers cannot change the fact that a certain proportion of a drug is necessarily going to be expelled again by the human body, either in its original form or as a metabolite, they can however exert influence on factors such as which pharmaceutical product will be used, or what will be the fate of unused or expired pharmaceutical products. Both communication and education are crucial for informing prescribers and consumers of the consequences of their decisions, for allowing them to e.g. choose among several adequate products the most environmentally friendly one, and for letting them dispose their residual pharmaceuticals in a manner that will not harm the environment.

This chapter begins by exploring the lessons that can be derived of existing communication and education initiatives, presenting in some detail their characteristics. It then proceeds to cover possibilities for additional initiatives based on communication and education which could help reduce the input of pharmaceutical products into the environment.

9.2 LESSONS LEARNED FROM EXISTING INITIATIVES

Of the multiple possible approaches that can contribute to reducing the presence of pharmaceuticals in the environment, source control is widely recognised as having a large potential for reduction. Source control includes targeting "hot spots" of pharmaceutical input into the environment, such as hospitals and nursing homes, but also refers to actively influencing everyday consumption volume and compound choice (Ternes and Joss, 2006). Particularly compound choice, i.e., opting for the environmentally friendly choice in case of alternative compounds being available with comparable effectiveness, can be addressed through classification and labelling schemes. Section 9.2.1 illustrates the most comprehensive existing classification and labelling scheme of pharmaceuticals, currently implemented in Sweden.

Another approach addresses the fraction of pharmaceutical products that does not end up being consumed. Whereas the usual recommendation for dealing with this fraction used to be (and in some places still is) disposal via toilet, the last years have seen widespread efforts targeting the development and set up of take-back schemes of unused/expired

©2010 IWA Publishing. *Pharmaceuticals in the Environment: Current knowledge and need assessment to reduce presence and impact*. Edited by Benoit Roig. ISBN: 9781843393146. Published by IWA Publishing, London, UK.

medication. This medication is then, ideally, disposed of in a manner that ensures its not making its way into the environment, e.g. high-temperature combustion.

Both take-back schemes and classification and labelling schemes are heavily reliant on information and communication to the different actors. In the case of take-back schemes, day-to-day users have to be aware both of the environmental problem posed by pharmaceutical products in the environment, and of the possibility of their adequate disposal in the respective take-back scheme. These schemes as a measure to reduce pharmaceutical products in the environment are thus seen, not surprisingly, as best coupled with public awareness and education on the subject (Doerr-MacEwen and Haight, 2006). The Swedish classification and labelling scheme even provides education, in the form of short training, for medical practitioners, so that they count with a complete understanding of the none-too-simple scheme.

9.2.1 Classification and labelling schemes for pharmaceuticals: The Swedish example

The most comprehensive existing classification and labelling scheme of pharmaceuticals is currently that implemented in Sweden. This scheme targets specific active pharmaceutical substances and was first initiated by the Stockholm County Council and Apoteket (National Corporation of Swedish Pharmacies). The scheme is voluntary (the Swedish Medical Products Agency concluded in 2003 that, due to European rules, it was not possible to implement a mandatory classification and labelling scheme in Sweden (Matsson *et al.*, 2007)). In this scheme, information on the environmental impacts of pharmaceutical substances is made public on websites and information booklets but not via labels on drug packaging.

9.2.1.1 The "Stockholm model"

The first approach implemented in Sweden was the one developed in cooperation between the Stockholm County Council and Apoteket in 2003; it is known as the "Stockholm model". Its focus is on the environmental hazard of a pharmaceutical product, i.e., on its inherent ability to affect the environment. Due to the lack of official standards, a working model was developed, based on biodegradability (i.e., related to persistence in the environment), potential for bioaccumulation, and toxicity to aquatic organisms (called "PBT assessment"). The evaluation is based on data provided by the manufacturers. Every year, the evaluation of new active substances is made public (around 50 substances were added to the list in 2007).

The outcome of the classification is presented to the pharmaceutical expert groups of the Stockholm County Council. These experts use this information, as well as considerations regarding therapeutic efficiency, side effects, price, etc., in selecting the drugs that are recommended for use in the County Council health care. According to the Stockholm County Council, doctors displayed a high obedience (around 75%) to the recommendations of these pharmaceutical expert groups. The presumed achievements of the model were that doctors, health care staff and patients increase their awareness of possible environmental impacts of pharmaceuticals, and that the industry increases its efforts to design new biodegradable and not bioaccumulating drugs (Wennmalm, 2004).

9.2.1.2 The "Swedish model"

In 2005, a complementary approach ("Swedish model") was developed at the initiative of the Swedish Association of Pharmacy Industries (LIF), and in collaboration with the Swedish Medical Products Agency, Apoteket, the Swedish Association of Local Authorities and Regions, and Stockholm County Council. In fact, LIF and member companies were not very happy with the layout of the initial Stockholm model. For example, the level of metabolism in any given substance was not taken into account, so the amounts of the original substance that eventually reached the water supply would have been misleading (Matsson *etal.*, 2007).

There are two major differences between the Swedish and the original Stockholm model:

(i) First, the evaluation of active substances is now not only based on the environmental hazard of the product, but an evaluation of both the hazard and the associated risk, i.e., an assessment of the *probability* that adverse effects will occur and of their possible extent, based on the current use of the pharmaceutical product (Stockholm County Council, 2007; Apoteket, 2006). The environmental risk assessment is carried out by the Swedish Association of the Pharmaceutical Industry; it is based on the EU guidelines for environmental risk assessment of pharmaceuticals. The ambition of LIF is to have an environmental classification published on all substances available on the Swedish market within 5 years from the system launch in October 2005 (Matsson *et al.*, 2007).

(ii) Second, the information on risk/hazard is presented differently for three target groups with different levels of expertise: patients, prescribers, and specialists (scientists, experts, etc.) (Table 9.1). The warning labels are not placed on the packaging due to European restrictions on the information that can be added to packaging of medication. The relevant information/labels are placed on websites; Stockholm County Council provides both a detailed book on the subject background, and booklets with the latest classifications of substances.

Table 9.1 Information on risk/hazard for three target groups with different levels of expertise: patients, prescribers, and specialists (scientists, experts etc.).

Patient level: environmental risk information is based on the PEC/PNEC ratio	
PEC/PNEC < 0.1	"Use of the medicine has been considered to result in insignificant environmental risk."
0.1 < PEC/PNEC < 1	"Use of the medicine has been considered to result in low environmental risk."
1 < PEC/PNEC < 10	"Use of the medicine has been considered to result in moderate environmental risk."
PEC/PNEC < 1, but potentialally PBT or vPvB[1]	"Hazardous environmental properties."
PEC/PNEC > 10	"Use of the medicine has been considered to result in high environmental risk."
Insufficient data to characterize PEC/PNEC ratio	"Risk of environmental impact cannot be excluded due to lack of data."
Prescriber level: environmental risk information given to patients is repeated, but it also includes information on the environmental persistence (degradation) and bioaccumulation of the active substance.	
Degradation level	"The medicine is degraded in the environment", or "The medicine is slowly degraded in the environment", or "The medicine is potentially persistent".
Bioaccumulation	"No significant bioaccumulation potential", or "Potential to bioaccumulate in aquatic organisms".
If the pharmaceutical fulfills the criteria for PBT and/or vPvB	"According to the established EU criteria, the compound should be regarded as a PBT/vPvB substance."
Insufficient data to characterize the potential for degradation	"The potential for persistence cannot be excluded due to lack of data."
If there is insufficient data to characterize the potential for bioaccumulation	"The potential for bioaccumulation cannot be excluded due to lack of data"
Specialist level: all environmental information available is provided such as	
– Risk assessment, i.e. PEC/PNEC, calculations as well as the specific PEC and PNEC calculation, given in microgram/l, where applicable – Total sold amount in kilograms of the active substance on the market (including all products and enantiomers containing the same active substance) in the most recent year for which data are available. – Results from ecotoxicity tests (given in microgram/l). – Test guidelines used (e.g. OECD, FDA). – Information about which forms the pharmaceutical is excreted as, parent compound as well as metabolites, and the percentages thereof. – Data interpretation in the context of risk and hazard assessment	

[1]very Persistent, very Bioaccumulative)

The possibilities of the Swedish model for limiting the discharge of pharmaceuticals into environmental waters seem considerable, particularly when considering the high compliance with expert groups' recommendations, the informing of both doctors and patients on the environmental aspects of their prescription practices, and the general awareness-raising regarding this issue.

In this context, it should be noted that the healthcare system in Sweden is highly particular, which has incidence on this Swedish scheme. Until January 2010, all pharmacies in Sweden belonged to the national retail monopoly Apoteket (after a change in the law, private pharmacies are now returning to Sweden). Additionally, the Swedish health system is very public, with County Councils providing the health care for its inhabitants. The decisions of these public institutions thus have great influence, both on patients and practitioners, which need not be the case in other national healthcare systems that are not as centralised. In addition, the issue of pharmaceuticals in the environment was taken up as one of the main environmental issues by both Stockholm County Council and Apoteket, and County Councils also write tenders for pharmaceuticals: including environmental criteria in these tenders is a possibility for the future (Wennmalm, 2004).

All these points seem to have posed a significant incentive for the industry to participate in an environmental classification scheme of pharmaceuticals. In spite of the approach being voluntary, it seems to be the significant clout of the Apoteket and the County Councils, and the price of not participating in the scheme (i.e., not being able to influence the process), that created the incentive for industry to participate in it.

The initiative has overall allowed the industry to be regarded as a serious and collaborative partner by the stakeholders in environmental discussions. The industry has gained trust because of the transparency and openness of their knowledge and in some cases lack of data and knowledge. It is also likely that the model may spread to other countries. Mattson *et al.* (2007) point out three things worth considering before implementing the system abroad:

(i) Several developments and refinements to the model are ongoing (e.g., regarding the classification on persistence) and will not be fully implemented until after fall 2006.
(ii) The availability of a specific internet portal (www.fass.se) for the information and classification has been critical for the launch in Sweden. Other countries must identify their most appropriate communication channel before their implementation.
(iii) The PEC values have to be recalculated to fit the specific market in which the model is implemented.

9.2.2 Drug take-back schemes

The development and set up of take-back schemes of unused/expired medication is a quite widespread post-pharmacy stewardship approach for reducing the discharge of pharmaceutical products into waters.

In a survey of expert stakeholders' views (including government, academia, pharmaceutical and consulting industries) (Doerr-MacEwen and Haight, 2006), the implementation of take-back schemes was put forward as the most popular risk management strategy for pharmaceuticals in the environment. In this context, take-back schemes were seen best coupled with public education. Although the contribution of improper pharmaceutical disposal to the environmental loading of pharmaceutical products is not well understood and is generally believed to be minor, one of the interviewed stakeholders explained that take-back schemes are seen as a positive management response because disposal of unused and expired medicine seems like a logical, easy portion to remove from the environmental loading.

In Europe, take-back schemes for unused or expired medicine are required since 2004 by EU legislation. Directive 2004/27/EC requires Member States to "ensure that appropriate collection systems are in place for human medicinal products that are unused or have expired" (Article 1.87). Reference to these collection systems is to be made on the labelling or package leaflet. These systems had to be set up by the end of October 2005. As concerns veterinary products, the requirement to have appropriate collection systems dates from 2001 and was set out by Directive 2001/182/EC. The following focuses on take-back schemes for human medication.

According to the European Federation of Pharmaceutical Industries and Associations (Taylor and Poulmair, 2007), 20 European countries (out of 28 countries surveyed) have a drug take-back scheme in place. Seven countries (Bulgaria, Cyprus, Greece, Latvia, Malta, Romania and Slovenia) however have no scheme, and no information was available for Luxembourg. Eleven of the schemes in place are managed by pharmacies, six are co-ordinated nationally and five locally. The majority of the returned material is incinerated.

Quantitative information on the performance of the existing drug take-back schemes is not readily available for many countries. In fact, in the recent survey of 28 countries mentioned above, only ca. 30% of respondents were able to provide actual data. According to these responses, 19 kilo tonnes of medicine have been recovered from 8 countries (figures may also include packaging). The following gives some more detailed information on the national take-back schemes in France and Sweden, which are considered to have the most comprehensive human drug take-back schemes operating at present.

Some regions outside of the EU have taken an approach of sporadic take-back systems, e.g. calling for all consumers to return unused or expired medications during a period of a few weeks every year. This kind of campaigns can be combined with a permanent take-back scheme, and has the advantage of high visibility and topicality.

9.2.2.1 France

In 1992, France issued a decree specifying that all manufacturers of products for the general public had to contribute to the elimination of the waste resulting from the packaging of these products (Décret N° 92-377, 01.04.1992). The manufacturers could choose one of two forms for the organisation of the elimination: participation in a common system, *Éco-emballages*, or setting up their own system. The French pharmaceutical industry chose the latter option, setting up CYCLAMED in 1993. CYCLAMED originally had two main functions: It organised and financed the incineration of medicinal waste and packaging, and sorted unused medicine to distribute it to NGOs for reuse in Eastern Europe, the global South, or among people in France without social protection. Consumers were to bring their pharmaceutical wastes and packages to their chemist, who was in charge of collecting and sorting it.

Following press attention on cases of deceptive use and resale of unused medicine by certain chemists, the Minister for Health and Social Protection demanded in 2004 a review of the system. The resulting report (Grass and Lalande, 2005) pointed out several weaknesses and problems in the system, among others the low rate of packaging and medicine collecting, the unsuitability of unused medicine for redistribution, especially in the global South, and the risk of fraud linked to the reuse of medicine.

A law was eventually passed in early 2007 stating that pharmacists were to collect, free of charge, unused medicine brought to them by the consumers (Loi n° 2007-248, Art. 32) (Republique Française, 2007). From the end of August 2008 and onwards, the redistribution and reuse of collected medicine will be forbidden. The medicine is to be destroyed "under secured conditions".

CYCLAMED collects 5.7% of the medicine sold yearly. At the same time, Grass and Lalande (2005) estimate that on average, one out of two prescribed medicines is not consumed (i.e., ca. 50% of medicine sold). Therefore, the collection rate of medicine sold but not consumed seems not to be very high. Having said that, there are different and contradicting estimations of the effectiveness of the French drug take-back scheme. For instance, Taylor and Poulmair (2007), of the European Federation of Pharmaceutical Industries and Associations, mention an estimated recovery rate of up to 80% for the take-back scheme of France.

9.2.2.2 Sweden

The Swedish take-back scheme is strongly determined by the structure of its healthcare system. All pharmacies in Sweden used to belong to the state-owned Apoteket (National Corporation of Swedish Pharmacies), which provides this company with significant powers in its dealing with other stakeholders[2]. (Changes to the law have meant the return of private pharmacies to Sweden, the first of which opened in January 2010.) Apoteket runs a pharmaceutical take-back scheme (since the 1990s), in which consumers can return their unused or expired medication to any of the chain's pharmacies. These pharmaceuticals are then collected and incinerated in approved incineration plants.

There is a high level of awareness of Apoteket's take-back scheme among the Swedish public. Apoteket claimed in 2004 that approximately 65% of the Swedish public disposed their unused medications by returning them at pharmacies, whereas in 2006 the figure had increased to 73%. In the year 2006, the amount of pharmaceuticals returned increased by 4% compared to the previous year (Apoteket, 2006).

The company has invested significant efforts in increasing the available information on the environmental risks of drugs[3], and in increasing the public's awareness of the environmental consequences of their disposal. Among the latter are campaigns that address the public's behaviour regarding surplus or leftover drugs (e.g. through advertising in newspapers and poster distribution).

Other approaches used by Apoteket and others for reducing the environmental impact of pharmaceuticals include reducing the amount of unused medicine, e.g. through the use of "starter packs", which reduce the amount of leftover medicine in the case of treatment being interrupted (Stockholm County Council, 2007).

Apoteket does not deliver numbers regarding the total amount or the percentage of dispensed medicines that are returned to pharmacies. Information derived from other studies seems to indicate that, in spite of 73% of the public claiming to return their medication, the proportion of unused medication that is returned may not be as large. According to studies cited in Ekedahl (2006) (Ekedahl, 2006), the proportion of dispensed medicines returned to pharmacies range

---

[2]This retailing monopoly will be broken up and by January 2009 new companies should be permitted to enter the business, according to the promises of the recently elected government

[3]See also section 9.1 on Sweden's classification and labelling schemes that inform the general public on environmental risks of pharmaceutical products

between 2.3 and 4.6% of the total volume dispensed (studies published in 1999 and 2003). In the book "Environment and Pharmaceuticals", published in collaboration between Apoteket, Stockholm County Council and Stockholm University, the proportion of drugs that never get used is placed at "about 5 per cent" (Apoteket 2006). This estimation may be optimistic, considering for example that for France, Grass and Lalande (2005) estimated that on average one out of two prescribed medications is not consumed (i.e., ca. 50% of medicine sold).

9.3 RECOMMENDATIONS GOING BEYOND EXISTING INITIATIVES: APPROACHES TO COMMUNICATING METHODS OF "GOOD PRACTICE"

Eco-pharmacostewardship may be employed to reduce the environmental impact of pharmaceutical products throughout their life cycle. In the following section, approaches to communicate methods of "good practice" to manufacturers, prescribers and users of pharmaceuticals are examined. Environmental considerations regarding drugs should be emphasised in order to include them throughout the life cycle of a drug. Measures to reduce drugs' potential effects on the environment must be based on current knowledge, particularly on acute and mainly long-term effects.

9.3.1 Informing the public

Strategies to enhance public awareness have to be developed in order to stimulate an environmental approach to pharmaceutical production and consumption. Society wants and needs information and training regarding the effects of drugs on the environment (Medical Product Agency, 2004).

What type of information can be provided and how?

1) When known, information on pharmaceutical substances' effects on the environment could be provided through products résumés and pack inserts.

The relevant information concerns mainly chemical and biological data on pharmaceutical substances, their metabolites and excipients. This information has to be based on scientific data and not delivered in the form of an advertising message. The knowledge on existing environmental data has to be continually updated and needs to be turned into relevant and practical information. News on drugs and their effects on the environment should be presented in the same way as news on side effects and drug interaction, and made available on the websites of relevant agencies. European legislation currently provides the opportunity for the product résumé to include information on veterinary drugs' effects on the environment. The possibility of establishing an equivalent facility for human drugs should be explored.

2) Information on source-based sorting of packs, handling of superfluous drugs, and handling of used drugs that contain substance residues can be issued through product résumés and pack inserts.

3) Other resources

Communication campaigns (e.g. texts such as brochures and posters) can be used to target the general public regarding the return of unused/expired drugs to pharmacies. Brochures and posters could provide directions on what is accepted and what is not, how to return the unused or expired medications, and the closest and most accessible points of return. These brochures could be made available in pharmacies; their distribution could be extended to municipalities, and could be inserted in local newspapers and bulletins of healthcare communities (doctors, vets, etc.). Information should be disseminated as largely as possible through relevant websites and networks.

Awareness campaigns can be established inside chemists and pharmacies (e.g. one-week long campaign), during which the pharmacist could for example provide a brochure to everyone bringing a prescription. The brochure could show, via a simple scheme, the route that unwanted or out of date medicines follow when thrown into the toilet/sink, eventually reaching the environment.

9.3.2 Ecolabels for pharmaceuticals

Ecolabels are important instruments in an environmental policy mix. Compared to command and control mechanisms, they are more flexible and their introduction is easier, if supported by stakeholders (Klashka *et al.*, 2004). Participation in the schemes is usually voluntary. To our knowledge, no ecolabels of the classical kind exist which address the issue of pharmaceutical products in the environment. To date, proposals have only been made which address the related issue of personal care products (e.g. cosmetics, sun-screen products). Considerations for introducing ecolabels for pharmaceutical products remain so far on the theoretical level.

In contrast to other initiatives such as the classification and labelling approach implemented in Sweden, which targets individual active substances and whose results are not to be found on the packaging (see Section 9.2.1), ecolabels target whole products (thus being product-specific) and the labelling information usually appears on product packaging, targeting the end consumer. It is often the case that chemicals of very different origin and purpose (e.g. a number of active substances plus other chemical compounds necessary for stabilising the intake of the drug) are mixed in a product, and as a consequence released jointly into the environment. By addressing the product as a whole, an ecolabel could be a step towards more sustainable chemistry in pharmaceuticals, often termed "green pharmacy" (Daughton, 2003).

A possible drawback of ecolabels is that voluntary participation is limited to reaching only part of the manufacturers. Additionally, ecolabels appeal to a limited part of market consumers. Especially, in the case of pharmaceuticals, market consumers choose themselves only over-the-counter drugs but not drugs prescribed by the doctor. Experience with ecolabels in other sectors shows that there is a possibility of a larger number of consumers participating, if labelled products are priced competitively regarding non-labelled ones. In the case of pharmaceuticals, however, price is not considered to be as influential on consumption as in other sectors (e.g. food).

Furthermore, the type of label most often discussed when referring to pharmaceutical products is that under the control of independent third parties (Type I environmental labelling programmes, ISO, 1999). Such labels should be distinguished from labels that are set up by single companies (or, possibly, by industries) that often serve as advertising and where the criteria for labelling are often not made public. Examples of well-known third-party ecolabels are those of the Forest Stewardship Council (FSC), Marine Stewardship Council (MSC), and the German "Blue Angel". Voluntary approaches, not reviewed by third parties, can be deceptive, serving corporate interests instead of the public good (van Amstel *et al.*, 2006). In addition, the issue of what information a label provides, and how it is communicated, is also of crucial importance.

All in all, ecolabels could be a powerful communication and education tool to sensibilise the broader public on the environmental effects of pharmaceutical products, as well as providing strong incentives to the industry to improve, when possible, the environmental behaviour of their products. Ecolabels are seen as helpful in initiating public discussion on the topic of pharmaceutical products in the environment, particularly when linked to the issue of pharmaceuticals in drinking water, and a powerful means to increase public awareness.

9.3.3 Involvement of producers and distributors

In a survey on the household disposal of unused and expired pharmaceuticals as a source of pharmaceutical compounds in the environment, it emerged that a prominent disposal route of out-of-date or unwanted medicines may be via the sink/toilet or in household waste (Khetan and Collins, 2007).

In response to rising trends in prescription drug abuse and potential environmental concerns, the US Federal prescription disposal guidelines (2007) include flushing when it is instructed to do so and return of unused, unneeded or expired medicines to pharmaceutical take-back locations for safe disposal. Through pharmaceutical-return programs, residual medications can be collected from the public at take-back locations and disposed of in environmentally sound manner. In the same frame, the concept of "Extended Producer Responsibility" (EPR) for manufacturers and distributors has been implemented in British Columbia (2006). Those who produce, use or sell a product are responsible for the impacts of the product throughout the stages of its life cycle, including the end of life, waste management stage. The regulation require all brand owners of pharmaceutical products sold in British Columbia to take responsibility for the management of their products by providing a way for the public to dispose of their unused or expired products in an environmentally responsible manner.

9.3.4 Professional training – Taking up the subject in schools

Universities could include environmental knowledge in courses for professional categories (doctors, pharmacists) involving prescription/handling of pharmaceutical products. In the same way, information of the environmental effects of drugs could be introduced in environment-related courses. The corresponding teaching aids (e.g. Powerpoint files) could be made available on relevant websites of stakeholders (professional associations, environmental agencies, etc).

Awareness of secondary school students must be stimulated, too, as well as in primary education. Current research shows that information about recycling, given at primary school, influences the parents' behaviour; there is a clear opportunity to achieve the same effect of environmental stewardship of pharmaceuticals through the education system.

REFERENCES

Apoteket A.B, Environment and Pharmaceuticals, 2006, Second edition, 142 pages, available online at www.janusinfo.org/imcms/servlet/GetDoc?meta_id = 8643

Apoteket, Sustainability, Report 2006. Available online at www.apoteket.se/NR/rdonlyres/0930EB8B-C79A-4E62-8376-B6498EEC7AFC/14391/Apo_2006_Hållbarhet_eng.pdf

British Columbia (2006) Post-consumer Pharmaceutical Stewardship Association. Medication Return Program, July 2006, Available online at http://www.medicationsreturn.ca/pp2006/pp2006_program_plan.pdf

Daughton C.G. (2003) Environmental Stewardship of Pharmaceuticals: The *Green Pharmacy*. Available online at http://epa.gov/nerlesd1/bios/daughton/ngwa2003.pdf

Deutscher Bundestag (2007) Antwort der Bundesregierung auf die Kleine Anfrage der Abgeordneten Cornelia Behm, Sylvia Kotting-Uhl, Nicole Maisch, weiterer Abgeordneter und der Fraktion BÜNDNIS 90/DIE GRÜNEN. Drucksache 16/5205. 21.05.2007

Doerr-MacEwen, N.A. and Haight, M.E. (2006) Expert stakeholders' views on the management of human pharmaceuticals in the environment. *Environmental Management* **38**, 853–866.

Ekedahl, A. (2006) Reasons why medicines are returned to Swedish pharmacies unused. *Pharmacy world and science* **28**(6), 352–358.

European Parliament and the Council, Directive 2001/82/EC of the European Parliament and of the Council of 6 November 2001 on the Community code relating to veterinary medicinal products. OJ L 311, 28.11.2001, p. 1.

European Parliament and the Council, Directive 2004/27/EC of the European Parliament and of the Council amending Directive 2001/83/EC on the Community code relating to medicinal products for human use. OJ L 136, 30.4.2004, p. 34

Grass, E. and Lalande, F. (2005) Enquête sur le dispositif de recyclage des médicaments «Cyclamed». Rapport no 2005 001, Janvier 2005. IGAS. Available at http://lesrapports.ladocumentationfrancaise.fr/BRP/054000106/0000.pdf [08.10.07]

Innovations Report (2002) Selbstentsorger verfehlen gesetzliche Quoten für Arzneimittelverpackungen. 26.09.2002. http://www.innovations-report.de/html/berichte/studien/bericht-13195.html [08.10.2007].

Khetan, S.K. and Collins, T.J. (2007) Human pharmaceuticals in the aquatic environment: A challenge to green chemistry. *Chem. Rev.* **107**, 2319–2364.

Klashka, U., Liebig, M., Moltmann, J., Knacker, T. (2004) Potential Environmental Risks by Cleaning Hair and Skin – Ecolabel – A possibility to reduce exposure to personal care products. In *Pharmaceuticals in the Environment: Source, Fate, Effects and Risks*, (ed.K. Kümmerer), pp. 411–428, Springer, Heidelberg/New York.

Mattson, B., Nasman, I. and Strom, J. (2007) Sweden's Voluntary Environmental Drug Classification System, *Regulatory Affairs Journal (RAJ) Pharma*, 153–158, Available online at http://www.lif.se/cs/default.asp?id = 12620

Medical Products Agency (August 2004) Environmental effects of pharmaceuticals, cosmetics and hygiene products. Report

NHS – United Kingdom: The National Health Service (Pharmaceutical Services) Regulations. S.I. 2005/641. Dated 10 March 2005.

NHS Health and Social Care Information Centre, 2006: General Pharmaceutical Services in England and Wales 1995–96 to 2004–2005. Bulletin 2006/1.

République Française, Loi no 2007-248 du 26 février 2007 portant diverses dispositions d'adaptation au droit communautaire dans le domaine du médicament, J.O. 27.02.07.

SRU – Sachverständigenrat für Umweltfragen, 2007: Arzneimittel in der Umwelt. Stellungnahme. Nr. 12. April 2007.

Stockholm County Council, Environmentally Classified Pharmaceuticals, 2007, 28 pages, available online at http://www.janusinfo.se/imcms/servlet/GetDoc?meta_id = 9759

Taylor, D. and Poulmaire, M. Survey of European "Take-Back" schemes for unused medicines 2007. efpia. Powerpoint Presentation for the EHS ad hoc meeting, 3 July 2007.

Ternes, T.A. and Joss, A., *Human Pharmaceuticals, Hormones and Fragrances. The Challenge of Micropollutants in Urban Water Management.* 2006, IWA Publishing, London – Seattle.

van Amstel-van Saane, M., Driessen, P. and Glasbergen, P. (2006) Eco-labeling and information asymmetry: A comparison of five eco-labels in the Netherlands. Conference Paper, First Annual Conference on Institutional Mechanisms for Industry Self-Regulation, February 24–25, 2006, Hanover, USA. Available online at: http://mba.tuck.dartmouth.edu/mechanisms/pages/Papers/VanAmstel_paper.pdf

Wennmalm, A. (2004) Management of Environmental Risks Related to Use of Pharmaceutical Products, Powerpoint Presentation, Cleanmed Conference.

IWA Publishing's authorised EU representative for General Product Safety Regulations is Diane D'Arras, 15 rue Duret, 75116 Paris, France, e-mail: safety@iwap.co.uk.

Printed and bound by CPI Group (UK) Ltd, Croydon, CR0 4YY
17/09/2026
02218799-0020